Homology & Cohomology

**A Primer for Undergraduates
Through Applications**

Homology & Cohomology

A Primer for Undergraduates Through Applications

Luca Vitagliano

University of Salerno, Italy

World Scientific

NEW JERSEY · LONDON · SINGAPORE · BEIJING · SHANGHAI · TAIPEI · CHENNAI

Published by

World Scientific Publishing Co. Pte. Ltd.
5 Toh Tuck Link, Singapore 596224
USA office: 27 Warren Street, Suite 401-402, Hackensack, NJ 07601
UK office: 57 Shelton Street, Covent Garden, London WC2H 9HE

Library of Congress Control Number: 2025006454

British Library Cataloguing-in-Publication Data
A catalogue record for this book is available from the British Library.

HOMOLOGY & COHOMOLOGY
A Primer for Undergraduates Through Applications

ISBN 978-981-98-0582-2 (hardcover)
ISBN 978-981-98-0710-9 (paperback)
ISBN 978-981-98-0583-9 (ebook for institutions)
ISBN 978-981-98-0584-6 (ebook for individuals)

For any available supplementary material, please visit
https://www.worldscientific.com/worldscibooks/10.1142/14116#t=suppl

Desk Editors: Murali Appadurai/Angeline Husni

Typeset by Stallion Press
Email: enquiries@stallionpress.com

Preface

A *chain complex* is a sequence $C_\bullet = (C_i)_{i \in \mathbb{Z}}$ of abelian groups together with a sequence of group homomorphisms

$$\cdots \xleftarrow{d} C_{i-1} \xleftarrow{d} C_i \xleftarrow{d} C_{i+1} \longleftarrow \cdots$$

such that $d \circ d = 0$. This immediately implies that, for all i, the image of $d : C_{i+1} \to C_i$ is contained in the kernel of $d : C_i \to C_{i-1}$. Hence, we can form the quotient abelian group

$$H_i(C, d) := \frac{\ker \left(d : C_i \to C_{i-1} \right)}{\operatorname{im} \left(d : C_{i+1} \to C_i \right)}$$

which is called the *i-th homology* of $(C_\bullet, d)$. This apparently *ad hoc* notion turns out to pop up surprisingly often in mathematics, particularly in algebra and geometry, and also in mathematical logic, analysis, mathematical physics, and even numerical analysis and data science.

For instance, one can associate a chain complex $(C_\bullet(X), \partial)$ to a topological space X. The ith homology of $(C_\bullet(X), \partial)$ roughly computes how many *holes of dimension i* X has, and therefore it is an extremely useful tool to study (topological) properties of X. As $(C_\bullet(X), \partial)$ is a purely algebraic object, we passed in this way from the realm of topology to the realm of algebra opening the new hybrid world of *Algebraic Topology*. This is exactly the way how *Homology* was invented at the end of the 19th century by Poincaré in his studies on the topology of manifolds in the celebrated work *Analysis Situs* (Poincaré, 1895). It was later realized that similar structures do actually appear in several other branches of mathematics. The abstract theory of chain complexes (independently of how a chain complex arises here or there) is nowadays called *Homological Algebra*.

In this book, we introduce homology (and the dual concept of *cohomology*) from the scratch, together with some of its applications in algebra, topology and (differential) geometry. Our main aim is convincing the reader that homology is an important theory transversal to different (and partially unrelated) branches of mathematics, and it is worth to know at least the fundamentals of homology theory whatever ones primary interest is. This book is organized into two parts. In the first part, we discuss the algebraic preliminaries. We work in the general setting of *modules over a commutative ring with unit* and present the main constructions with them, including direct sums/products and tensor products. We also define (co)chain complexes and their (co)homology and discuss two basic tools to compute the latter: namely, *algebraic homotopies* and the *long homology exact sequence*. In the second part of the notes, we discuss applications. We first discuss applications in algebra. Specifically, we show how various algebraic structures give rise to (co)chain complexes and how the associated (co)homologies encode appropriate properties of those algebraic structures. We consider three cases: groups, associative algebras and Lie algebras. Later, we turn our attention to topological spaces and (one of) the associated homology theory: *singular homology*. We dedicate more space to this example as it is central in algebraic topology. After some preliminary work, we are able to compute the singular homology of spheres and this in turn has several interesting applications that, somewhat surprisingly, include a topological proof of the fundamental theorem of algebra. In the last chapter, we discuss de Rham cohomology which plays an important role in modern differential geometry.

We assume throughout this book that the reader is familiar with the fundamentals of algebra (in particular, groups and their homomorphisms, vector spaces and linear maps), set theoretic topology (topological spaces and continuous maps) and calculus (differentiable functions and vector valued maps, differentiation and integration). Any undergraduate textbook on those topics will do to recover the necessary prerequisites.

About the Author

Luca Vitagliano is a Full Professor of Geometry at the University of Salerno, Italy, where he regularly teaches Geometry courses to bachelor's, master's, and PhD students. His main research interests lie in Differential Geometry, with a slight emphasis on those aspects that are inspired by Mathematical Physics. He is the (co)author of around 40 scientific papers in the field published in distinguished international mathematical journals and has been invited to present his results to several international conferences.

Contents

Chapter 1

Multilinear Algebra

In this chapter, we concentrate on a specific algebraic structure: that of a *module over a ring*. Abelian groups and vector spaces are examples of modules. An ideal in a ring is also a module. Modules are particularly flexible objects appearing in several different situations both in algebra and in geometry, and they are the main building blocks in homological algebra. Here, we study their main properties and the first constructions with them. Such constructions will pop up every now and then along these lecture notes. For more information on modules and, more generally, on commutative algebra, the reader may consult (Atiyah and Macdonald, 2016).

1.1 Modules and Linear Maps

Let R be a commutative ring with unit.

Remark 1.1. Recall that a *ring* is a non-empty set R equipped with two composition laws

$$+ : R \times R \to R, \quad (a, b) \mapsto a + b \quad \textbf{(sum)},$$
$$\cdot : R \times R \to R, \quad (a, b) \mapsto a \cdot b \quad \textbf{(product)},$$

such that $(R, +)$ is an abelian group and, additionally, the product $\cdot$

- is associative,
- is both left and right distributive with respect to the sum.

The neutral element with respect to $+$ is usually denoted 0. A ring R is *commutative* if the product is commutative and it is a *ring with unit* if there is a neutral element with respect to the product (called the *unit* and usually denoted 1). For instance, a *field* is a commutative ring with unit such that every non-zero element is invertible with respect to the product. $\diamond$

In the following, all rings will be commutative with unit, unless otherwise stated.

Definition 1.2 (Module over a Ring). A *module over* R (or, simply, an *R-module*) is a non-empty set M equipped with two additional structures: a composition law

$$+ : M \times M \to M, \quad (p, q) \mapsto p + q \quad \textbf{(sum)}$$

and an *action* of R

$$\cdot : R \times M \to M, \quad (a, p) \mapsto a \cdot p \quad \textbf{(product by a scalar)},$$

such that $(M, +)$ is an abelian group and, additionally, the product by a scalar satisfies

- $a \cdot (b \cdot p) = (a \cdot b) \cdot p$,
- $a \cdot (p + q) = a \cdot p + a \cdot q$,
- $(a + b) \cdot p = a \cdot p + b \cdot p$,
- $1 \cdot p = p$,

for all $a, b \in R$ and $p, q \in M$. When working with R-modules, the elements of the ring R are called *scalars*. A subset $N \subseteq M$ in an R-module M is a *submodule* if it contains 0 and it is closed under both the sum and product by a scalar.

So, a module looks very much like a vector space, the only difference with the latter definition being that the scalars form a commutative ring with unit rather than a field. As customary for vector spaces, we often omit the symbol $\cdot$ in a product by a scalar and write, e.g., ap instead of $a \cdot p$, where $a \in R$, $p \in M$. We also omit the round brackets when associativity permits. For instance, we simply write abp for $a(bp) = (ab)p$, where $a, b \in R$, $p \in M$. It is clear that, with the restricted operations, any submodule is a module itself.

Proposition 1.3. *Let M be an R module. Then, for every $a \in R$ and every $p \in M$, we have*

$$a \cdot 0 = 0 \cdot p = 0$$

(where the 0 in the first and the last term is the zero in M, while the 0 in the second term is the zero in R).

Proof. Left as Exercise 1.1. □

Exercise 1.1. Prove Proposition 1.1.

Example 1.4. The trivial module is the module 0 containing only 1 element, necessarily the zero element 0. ◆

Example 1.5 (Vector Spaces are Modules). Let $\mathbb{K}$ be a field. In particular, $\mathbb{K}$ is a (commutative) ring (with unit) and $\mathbb{K}$-modules are precisely $\mathbb{K}$-vector spaces. ◆

Example 1.6 (Abelian Groups are $\mathbb{Z}$-modules). Denote by $\mathbb{Z}$ the ring of integers. Any abelian group G can be seen as a $\mathbb{Z}$-module as follows. The sum in G is just the pre-existing group sum (we always adopt the additive notation for abelian groups unless otherwise stated). The product by a scalar is defined as follows:

$$\mathbb{Z} \times G \to G, \quad (n,g) \mapsto ng := \begin{cases} \underbrace{g + \cdots + g}_{n \text{ times}} & \text{if } n > 0 \\ \underbrace{-g - \cdots - g}_{-n \text{ times}} & \text{if } n < 0 \\ 0 & \text{if } n = 0 \end{cases} . \tag{1.1}$$

It is easy to see that, with this two operations, G is indeed a $\mathbb{Z}$-module (see Exercise 1.2). Conversely, any $\mathbb{Z}$-module is, in particular, an abelian group and the product by a scalar is completely determined by the sum via Formula (1.1). In other words, if two $\mathbb{Z}$-module structures on the same set G share the same sum, then they also share the same product by a scalar (see Exercise 1.2 again). This shows that abelian groups are *one and the same thing* with $\mathbb{Z}$-modules, and talking about abelian groups or $\mathbb{Z}$-modules makes no difference. ◆

Exercise 1.2. Show that abelian groups are the same as $\mathbb{Z}$-modules proving that

(1) the pre-existing sum in an abelian group, together with the product by a scalar (1.1), equips G with a $\mathbb{Z}$-module structure and
(2) in a $\mathbb{Z}$-module, Formula (1.1) holds, hence the product by a scalar is actually determined by the sum.

Example 1.7 (Ideals are Modules). Let $I \subseteq R$ be an *ideal* in a ring R. Recall that this means that I is an abelian subgroup (with respect to the sum) and, additionally, for any $a \in R$ and any $b \in I$, the product ab is in I again. In other words, the product in R restricts to a product

$$R \times I \to I, \quad (a, b) \mapsto ab.$$

It is easy to see that, with the restricted operations, I is an R-module (Exercise 1.3). For instance, R itself is an R-module.   ◆

> **Exercise 1.3.** Show that, with the restricted operations, any ideal I in a ring R is an R-module.

Example 1.8 (The Module of n-tuples). Let n be a positive integer, and denote by R^n the set consisting of n-tuples of elements in R:

$$(a_1, \ldots, a_n), \quad a_1, \ldots, a_n \in R.$$

With the *entrywise sum*

$$(a_1, \ldots, a_n) + (b_1, \ldots, b_n) := (a_1 + b_1, \ldots, a_n + b_n),$$

$$(a_1, \ldots, a_n), (b_1, \ldots, b_n) \in R^n,$$

and the *entrywise product by a scalar*

$$a \cdot (a_1, \ldots, a_n) = (aa_1, \ldots, aa_n), \quad a \in R, \quad (a_1, \ldots, a_n) \in R^n.$$

Then R^n is an R-module (Exercise 1.4). The zero element in the R-module R^n is the *zero n-tuple* $(0, \ldots, 0)$.   ◆

> **Exercise 1.4.** Show that, with the entrywise operations, the space R^n of n-tuples of elements in a ring R is an R-module.

Example 1.9 (Function Ring and Function Module). Let X be a set. Denote by R^X the space of all R-valued functions on X:

$$f : X \to R.$$

Such space carries the structure of an R-module. The operations are *pointwise*, i.e., for any two functions $f, g \in R^X$ and any scalar $a \in R$, we define

$$f + g : X \to R, \quad x \mapsto (f + g)(x) := f(x) + g(x),$$

and

$$af : X \to R, \quad x \mapsto (af)(x) := af(x).$$

As announced, with this two operations, R^X is an R-module. For instance, when $X = X_n := \{1, \ldots, n\}$ is the set of the first n positive integers, then the R-module R^X identifies with the module R^n via the assignment

$$f \mapsto (f(1), \ldots, f(n)).$$

In other words, we could have defined R^n simply as R^{X_n} (do you see this?).

Now, let X be again any set. Not only R^X is an R-module, but it is also a ring with the (already defined) pointwise sum, and the *pointwise product* given by

$$fg : X \to R, \quad x \mapsto (fg)(x) := f(x)g(x)$$

for all $f, g \in R^X$. When interpreted as a ring, we denote R^X by $\mathcal{F}(X, R)$. The zero element in $\mathcal{F}(X, R)$ is the constant function 0, while the unit is the constant function 1 (do you see it?). Note that the ring $\mathcal{F}(X, R)$ is never a field (unless X consists of just one point $*$ and R is a field itself, in which case the assignment $f \mapsto f(*)$ identifies $\mathcal{F}(X, R)$ with the field R). To see this, let x_1, x_2 be distinct points in X. Take the functions $\chi_1, \chi_2 : X \to R$ defined by

$$\chi_i(x) = \begin{cases} 1 & \text{if } x = x_i \\ 0 & \text{otherwise} \end{cases}, \quad i = 1, 2.$$

As $x_1 \neq x_2$, we clearly have $\chi_1 \chi_2 = 0$ but nor χ_1 nor χ_2 is the zero function, hence $\mathcal{F}(X, R)$ is not an integral domain. We leave it to the reader to discuss the only remaining case, when X consists of just one point but R is *not* a field.

Finally, let M be an R module. Consider the space $M^X = \mathcal{F}(X, M)$ of M-valued maps on X:

$$F : X \to M.$$

The space $\mathcal{F}(X, M)$ is an $\mathcal{F}(X, R)$-module. The operations in $\mathcal{F}(X, M)$ are again pointwise:

$$F + G : X \to M, \quad x \mapsto (F + G)(x) := F(x) + G(x)$$

and

$$fF : X \to M, \quad x \mapsto (fF)(x) := f(x)F(x)$$

for all $F, G \in \mathcal{F}(X, M)$ and all $f \in \mathcal{F}(X, R)$. The zero element in $\mathcal{F}(X, M)$ is the constant map 0.   ◆

Exercise 1.5. Prove all the unproven claims in Example 1.9.

Similarly as for vector spaces, modules over the same ring R can be compared via suitable maps that we now discuss. Let M, N be R-modules.

Definition 1.10 (Module Homomorphism). An *R-module homomorphism*, or an *R-linear map* (or simply, a *linear map*), between M and N is a map

$$f : M \to N$$

such that

(1) $f(p+q) = f(p) + f(q)$ and
(2) $f(ap) = af(p)$,

for all $p, q \in M$ and all $a \in R$. An injective linear map is called a(n *R*-module) *monomorphism*. A surjective linear map is called an *epimorphism*. A bijective linear map is called an *isomorphism*. Two *R*-modules M, M' are said to be *isomorphic* if there exists an isomorphism $\Phi : M \to M'$ connecting them. In this case, we also write $M \cong M'$.

If R is a field, then Definition 1.10 agrees with the definition of a vector space homomorphism.

Example 1.11. Let G, H be abelian groups. It should be clear that a map $f : G \to H$ is a $\mathbb{Z}$-module homomorphism if and only if it is a group homomorphism (do you see it?).   ◆

Example 1.12. Let M be an R-module and let 0 be the zero module. There exists exactly 1 linear map $0 \to M$, namely the zero map. The only map $M \to 0$ is also R-linear.   ◆

Actually, R-module homomorphisms share several properties with vector space homomorphisms. We summarize some of them in a series of propositions whose proofs are straightforward and are left as exercise.

Proposition 1.13.

(1) *For any R-module M, the identity map $\mathrm{id}_M : M \to M$ is an R-module homomorphism (actually an isomorphism).*
(2) *The composition of R-module homomorphisms (resp. monomorphisms, epimorphisms, isomorphisms) is an R-module homomorphism (resp. monomorphism, epimorphism, isomorphism).*
(3) *The inverse of an R-module isomorphism is an R-module isomorphism.*

Proof.   Left as Exercise 1.6.   □

Exercise 1.6. Prove Proposition 1.13.

The kernel of an R-module homomorphism is defined exactly as for a vector space homomorphism. Let M, N be R-modules, and let $f : M \to N$ be a linear map.

Definition 1.14 (Kernel of a Module Homomorphism). The *kernel* of f is the subset

$$\ker f := \{p \in M : f(p) = 0\} \subseteq M.$$

If R is a field, then Definition 1.14 agrees with that of *kernel* of a vector space homomorphism.

Example 1.15. Let G, H be abelian groups and let $f : G \to H$ be a group homomorphism, hence a $\mathbb{Z}$-module homomorphism. In this case, Definition 1.14 agrees with that of *kernel* of a group homomorphism. ♦

Proposition 1.16. *Let $f : M \to N$ be a linear map of R-modules. Then both the kernel* $\ker f \subseteq M$ *and the image* $\operatorname{im} f \subseteq N$ *of f are submodules.*

Proof. Left as Exercise 1.7. □

Exercise 1.7. Prove Proposition 1.16.

Proposition 1.17 (Kernel Criterion). *A linear map $f : M \to N$ is injective if and only if the kernel* $\ker f$ *is trivial, i.e., $\ker f = 0$.*

Proof. Left as Exercise 1.8. □

Exercise 1.8. Prove Proposition 1.17.

We conclude this section discussing *quotient modules* which play an important role throughout this book. Begin with an R-module M. Similarly as for groups (and for vector spaces), a submodule $N \subseteq M$ determines an equivalence relations $\sim$ on M defined by

$$p \sim q \quad \text{if } p - q \in N$$

(can you prove in details that $\sim$ is reflexive, symmetric and transitive?). The space $M/\sim$ of equivalence classes of M under this relation is also denoted M/N. The equivalence class of an element $p \in M$ is denoted $p \bmod N$. For instance, the equivalence class $0 \bmod N$ is exactly N (do you see it?).

Proposition 1.18. *There exists a unique R-module structure on M/N such that the natural projection*

$$\pi : M \to M/N, \quad p \mapsto p \bmod N$$

is a linear map.

Proof. We sketch the proof, leaving the details as an exercise for the reader. We begin defining an R-module structure on M/N. The sum is defined by

$$(p \bmod N) + (q \bmod N) = p + q \bmod N, \tag{1.2}$$

and the product by a scalar is defined by

$$a(p \bmod N) = ap \bmod N, \tag{1.3}$$

for all $p, q \in M$ and $a \in R$. The reader is invited to check that these operations are well defined (i.e., they are independent of the chosen representatives in the equivalence classes involved), and they equip M/N with an R-module structure. It immediately follows from this definition that $\pi : M \to M/N$ is a linear map, indeed, for all $p, q \in M$,

$$\pi(p + q) = p + q \bmod N = (p \bmod N) + (q \bmod N) = \pi(p) + \pi(q),$$

and similarly for the product by a scalar. Uniqueness also follows immediately: if π is a linear map, the operations in M/N cannot be defined in a way other than (1.2). Indeed, in this case, for all $p, q \in M$,

$$(p \bmod N) + (q \bmod N) = \pi(p) + \pi(q) = \pi(p + q) = p + q \bmod N,$$

and similarly for the product by a scalar. $\qquad\square$

Definition 1.19 (Quotient Module). The module M/N is called the *quotient module* of M over the submodule N.

Example 1.20. Let G be an abelian group and let $H \subseteq G$ be a subgroup. As G is abelian, H is a normal subgroup. It is also clear that H is a submodule of the $\mathbb{Z}$-module G (do you see it?). In this case, Definition 1.19 agrees with that of *quotient group* (over a normal subgroup). ◆

Note that the kernel of the projection $\pi : M \to M/N$ is exactly the submodule N.

Example 1.21. Let M be an R-module and let $N \subseteq M$ be a submodule. Then the quotient M/N is the zero module 0 if and only if $N = M$. At the other extreme, the projection $\pi : M \to M/N$ is a module isomorphism if and only if $N = 0$. In this case, we often use π to identify M with M/N and simply write $M/0 = M$. ◆

Proposition 1.22 (Homomorphism Theorem). *Let M, Q be R-modules, let $N \subseteq M$ be a submodule and let $f : M \to Q$ be a linear map. The following two conditions are equivalent:*

(1) $N \subseteq \ker f$;
(2) *the map f factorizes as the composition $f_{M/N} \circ \pi$ of the projection $\pi : M \to M/N$ followed by a linear map $f_{M/N} : M/N \to Q$, i.e., there exists a linear map $f_{M/N} : M/N \to Q$ such that the diagram*

$$\begin{array}{ccc} M & \xrightarrow{f} & Q \\ {\scriptstyle \pi}\downarrow & \nearrow{\scriptstyle f_{M/N}} & \\ M/N & & \end{array} \tag{1.4}$$

commutes.

In this situation, the linear map $f_{M/N}$ making Diagram (1.4) commutative is unique.

Proof. Let $\ker f \supseteq N$. Then we can define a map $f_{M/N} : M/N \to Q$ by putting

$$f_{M/N}(p \bmod N) = f(p)$$

for all $p \in M$. If p' is another representative for the same class $p \bmod N$, i.e., $p' - p \in N$, then we have

$$f(p') = f(p + p' - p) = f(p) + f(p' - p) = f(p),$$

where, in the last step, we used that $N \subseteq \ker f$. This shows that $f_{M/N}$ is well defined. Additionally, for all $p \in M$,

$$(f_{M/N} \circ \pi)(p) = f_{M/N}(\pi(p)) = f_{M/N}(p \bmod N) = f(p),$$

i.e., Diagram (1.4) commutes. So, (1) $\Rightarrow$ (2). That (2) $\Rightarrow$ (1) is straightforward. Finally, if $g : M/N \to Q$ is another linear map such that $f = g \circ \pi$, then necessarily $g = f_{M/N}$ indeed, for all $p \in M$, $g(p \bmod N) = (g \circ \pi)(p) = f(p) = f_{M/N}(p \bmod N)$. $\qquad\square$

Corollary 1.23. *Let $f : M \to Q$ be a linear map. Then there is a unique isomorphism $\overline{f} : M/\ker f \to \mathrm{im}\, f$ such that the diagram*

$$\begin{array}{ccc} M & \xrightarrow{f} & \mathrm{im}\, f \\ {\scriptstyle \pi}\downarrow & \nearrow{\scriptstyle \overline{f}} & \\ M/\ker f & & \end{array} \tag{1.5}$$

commutes. In particular, if f is surjective, then $M/\ker f \cong Q$.

Proof.   From Proposition 1.22, there exists a unique linear map $\overline{f}$: $M/\ker f \to Q$ such that $f = \overline{f} \circ \pi$. It is clear that $\overline{f}$ takes values in $\operatorname{im} f$, hence it restricts to a linear map $M/\ker f \to \operatorname{im} f$ which, abusing the notation, we call $\overline{f}$ again. It remains to show that $\overline{f} : M/\ker f \to \operatorname{im} f$ is a module isomorphism. As the restriction $f : M \to \operatorname{im} f$ to $\operatorname{im} f$ in the codomain is surjective by construction, and $f = \overline{f} \circ \pi$, it immediately follows that $\overline{f} : M/\ker f \to \operatorname{im} f$ is surjective as well. In order to show that it is also injective, we use the *kernel criterion*. So, let $p \in M$ be such that $\overline{f}(p \bmod \ker f) = 0$. This means that

$$0 = \overline{f}(p \bmod \ker f) = \overline{f}(\pi(p)) = f(p),$$

i.e., $p \in \ker f$, hence $p \bmod \ker f = 0$. We conclude that $\overline{f} : M/\ker f \to \operatorname{im} f$ is also injective.   □

Corollary 1.24. *Let M_1 and M_2 be R-modules, and let $N_1 \subseteq M_1$ and $N_2 \subseteq M_2$ be submodules. Additionally, let $f : M_1 \to M_2$ be a linear map such that $f(N_1) \subseteq N_2$. Then there exists a unique linear map $\overline{f} : M_1/N_1 \to M_2/N_2$ such that the diagram*

$$
\begin{array}{ccc}
M_1 & \xrightarrow{\ f\ } & M_2 \\
{\scriptstyle \pi}\downarrow & & \downarrow{\scriptstyle \pi} \\
M_1/N_1 & \xrightarrow{\ \overline{f}\ } & M_2/N_2
\end{array}
$$

commutes, i.e., $\overline{f}(p \bmod N_1) = f(p) \bmod N_2$ for all $p \in M_1$.

Proof.   Left as Exercise 1.9.   □

Exercise 1.9. Prove Corollary 1.24.

A sequence of R-module homomorphisms of the type

$$0 \longrightarrow N \xrightarrow{\ \alpha\ } M \xrightarrow{\ \beta\ } Q \longrightarrow 0 \tag{1.6}$$

is called a *short exact sequence* (of modules) if $\alpha : N \to M$ is injective, $\beta : M \to Q$ is surjective and, additionally, $\operatorname{im}\alpha = \ker\beta$. Note that, in this situation, the kernel of any arrow in (1.6) coincides with the image of the preceding arrow, indeed $\ker\alpha = 0 = \operatorname{im}(0 \to N)$ and $\ker(Q \to 0) = Q = \operatorname{im}\beta$. It is clear that the restriction $\alpha : N \to \operatorname{im}\alpha$ of α to its image in the codomain is an R-module isomorphism. Hence, the map α

identifies N with im $\alpha = \ker \beta$. On the other hand, from Corollary 1.23, the map $\bar{\beta}$ identifies Q with $M/\ker \beta = M/\operatorname{im}\alpha$.

Example 1.25. Let $2\mathbb{Z} \subseteq \mathbb{Z}$ be the subgroup of even integers. Then $2\mathbb{Z}$ is also a submodule and the inclusion map $i : 2\mathbb{Z} \to \mathbb{Z}$, $n \to n$, is a $\mathbb{Z}$-module homomorphism. The quotient $\mathbb{Z}/2\mathbb{Z}$ is the abelian group $\mathbb{Z}_2$ of integers modulo 2 and the sequence

$$0 \longrightarrow 2\mathbb{Z} \xrightarrow{i} \mathbb{Z} \xrightarrow{\pi} \mathbb{Z}_2 \longrightarrow 0$$

is a short exact sequence of $\mathbb{Z}$-modules. ♦

Example 1.26. More generally, let M be an R-module, and let $N \subseteq M$ be a submodule. Denote by $i_N : N \to M$, $p \mapsto p$ the *inclusion*. It is clear that i_N is a linear map. The sequence

$$0 \longrightarrow N \xrightarrow{i_N} M \xrightarrow{\pi} M/N \longrightarrow 0$$

is a short exact sequence and the discussion above shows that every short exact sequence is of this type up to appropriate identifications. ♦

Example 1.27. For any R-module homomorphism $f : M \to Q$, the sequence

$$0 \longrightarrow \ker f \longrightarrow M \xrightarrow{f} \operatorname{im} f \longrightarrow 0$$

is a short exact sequence. Here, the second arrow is the inclusion. ♦

1.2 Free Modules

There is a class of modules, called *free modules*, which are similar to vector spaces in many respects (but not all respects), even if their ring of scalars is not necessarily a field. We discuss this class in this section.

Let R be a ring, and let M be an R-module. Consider a family $S = (p_i)_{i \in I} \subseteq M$ of (non-necessarily distinct) elements of M, parameterized by a (possibly infinite) index set I. A finite linear combination of elements of S (with coefficients in R) will be also denoted

$$\sum_{i \in I} a_i p_i, \quad a_i \in R,$$

where we tacitly assume that the scalars a_i vanish for all but finitely many $i \in I$. The subset

$$\operatorname{Span}(S) := \left\{ \text{finite linear combinations of elements of } S \right\} \subseteq M$$

is a submodule (do you see it?), called the *submodule spanned* (or, *generated*) *by S*. If $N \subseteq M$ is a submodule and $S \subseteq M$ is a subset such that $N = \text{Span}(S)$, we also say that S is a *family* or a *set of generators* of N (or that S *generates* N). For instance, a family $S = (p_i)_{i \in I} \subseteq M$ is a set of generators of M itself if and only if every element of M can be written as a (finite) linear combination of elements of S. A (sub)module N is *finitely generated* if it is spanned by a finite family, i.e., a family $(p_i)_{i \in I} \subseteq M$ with I a finite set.

A family $S = (p_i)_{i \in I} \subseteq M$ is said to be *independent* if a finite linear combination

$$\sum_{i \in I} a_i p_i, \quad a_i \in R,$$

vanishes only if the coefficients a_i all vanish. For instance, if there are repetitions in S, then S is *not* independent. A *basis* for M is a family $B = (q_i)_{i \in I} \subseteq M$ which is both independent and a family of generators.

Exercise 1.10. Let $f : M \to N$ be an R-module homomorphism. Prove that

(1) if f is injective, then it transforms independent families into independent families (the converse might not be true);
(2) f is surjective if and only if it transforms one (hence any) family of generators into a family of generators;
(3) if f is bijective, then it transforms bases into bases (the converse might not be true).

Example 1.28. Consider the R module R^n of n-tuples of scalars. Put

$$E_1 := (1,0,0,\ldots,0),\ E_2 := (0,1,0,\ldots,0),\ \ldots\ ,\ E_n = (0,0,0,\ldots,1).$$

It is easy to see that the family $(E_1, \ldots, E_n)$ is both independent and a family of generators for R^n (do you see this?). Hence, it is a basis for R^n. ◆

Example 1.29. More generally, let X be any set and consider the function module R^X. Consider the submodule $RX \subseteq R^X$ consisting of functions $a : X \to R$ such that $a(x) \neq 0$ for finitely many $x \in X$ only (do you see that RX is a submodule?). When X is a finite set, clearly $RX = R^X$.

We want to show that RX possesses a basis with the same cardinality as X. To do this, for any $x_0 \in X$ consider the *characteristic function* $\chi_{x_0} : X \to R$

defined by

$$\chi_{x_0}(x) = \begin{cases} 1 & \text{if } x = x_0 \\ 0 & \text{otherwise} \end{cases}.$$

It is clear that $\chi_{x_0} \in RX$ for all $x_0 \in X$. We want to show that the family

$$B := (\chi_x)_{x \in X} \subseteq RX$$

is a basis of RX. To do this, we adopt a slight change of notation, which will prove to be rather convenient in the following. We fix once for all another set I with the same cardinality as X, and a bijection $I \to X$, denoted $i \mapsto x_i$. We can now interpret X as a family $(x_i)_{i \in I}$ (without repetitions) indexed by I (e.g., when X is a finite set of cardinality $n \in \mathbb{N}$, we can choose I to be X_n and interpret X as a string $(x_1, \ldots, x_n)$). In practice, we are relabeling the elements of X. This is not strictly necessary but makes the final formulas slightly more handy for the beginners. With this choice, B can be thought of as a family indexed by I (rather than by X):

$$B = (\chi_{x_i})_{i \in I}.$$

Consider a zero finite linear combination

$$\sum_{i \in I} a_i \chi_{x_i} = 0, \quad a_i \in R \text{ vanishing for all but finitely many } i \in I,$$

i.e., the left-hand side is the constant zero function $0 : X \to R$. In particular, for every $j \in I$, we have

$$0 = \left(\sum_{i \in I} a_i \chi_{x_i} \right)(x_j) = \sum_{i \in I} a_i \chi_{x_i}(x_j) = \sum_{i \in I} a_i \delta_{ij} = a_j,$$

where we denoted

$$\delta_{ij} = \begin{cases} 1 & \text{if } i = j \\ 0 & \text{otherwise} \end{cases}, \quad i, j \in I.$$

This shows that B is independent. Now, let $a : X \to R$ be a function in RX and denote $a_i := a(x_i) \in R$. By definition of RX, all a_i vanish but finitely many. Hence,

$$a' := \sum_{i \in I} a_i \chi_{x_i}$$

is a finite linear combination of elements of B. We have $a = a'$, indeed the same computation as above shows that, for any $j \in I$,

$$a'(x_j) = a_j = a(x_j),$$

whence B generates RX.

The module RX will play an important role in the following. For this reason, we provide for its elements an alternative interpretation that is often useful. First of all, note that the map $\chi : X \to RX, x \mapsto \chi_x$ is injective. Hence, we can use it to identify $x_i \in X$ with $\chi_{x_i} \in RX$, for all $i \in I$. If we do so, we can interpret an element $a \in RX$ as a (formal) finite linear combination of elements of X and write

$$a = \sum_{i \in I} a_i x_i \tag{1.7}$$

(instead of $a = \sum_{i \in I} a_i \chi_{x_i}$) with the $a_i \in R$ all vanishing but finitely many as usual. For this reason, RX is also called the *module of formal linear combinations* of elements in X. Note that the usual computational rules hold for such formal linear combinations.

Finally, we recover Example 1.28 in the case $X = X_n$ (hence, $RX = R^X = R^{X_n} = R^n$). ◆

Although modules are very similar to vector spaces in the definition, they can differ significantly from the latter in practice. The main difference is that, in general, modules do not possess bases. Indeed, as the ring of scalars R is not a field, it is impossible to provide a proof of the Linear Dependence Lemma (Axler, 2024, Statement 2.19) (hence the existence of bases) for modules.

Example 1.30. Denote by $\mathbb{Z}_2 = \{\bar{0}, \bar{1}\}$ the quotient of the ring $\mathbb{Z}$ under the congruence mod 2. Forget about the product in $\mathbb{Z}_2$ and regard it as a plain abelian group, hence a $\mathbb{Z}$-module. Bases do not have repeated elements. There are exactly 4 families of elements of $\mathbb{Z}_2$ without repetitions (a family without repetitions is basically a subset)

$$\varnothing, \ (\bar{0}), \ (\bar{1}), \ (\bar{0}, \bar{1})$$

(the order of the elements in the family is not relevant). The empty family does not generate $\mathbb{Z}_2$. The second and the last families are not independent because they contain the zero vector which is never independent (do you see it?). Finally, the one element family $(\bar{1})$, while generating $\mathbb{Z}_2$, is not independent. Indeed,

$$2 \cdot \bar{1} = \bar{1} + \bar{1} = \bar{0}$$

is a zero linear combination with non-trivial coefficient 2. ◆

Definition 1.31 (Free Module). An R-module M is *free* if there exists a basis of M.

Example 1.32. If R is a field, then any R-vector space is a free R-module (there are no non-free R-modules in this case). ◆

Example 1.30 shows that not all modules are free. On the other hand, if X is any set, then Example 1.29 shows that there exists a free module RX together with an injection $\chi : X \to RX$, $x \mapsto \chi_x$ such that $(\chi_x)_{x \in X} \subseteq RX$ is a basis for RX. In particular, X identifies (via χ) with a basis of RX.

Similarly as for finite dimensional vector spaces, in a free module, it makes sense the notion of *components* (or *coordinates*) of an element in a basis. To see this, let R be any ring, and let M be a free R-module. Fix a basis $B = (q_i)_{i \in I} \subseteq M$ of M. In the case when B is infinite, formalizing the idea of coordinates is a little bit harder than for finite bases in a vector space. Let $p \in M$. As B is a set of generators for M, then p can be written as a (finite) linear combination of elements of B. In other words,

$$p = \sum_{i \in I} a_i q_i \qquad (1.8)$$

for some family $(a_i)_{i \in I} \subseteq R$ of scalars such that $a_i = 0$ for all but finitely many $i \in I$. The linear combination (1.8) defines a map $a : B \to R$, $q_i \mapsto a_i$. Notice that $a \in RB$ and it is uniquely defined by p, i.e., if $a' : B \to R$, $q_i \mapsto a_i'$ is another map in RB such that

$$p = \sum_{i \in I} a_i' q_i,$$

then necessarily $a = a'$, indeed

$$0 = \sum_{i \in I} a_i q_i - \sum_{i \in I} a_i' q_i = \sum_{i \in I} (a_i - a_i') q_i$$

and, from the independence, $a_i - a_i' = 0$, i.e., $a_i = a_i'$, for all $i \in I$. The a_i are the *components* of p in the basis B. In this way, we constructed a map

$$c_B : M \to RB, \quad p \mapsto c_B(p) := a \qquad (1.9)$$

called the *coordinate map*. The coordinate map is injective, indeed if two elements p_1, p_2 have the same components, they clearly agree. It is also surjective as, given a function $a \in RB$, $a : q_i \mapsto a_i$, the linear combination $p = \sum_{i \in I} a_i q_i$ is well defined and, by construction, its components are exactly the a_i. Finally, it is easy to see that the coordinate map is R-linear (do you see it?). So, it is an isomorphism of R-modules. This shows that every free module M with a basis B is (canonically) isomorphic to the free module RB.

Proposition 1.33. *Every free and finitely generated module possesses a finite basis.*

Proof.   Let M be a free R-module and let $S = (p_1, \ldots, p_n)$ be a finite set of generators. Pick any basis $\overline{B} = (q_i)_{i \in I}$ of M. Every p_α is a linear combination of finitely many q_i, i.e., for every $\alpha = 1, \ldots, n$, there is a finite subset $I_\alpha \subseteq I$ such that p_α is a linear combination of the q_j with $j \in I_\alpha$. The set $J = \bigcup_{\alpha=1}^n I_\alpha \subseteq I$ is a finite subset, hence the family $B := (q_j)_{j \in J} \subseteq \overline{B}$ is a finite subfamily. As $\overline{B}$ is independent, B is independent as well, and as S is a set of generators, $\overline{B}$ is a set of generators (do you see it?). We conclude that B is a finite basis as desired. $\square$

Free modules are characterized by a *universal property* that we now discuss. We begin axiomatizing this universal property. Let X be a set.

Definition 1.34 (Free Module Spanned by a Set). A *free module spanned by* X is a pair (M, χ) consisting of an R-module M and a map $\chi : X \to M$ with the following *universal property*: for every R-module N and every map $\phi : X \to N$, there exists a unique R-module homomorphism $f : M \to N$ such that $\phi = f \circ \chi$, i.e., the diagram

$$
\begin{array}{ccc}
X & \xrightarrow{\ \phi\ } & N \\
{\scriptstyle \chi}\big\downarrow & \diagup & \\
M & {\scriptstyle f} &
\end{array}
$$

commutes.

Example 1.35. Let $R = \mathbb{K}$ be a field, and let V be a $\mathbb{K}$-vector space of finite dimension n. Moreover, let $B = (q_1, \ldots, q_n) \subseteq V$ be a basis and let $i_B : B \to V, q_i \mapsto q_i$ be the inclusion. The *Linear Extension Theorem* (Axler, 2024, Statement 3.4) then shows that (V, i_B) is a free $\mathbb{K}$-module spanned by B.   ♦

Theorem 1.36 (Universal Property of Free Modules). *Let R be a ring and let X be a set. Then*

(1) *there exists a free module spanned by X;*
(2) *the free module spanned by X is unique up to unique isomorphisms, i.e., if $(M_1, \chi_1), (M_2, \chi_2)$ are free modules spanned by X, then there exists a unique R-module isomorphism $\Phi : M_1 \to M_2$ such that the diagram*

$$
\begin{array}{ccc}
 & X & \\
{\scriptstyle \chi_1}\diagup & & \diagdown{\scriptstyle \chi_2} \\
M_1 & \xrightarrow{\ \ \Phi\ \ } & M_2
\end{array}
\tag{1.10}
$$

commutes.

Proof. For item (1), consider the module RX and the map $\chi : X \to RX$, $x \mapsto \chi_x$ that maps x to the corresponding characteristic function. We want to show that (RX, χ) is a free module spanned by X. As we did in Example 1.29, we fix once for all an index set I with the same cardinality as X and a bijection $I \to X$, $i \mapsto x_i$. In this way, we interpret X as a family $(x_i)_{i \in I}$. As suggested at the end of Example 1.29, we also denote χ_{x_i} simply by x_i, for all $i \in I$. With this notation, X becomes just a subset in RX and χ is just the inclusion $X \to RX$. Now, let N be another R-module and let $\phi : X \to N$ be a map. Define a map $f : RX \to N$ as follows. For $a = \sum_{i \in I} a_i x_i \in RX$, put

$$f(a) = \sum_{i \in I} a_i \phi(x_i). \tag{1.11}$$

As only finitely many of the a_i are non-zero, $f(a)$ is well defined. We want to show that

(i) f is an R-linear map;
(ii) $f \circ \chi = \phi$;
(iii) f is uniquely determined by conditions (i), (ii).

We leave item (i) as Exercise 1.11 and prove items (ii) and (iii) here. So, let $i \in I$ and compute

$$(f \circ \chi)(x_i) = f(x_i) = \phi(x_i),$$

and (ii) follows from the arbitrariness of i. For (iii), let $f' : RX \to N$ be another linear map such that $f' \circ \chi = \phi$. Take $a \in RX$. Then,

$$a = \sum_{i \in I} a_i x_i$$

for some family $(a_i)_{i \in I} \subseteq R$ such that $a_i = 0$ for all but finitely many i. Hence,

$$f'(a) = f'\left(\sum_{i \in I} a_i x_i\right) = \sum_{i \in I} a_i f'(x_i) = \sum_{i \in I} a_i (f' \circ \chi)(x_i) = \sum_{i \in I} a_i \phi(x_i) = f(a),$$

and (iii) follows from the arbitrariness of a. This concludes the proof of item (1).

For item (2), let $(M_1, \chi_1), (M_2, \chi_2)$ be free modules spanned by X. In particular, $\chi_2 : X \to M_2$ is a map and, as (M_1, χ_1) satisfies the universal property of free modules spanned by X, there exists a unique linear map $\Phi : M_1 \to M_2$ such that diagram (1.10) commutes. We want to show that Φ is an R-module isomorphism. To do this, note that, exchanging the roles of (M_1, χ_1) and (M_2, χ_2) we find another linear map $\Psi : M_2 \to M_1$ such that $\Psi \circ \chi_2 = \chi_1$. It is easy to see that Ψ inverts Φ. Indeed, consider the linear map $\Psi \circ \Phi : M_1 \to M_1$. It satisfies

$$\Psi \circ \Phi \circ \chi_1 = \Psi \circ \chi_2 = \chi_1.$$

However, there is only one linear map $I : M_1 \to M_1$ such that $I \circ \chi_1 = \chi_1$. As $\mathrm{id}_{M_1} : M_1 \to M_1$ is another such linear map, we necessarily have $\Psi \circ \Phi = \mathrm{id}_{M_1}$. Similarly, $\Phi \circ \Psi = \mathrm{id}_{M_2}$ and this concludes the proof. $\square$

Exercise 1.11. Prove that the map f defined by (1.11) in the proof of Theorem 1.36 is R-linear.

Theorem 1.36 says that, for any set X, a free module spanned by X exists and it is unique up to (unique) isomorphisms. The proof shows that (RX, χ) is a canonical choice of a free module spanned by X. For this reason, we also say that (RX, χ) (or simply RX) is *the* free module spanned by X.

The following corollary shows that free modules are completely characterized by the defining property of the free module spanned by a set and motivates the terminology used in Definition 1.34.

Corollary 1.37. *Let M be a free module with basis $B = (q_i)_{i \in I} \subseteq M$, and let $i_B : B \to M, q_i \mapsto q_i$ be the inclusion. Then, (M, i_B) is a free module spanned by B. Conversely, if $X = (x_i)_{i \in I}$ is a set and (M, χ) is a free module spanned by X, then $\chi : X \to M$ is an injective map, and M is a free module with basis $(\chi_{x_i})_{i \in I}$.*

Proof. For the first part of the statement, consider the coordinate map $c_B : M \to RB$. It is an R-module isomorphism that identifies the inclusion $i_B : B \to M$ with the canonical injection $\chi : B \to RB$ (do you see it?). As (RB, χ) is a free module spanned by B, it easily follows that (M, i_B) is also a free module spanned by B (check the details as an exercise). For the second part of the statement, note that, from uniqueness, there is an R-module isomorphism $\Phi : RX \to M$ that identifies $\chi : X \to M$ with the canonical injection $X \to RX$. Hence, $\chi : X \to M$ is also an injection. As X is a basis in RX and R-module isomorphisms map bases to bases (see Exercise 1.10), then $(\chi_{x_i})_{i \in I}$ is a basis in M. $\square$

Remark 1.38. The proof of Corollary 1.37 was very quick. We invite the reader to check all the details and fill in the possible gaps. $\Diamond$

Proposition 1.39. *Every module M is (isomorphic to) the quotient of a free module P (over an appropriate submodule). If M is finitely generated, then P can be chosen to have a finite basis.*

Proof. Let $S = (p_i)_{i \in I} \subseteq M$ be a family of generators (there is always a family of generators: at the worse, one can choose $S = M$). Consider the free module RS spanned by S. By the universal property of free modules, there is a linear map $f : RS \to M$ such that $f \circ \chi : S \to M$ is the inclusion of S into M. In other words, f maps χ_{p_i} to p_i. As RS is spanned by $(\chi_{p_i})_{i \in I}$, this shows that f maps a family of generators to a family of generators. It follows that it is a surjective linear map (Exercise 1.10). Hence, from Corollary 1.23, it induces an R-module isomorphism $RS/\ker f \cong M$. If M is finitely generated, then S can be chosen finite so that RS has a finite basis. □

1.3 Direct Sums and Direct Products

Let R be a ring and let $(M_i)_{i \in I}$ be a family of R-modules parameterized by a possibly infinite index set I. Roughly, the direct sum of the M_i is the "smallest module containing the M_i as *independent* submodules". The correct way to formalize this idea is via a universal property similar to that of the free module spanned by a set.

Definition 1.40 (Direct Sum). A *direct sum* of the modules $(M_i)_{i \in I}$ is a pair (D, ι) consisting of an R-module D and a family $\iota = (\iota_i : M_i \to D)_{i \in I}$ of R-linear maps with the following *universal property*: for every R-module M and every family $\lambda = (\lambda_i : M_i \to M)_{i \in I}$ of linear maps, there exists a unique R-module homomorphism $\lambda_D : D \to M$ such that $\lambda_i = \lambda_D \circ \iota_i$ for all $i \in I$, i.e., the diagram

$$
\begin{array}{ccc}
M_i & \xrightarrow{\lambda_i} & M \\
{\scriptstyle \iota_i} \downarrow & \nearrow {\scriptstyle \lambda_D} & \\
D & &
\end{array}
$$

commutes for all $i \in I$.

Direct sums are sometimes called *exterior direct sums* to distinguish them from the *direct sum of submodules* in a given modules (which is a closely related construction generalizing the direct sum of vector subspaces (Axler, 2024, Definition 1.41) in a vector space in the obvious way).

Theorem 1.41. *Let $(M_i)_{i \in I}$ be a family of R-modules. Then*

(1) *there exists a direct sum (D, ι) of $(M_i)_{i \in I}$;*
(2) *direct sums are* unique up to unique isomorphisms, *i.e., if $(D_1, \iota_1 = (\iota_{1i})_{i \in I}), (D_2, \iota_2 = (\iota_{2i})_{i \in I})$ are two direct sums of $(M_i)_{i \in I}$, then there*

exists a unique R-module isomorphism $\Phi : D_1 \to D_2$ such that the diagram

$$(1.12)$$

commutes for all $i \in I$.

Proof. For item (1), define D as the set of families $p = (p_i)_{i \in I}$ such that $p_i \in M_i$ for all $i \in I$, with the additional (extremely important!) assumption that $p_i = 0$ for all but finitely many i. There is a natural R-module structure on D: for all $p = (p_i)_{i \in I}, p' = (p'_i)_{i \in I}, q = (q_i)_{i \in I} \in D$, and $a \in R$, we put

$$p + p' := (p_i + p'_i)_{i \in I} \quad \text{and} \quad aq := (aq_i)_{i \in I}.$$

We leave it to the reader to check that with this two operations D is indeed an R-module. For each $i \in I$, there is a map

$$\iota_i : M_i \to D, \quad p \mapsto \iota_i(p) := (p_j)_{j \in I}, \quad \text{with } p_j = \begin{cases} p & \text{if } j = i \\ 0 & \text{otherwise} \end{cases}.$$

This map is obviously R-linear (do you see it? If not, check the necessary details). Put $\iota = (\iota_i : M_i \to D)_{i \in I}$. We want to check that (D, ι) satisfies the universal property of direct sums. So, let M be another R-module, and let $\lambda = (\lambda_i : M_i \to M)_{i \in I}$ be a family of linear maps. Define a map $\lambda_D : D \to M$ by putting

$$\lambda_D(p) := \sum_{i \in I} \lambda_i(p_i), \quad \text{for all } p = (p_i)_{i \in I} \in D. \qquad (1.13)$$

Note that the sum in (1.13) is well defined because the p_i are all zero but finitely many. It is easy to see that λ_D is a linear map (check the details as an exercise). Now, take $j \in I$ and $p \in M_j$, then all entries of $\iota_j(p)$ vanish except the jth one which is equal to p. Hence, we have

$$\lambda_D(\iota_j(p)) = \sum_{i \in I} \lambda_i \,(i\text{th entry of } \iota_j(p)) = \lambda_j(p)$$

as desired. To conclude with item (1), we have to show that λ_D is uniquely determined by R-linearity and the condition $\lambda_D \circ \iota_i = \lambda_i$ for all i. So, let $\lambda'_D : D \to M$ be another linear map such that $\lambda'_D \circ \iota_i = \lambda_i$, then $\lambda'_D = \lambda_D$.

Indeed, for any $p = (p_i)_{i \in I} \in D$, there are only finitely many i such that $p_i \neq 0$. Denote them $i_1, \ldots, i_r$. It should be clear that

$$p = \sum_{k=1}^{r} \iota_{i_k}(p_{i_k}). \tag{1.14}$$

Hence,

$$\lambda'_D(p) = \lambda'_D \left(\sum_{k=1}^{r} \iota_{i_k}(p_{i_k}) \right) = \sum_{k=1}^{r} \lambda'_D \left(\iota_{i_k}(p_{i_k}) \right) = \sum_{k=1}^{r} \lambda_{i_k}(p_{i_k})$$

$$= \sum_{k=1}^{r} \lambda_D \left(\iota_{i_k}(p_{i_k}) \right) = \lambda_D \left(\sum_{k=1}^{r} \iota_{i_k}(p_{i_k}) \right) = \lambda_D(p),$$

where we used the linearity of both λ_D, λ'_D. This concludes the proof of item (1).

For item (2), let $(D_1, \iota_1), (D_2, \iota_2)$ be direct sums of $(M_i)_{i \in I}$. In particular, $\iota_2 = (\iota_{2i} : M_i \to D_2)_{i \in I}$ is a family of linear maps and, as (D_1, ι_1) satisfies the universal property of the direct sum, there exists a unique linear map $\Phi : D_1 \to D_2$ such that Diagram (1.12) commutes for all $i \in I$. We want to show that Φ is an R-module isomorphism. To do this, note that, exchanging the roles of (D_1, ι_1) and (D_2, ι_2), we find another linear map $\Psi : D_2 \to D_1$ such that $\Psi \circ \iota_{2i} = \iota_{1i}$ for all $i \in I$. It is easy to see that Ψ inverts Φ and we leave the details to the reader as Exercise 1.12. This concludes the proof. □

> **Exercise 1.12.** Fill all the gaps in the proof of Theorem 1.41. In particular, show that the homomorphisms Φ, Ψ in the end of the proof are mutual inverses (**Hint:** *Use the same exact argument as in the end of the proof of Theorem* 1.36).

Theorem 1.41 shows that direct sums exist and are unique up to unique isomorphisms. The direct sum (D, ι) constructed in the proof is a canonical choice. For this reason, we call it *the direct sum* of $(M_i)_{i \in I}$ and denote it by

$$\bigoplus_{i \in I} M_i$$

(or also $M_1 \oplus \cdots \oplus M_k$, if the M_i are finitely many). Note from the proof of Theorem 1.41 that the maps $\iota_j : M_j \to \bigoplus_{i \in I} M_i$ are injective, and we often use them to identify the M_j with their images in $\bigoplus_{i \in I} M_i$. If we do this, then, in view of (1.14), any element in $\bigoplus_{i \in I} M_i$ can be seen as a finite sum of the type

$$\sum_{k=1}^{r} p_{i_k}$$

with $p_{i_k} \in M_{i_k}$ for some $i_1, \ldots, i_r \in I$.

It is some times convenient to consider direct sums different from the canonical choice.

Example 1.42 (Free Modules as Direct Sums). Let $X = (x_i)_{i \in I}$ be a set (interpreted as a family as usual). The free module RX spanned by X can also be seen as the direct sum

$$\bigoplus_{i \in I} R$$

of a family $(M_i = R)_{i \in I}$ of copies of R. Indeed, the map

$$\bigoplus_{i \in I} R \to RX, \quad (a_i)_{i \in I} \mapsto \sum_{i \in I} a_i x_i$$

is clearly an R-module isomorphism (a canonical one). So, we have already three interpretations of elements of RX:

(1) as functions $a : X \to R$, $x_i \mapsto a_i$,
(2) as formal linear combinations $\sum_{i \in I} a_i x_i$,
(3) as families $(a_i)_{i \in I}$,

with $x_i \in X$, and $a_i \in R$ such that $a_i = 0$ for all but finitely many $i \in I$. This is not surprising: each of these interpretations is just a different way of encoding the same piece of information. What the best interpretation is might depend on the concrete situation at hand. ◆

Example 1.43 (Split Short Exact Sequence). Consider a short exact sequence of R-modules:

$$0 \longrightarrow N \xrightarrow{\alpha} M \xrightarrow{\beta} Q \longrightarrow 0. \tag{1.15}$$

We say that Sequence (1.15) *splits* if there is an R-linear map $s : Q \to M$ inverting β on the right: $\beta \circ s = \mathrm{id}_Q$. In this case, s is called a *splitting* of the sequence.

Not all short exact sequences split. For instance, let $R = \mathbb{Z}$ and consider the short exact sequence of abelian groups:

$$0 \longrightarrow \mathbb{Z} \longrightarrow \mathbb{Z} \longrightarrow \mathbb{Z}_2 \longrightarrow 0 \tag{1.16}$$
$$n \longmapsto 2n,$$

where the arrow $\mathbb{Z} \to \mathbb{Z}_2$ is the canonical projection. As there is no non-trivial abelian group homomorphism $\mathbb{Z}_2 \to \mathbb{Z}$ (let $f : \mathbb{Z}_2 \to \mathbb{Z}$ be a $\mathbb{Z}$-linear map, then $0 = f(\overline{0}) = f(\overline{1} + \overline{1}) = f(\overline{1}) + f(\overline{1}) = 2f(\overline{1})$, hence $f(\overline{1}) = f(\overline{0}) = 0$), the short exact sequence (1.16) cannot split.

However, some short exact sequences split. For instance, if $R = \mathbb{K}$ is a field, then any short exact sequence of R-modules, i.e., $\mathbb{K}$-vector spaces, splits. Indeed, let

$$0 \longrightarrow W \xrightarrow{\alpha} V \xrightarrow{\beta} U \longrightarrow 0$$

be such a short exact sequence. Consider the image $\alpha(W) \subseteq V$ of W in V. It is a vector subspace. Hence, there is another vector subspace $U' \subseteq V$ such that $\alpha(V) + U' = \alpha(V) \oplus U' = V$ (usual direct sum of vector subspaces). To see this, choose a basis B_W of W, then $\alpha(B_W)$ is a basis of $\alpha(W)$. Complete it to a basis B of V adding an appropriate subset $B' \subseteq V$ of vectors (this is always possible), and let $U' \subseteq V$ be the subspace spanned by B' (if you are not following, check all the details in the finite dimensional case). The restriction $\beta|_{U'} : U' \to U$ is a vector space isomorphism. Indeed, it is injective: let $u' \in U'$ be such that $0 = \beta|_{U'}(u') = \beta(u')$. This means that $u' \in \ker \beta = \operatorname{im} \alpha = \alpha(W)$. Hence, $u' \in U' \cap \alpha(W) = 0$. It is also surjective: indeed, let $u \in U$. From the surjectivity of β there exists $v \in V$ such that $u = \beta(v)$. As $\alpha(W) + U' = V$, there exists $w \in W$ and $u' \in U'$ such that $v = \alpha(w) + u'$. Hence, $u = \beta(v) = \beta(\alpha(w) + u') = (\beta \circ \alpha)(w) + \beta(u') = \beta|_{U'}(u')$, where we used that $\beta \circ \alpha = 0$. Finally, put $s = i_{U'} \circ \beta|_{U'}^{-1} : U \to V$, where $i_{U'} : U' \to V$ is the inclusion. Then s is the splitting we were looking for (see Figure 1.1). Indeed,

$$\beta \circ s = \beta \circ i_{U'} \circ \beta|_{U'}^{-1} = \beta|_{U'} \circ \beta|_{U'}^{-1} = \operatorname{id}_U$$

as desired.

The latter example suggests that, given a short exact sequence of R-modules together with a splitting

$$0 \longrightarrow N \xrightarrow{\alpha} M \underset{s}{\overset{\beta}{\longrightarrow}} Q \longrightarrow 0 \,, \tag{1.17}$$

the middle module M might be a direct sum of the other two $M \cong N \oplus Q$. This is indeed the case. More precisely, the pair (M, ι) is a direct sum of

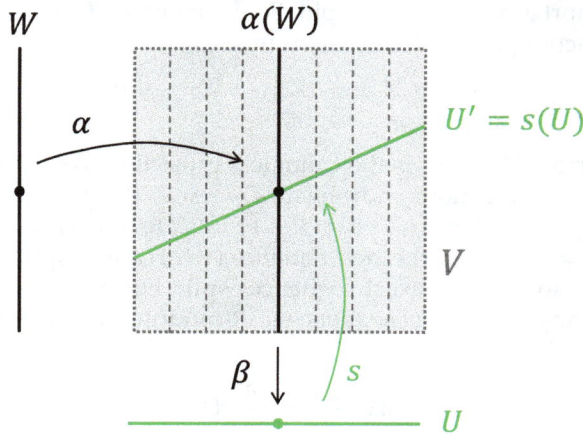

Figure 1.1.   A splitting of a short exact sequence of vector spaces.

the pair of modules (N, Q) with ι being the pair of linear maps

$$\iota = \Big(\alpha : N \to M, \; s : Q \to M\Big).$$

To see this, it is enough to check that the linear map

$$\Phi : N \oplus Q \to M, \quad (p, q) \mapsto \alpha(p) + s(q), \tag{1.18}$$

is a module isomorphism (such that $\Phi \circ \iota_N = \alpha$ and $\Phi \circ \iota_Q = s$, where $\iota_N : N \to N \oplus Q$ and $\iota_Q : Q \to N \oplus Q$ are the canonical monomorphisms). We leave the details to the reader as Exercise 1.13. ◆

> **Exercise 1.13.** Prove that, given a short exact sequence of R-modules together with a splitting as in (1.17), the middle module M together with the linear maps (α, s) is a direct sum of (N, Q) (**Hint:** *Prove that the map* (1.18) *is a module isomorphism such that* $\Phi \circ \iota_N = \alpha$ *and* $\Phi \circ \iota_Q = s$. *Why is this enough to solve the exercise?*).

There is a construction somehow "dual" to direct sums, called *direct product*. The notion of direct product is obtained from that of direct sum by inverting all the arrows in the definition. Namely, let $(M_i)_{i \in I}$ be a family of R-modules as above.

Definition 1.44 (Direct Product). A *direct product* of the modules $(M_i)_{i \in I}$ is a pair (P, π) consisting of an R-module P and a family $\pi = (\pi_i : P \to M_i)_{i \in I}$ of R-linear maps with the following *universal property*: for every

R-module M and every family $\mu = (\mu_i : M \to M_i)_{i \in I}$ of linear maps, there exists a unique R-module homomorphism $\mu_P : M \to P$ such that $\mu_i = \pi_i \circ \mu_P$ for all $i \in I$, i.e., the diagram

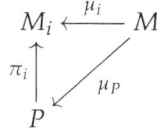

$$
\begin{array}{ccc}
M_i & \xleftarrow{\ \mu_i\ } & M \\
\pi_i \uparrow & \nearrow \mu_P & \\
P & &
\end{array}
$$

commutes for all $i \in I$.

Theorem 1.45. *Let $(M_i)_{i \in I}$ be a family of R-modules. Then*

(1) *there exists a direct product (P, π) of $(M_i)_{i \in I}$;*
(2) *direct products are unique up to unique isomorphisms, i.e., if $(P_1, \pi_1 = (\pi_{1i})_{i \in I})$, $(P_2, \pi_2 = (\pi_{2i})_{i \in I})$ are two direct products of $(M_i)_{i \in I}$, then there exists a unique R-module isomorphism $\Psi : P_1 \to P_2$ such that the diagram*

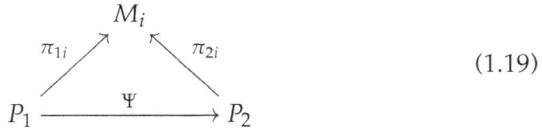

$$
\begin{array}{ccc}
 & M_i & \\
\pi_{1i} \nearrow & & \nwarrow \pi_{2i} \\
P_1 & \xrightarrow{\ \Psi\ } & P_2
\end{array}
$$

(1.19)

commutes for all $i \in I$.

Proof. For item (1), define P as the set of families $p = (p_i)_{i \in I}$ such that $p_i \in M_i$ for all $i \in I$, with no other assumption (beware the difference with the direct sum case!). There is a natural R-module structure on P: for all $p = (p_i)_{i \in I}, p' = (p'_i)_{i \in I}, q = (q_i)_{i \in I} \in P$, and $a \in R$, we put

$$p + p' := (p_i + p'_i)_{i \in I} \quad \text{and} \quad aq := (aq_i)_{i \in I}.$$

We leave it to the reader to check that with these two operations P is indeed an R-module (note that the module D constructed in the proof of Theorem 1.41 is then obviously a submodule of P. However, this fact will not play any role in what follows!). For each $i \in I$, there is a map

$$\pi_i : P \to M_i, \quad p = (p_j)_{j \in I} \mapsto \pi_i(p) := p_i.$$

In other words, $P = \times_{j \in I} M_j$, the cartesian product of the M_i, and π_i is the projection onto the ith factor. The map π_i is obviously R-linear for all i (do you see it?). Put $\pi = (\pi_i : P \to M_i)_{i \in I}$. We want to check that (P, π) satisfies the universal property of direct products. So, let M be another

R-module, and let $\mu = (\mu_i : M \to M_i)_{i \in I}$ be a family of linear maps. Define a map $\mu_P : M \to P$ by putting

$$\mu_P(q) := (\mu_i(q))_{i \in I}, \quad \text{for all } q \in M. \tag{1.20}$$

It is easy to see that μ_P is a linear map (check the details as an exercise). Now, take $q \in M$ and $j \in I$ and compute

$$\pi_j(\mu_P(q)) = \pi_j (\mu_i(q))_{i \in I} = \mu_j(q)$$

as desired. To conclude with item (1), we have to show that μ_P is uniquely determined by R-linearity and the condition $\pi_i \circ \mu_P = \mu_i$ for all i. We leave the easy check to the reader as part of Exercise 1.14.

For item (2), let $(P_1, \pi_1), (P_2, \pi_2)$ be direct products of $(M_i)_{i \in I}$. In particular, $\pi_1 = (\pi_{1i} : P_1 \to M_i)_{i \in I}$ is a family of linear maps and, as (P_2, π_2) satisfies the universal property of the direct sum, there exists a unique linear map $\Psi : P_1 \to P_2$ such that Diagram (1.19) commutes for all $i \in I$. We want to show that Φ is an R-module isomorphism. To do this, note that, exchanging the roles of (P_1, π_1) and (P_2, π_2), we find another linear map $\Phi : P_2 \to P_1$ such that $\pi_{1i} \circ \Phi = \pi_{2i}$ for all $i \in I$. It is easy to see that Ψ inverts Φ and we leave the details to the reader as the second part of Exercise 1.14. This concludes the proof. $\qquad\square$

Exercise 1.14. Fill all the gaps in the proof of Theorem 1.45. In particular, show that

(1) the map μ_P is uniquely determined by R-linearity and the condition $\pi_i \circ \mu_P = \mu_i$ for all i,
(2) the homomorphisms Φ, Ψ in the end of the proof are mutual inverses.

Theorem 1.45 shows that direct products exist and are unique up to unique isomorphisms. The direct product (P, π) constructed in the proof is a canonical choice. For this reason, we call it *the direct product* of $(M_i)_{i \in I}$ and denote it by

$$\prod_{i \in I} M_i.$$

Note from the proof of Theorem 1.45 that the maps $\pi_j : \prod_{i \in I} M_i \to M_j$ are surjective.

It is sometimes convenient to consider direct products different from the canonical choice.

Example 1.46 (Function Modules as Direct Products). Let $X = (x_i)_{i \in I}$ be a set (interpreted as a family as usual). The module R^X of functions $X \to R$ can also be seen as the direct product

$$\prod_{i \in I} R$$

of a family $(M_i = R)_{i \in I}$ of copies of R. Indeed, the map

$$R^X \to \prod_{i \in I} R, \quad f \mapsto (f(x_i))_{i \in I}$$

is clearly an R-module isomorphism (a canonical one). So, we have two interpretations of elements of R^X:

(1) as functions $f : X \to R$,
(2) as families $(f_i)_{i \in I}$,

with $f_i \in X$, and no other constraint. Yet another interpretation is provided in the following section (see Example 1.51). $\blacklozenge$

Example 1.47 (Finite Direct Products). Let $(M_1, \dots, M_k)$ be a finite family of R-modules. It is clear that, in this case, the direct sum and the direct product of the M_i are actually isomorphic as R-modules, and they are often denoted both by $M_1 \oplus \cdots \oplus M_k$. Note, however, that the direct sum D comes, by definition, with maps $M_i \to D$, while the direct product P comes with maps $P \to M_i$, so D and P, even in this case, should be thought of as two distinct mathematical objects. $\blacklozenge$

1.4 Tensor Products

In this section, we discuss a new construction with modules that plays an important role in algebra and geometry: the *tensor product*. Roughly, the tensor product of two modules M_1, M_2 is a new module $M_1 \otimes M_2$ with the key property that homomorphisms $M_1 \otimes M_2 \to N$ (to a third arbitrary module N) are "the same as" bilinear maps $M_1 \times M_2 \to N$. In this sense, the tensor product "represents" bilinear maps.

We begin with a discussion on *multilinear maps* in the setting of modules. So, let R be a ring, and let $M_1, \dots, M_k$ and N be R-modules.

Definition 1.48 (Multilinear Map). A *k-multilinear map* (defined on $M_1, \dots, M_k$ and with values in N) is a map

$$\mu : M_1 \times \cdots \times M_k \to N$$

which is R-linear in each entry, i.e., for all $i = 1, \ldots, k$,

$$\mu(\ldots, \underbrace{ap + bq}_{i\text{th place}}, \ldots)$$

$$= a\mu(\ldots, \underbrace{p}_{i\text{th place}}, \ldots) + b\mu(\ldots, \underbrace{q}_{i\text{th place}}, \ldots), \quad p, q \in M_i, \quad a, b \in R.$$

Example 1.49. Let $R = \mathbb{K}$ be a field and let n be a positive integer. The determinant

$$\det : \underbrace{\mathbb{K}^n \times \cdots \times \mathbb{K}^n}_{n\text{-times}} \to \mathbb{K}, \quad (A^{(1)}, \ldots, A^{(n)}) \mapsto \det(A^{(1)} \cdots A^{(n)})$$

is a multilinear map.   ◆

We have the following multilinear map analogue of Corollary 1.37.

Proposition 1.50 (Multilinear Extension Theorem). *Let $M_1, \ldots, M_k$ be free R-modules with bases $B_1, \ldots, B_k$, respectively. Then for any R-module N and any map $m : B_1 \times \cdots \times B_k \to N$, there exists a unique multilinear map $\mu : M_1 \times \cdots \times M_k \to N$ such that the diagram*

$$
\begin{array}{ccc}
M_1 \times \cdots \times M_k & \xrightarrow{\ \mu\ } & N \\
\uparrow & \nearrow{\scriptstyle m} & \\
B_1 \times \cdots \times B_k & &
\end{array}
$$

commutes, where the vertical arrow is the inclusion.

Proof.   Left as Exercise 1.15.   □

Exercise 1.15. Prove Proposition 1.50.

    Proposition 1.50 says two things: (1) in the existence, it says that if we fix a map on the bases then this can be extended multilinearly to the whole modules; (2) in the uniqueness, it says that a multilinear map is uniquely determined by its action on the basis elements.

    Multilinear maps $M_1 \times \cdots \times M_k \to N$ can be added and multiplied by a scalar as follows. For any two multilinear maps $\mu, \nu : M_1 \times \cdots \times M_k \to N$, we define

$$\mu + \nu : M_1 \times \cdots \times M_k \to N,$$

by putting

$$(\mu + \nu)(p_1, \ldots, p_k) := \mu(p_1, \ldots, p_k) + \nu(p_1, \ldots, p_k).$$

For any multilinear map $\mu : M_1 \times \cdots \times M_k \to N$ and any scalar $a \in R$, we define

$$a\mu : M_1 \times \cdots \times M_k \to N,$$

by putting

$$(a\mu)(p_1, \ldots, p_k) := a\mu(p_1, \ldots, p_k).$$

It is easy to see that $\mu + \nu$ and $a\mu$ are again multilinear maps. With these two operations, the space of multilinear maps $M_1 \times \cdots \times M_k \to N$ is an R-module that we denote by

$$\text{Mult}_R^k(M_1, \ldots, M_k; N),$$

or simply $\text{Mult}^k(M_1, \ldots, M_k; N)$ if it is clear which is the ring of scalars.

We also make sense of 0-multilinear maps by putting $\text{Mult}^0(N) := N$. When $k = 1$, a multilinear map $M_1 \to N$ is just a linear map. The R-module $\text{Mult}_R^1(M_1; N)$ is also denoted $\text{Hom}_R(M_1, N)$, or simply $\text{Hom}(M_1, N)$. Multilinear maps $M_1 \times M_2 \to N$ are also called *bilinear maps* and the R-module $\text{Mult}_R^2(M_1, M_2; N)$ is also denoted $\text{Bil}_R(M_1, M_2; N)$, or simply $\text{Bil}(M_1, M_2; N)$.

Example 1.51 (Dual of a Free Module). For an R-module M, denote $M^* := \text{Hom}(M, R)$ and call it the *dual module* of M. It is indeed a module with the operations on (multi)linear maps described above. Now let X be a set. We want to show that the dual $(RX)^*$ of the free module RX generated by X is canonically isomorphic to the function module R^X. To see this, interpret X as a family as usual: $X = (x_i)_{i \in I}$, and define a map

$$\iota : R^X \to (RX)^*$$

as follows. For every $f \in R^X$ and every

$$p = \sum_{i \in I} a_i x_i \in RX,$$

put

$$\iota(f)(p) = \sum_{i \in I} a_i f(x_i) \in R.$$

In other words, $\iota(f) : RX \to R$ is the unique linear map that acts as f on the basis $X \subseteq RX$. As all the a_i are zero but finitely many, $\iota(f)(p)$ is well-defined. It is easy to see that $\iota(f) : RX \to R$ is indeed a linear map for every $f \in R^X$. It is also easy to see that ι is a linear map. It remains to show that ι is bijective. It is injective. Indeed, if $f \in R^X$ is such that $\iota(f) = 0$, in particular $f(x_i) = 0$ for all $i \in I$, i.e., $f(x) = 0$ for all $x \in X$, i.e., $f = 0$. For the surjectivity, let $\varphi \in (RX)^*$ and define $f : X \to R$ by putting $f(x) = \varphi(x)$ for all $x \in X \subseteq RX$. It is now clear that $\iota(f) = \varphi$.

The isomorphism ι provides yet a third interpretation for elements in R^X, namely as linear forms on RX. ◆

Example 1.52. The present example simultaneously generalizes Examples 1.46 and 1.51. Let $(M_i)_{i \in I}$ be a family of R-modules, indexed by some (possibly infinite) set I. We want to show that the dual $(\oplus_{i \in I} M_i)^*$ of the direct sum $\oplus_{i \in I} M_i$ is canonically isomorphic to the direct product $\prod_{i \in I} M_i^*$:

$$\left(\bigoplus_{i \in I} M_i \right)^* \cong \prod_{i \in I} M_i^*.$$

To see this, define a map

$$\Phi : \left(\bigoplus_{i \in I} M_i \right)^* \to \prod_{i \in I} M_i^*$$

as follows. For every linear form

$$\varphi : \bigoplus_{i \in I} M_i \to R$$

and every $j \in I$, let $\varphi_j := \varphi|_{M_j} : M_j \to R$ be the linear form obtained by restricting φ to the submodule $M_j \subseteq \oplus_{i \in I} M_i$ (equivalently, $\varphi_j = \varphi \circ \iota_j$, where $\iota_j : M_j \to \oplus_{i \in I} M_i$ is the jth structure monomorphism of the direct sum). Now put

$$\Phi(\varphi) := (\varphi_i)_{i \in I} \in \prod_{i \in I} M_i^*.$$

We leave it to the reader to check that Φ is indeed an R-module isomorphism as Exercise 1.16. ◆

Exercise 1.16. Show that the map $\Phi : (\oplus_{i \in I} M_i)^* \to \prod_{i \in I} M_i^*$ defined in Example 1.52 is an R-modules isomorphism.

Example 1.53. Let M, N, P be R-modules. As the composition of R-module homomorphisms is an R-module homomorphism, we get a map

$$\circ : \mathrm{Hom}_R(N, P) \times \mathrm{Hom}_R(M, N) \to \mathrm{Hom}_R(M, P), \quad (f, g) \mapsto f \circ g.$$

It is easy to see that this map is bilinear, i.e., for all $f_1, f_2 \in \mathrm{Hom}_R(N, P)$ all $a_1, a_2 \in R$ and all $g \in \mathrm{Hom}_R(M, N)$, we have

$$(a_1 f_1 + a_2 f_2) \circ g = a_1(f_1 \circ g) + a_2(f_2 \circ g),$$

and likewise with respect to the second argument. We leave the details to the reader as Exercise 1.17. ♦

Exercise 1.17. Prove that composing linear maps is a bilinear operation (see Example 1.53 for a precise formulation).

Proposition 1.54. *Let R be a ring, let k be a non-negative integer, and let $M_1, \ldots, M_k, N$ be R-modules. Then*

(1) *for every $l \leq k$, there is a canonical R-module isomorphism*

$$\mathrm{Mult}_R^l(M_1, \ldots, M_l; \mathrm{Mult}_R^{k-l}(M_{l+1}, \ldots, M_k; N))$$

$$\cong \mathrm{Mult}_R^k(M_1, \ldots, M_k; N);$$

(2) *there is a canonical R-module isomorphism*

$$\mathrm{Mult}_R^{k+1}(M_1, \ldots, M_k, R; N) \cong \mathrm{Mult}_R^k(M_1, \ldots, M_k; N);$$

(3) *for every permutation $\sigma \in S_k$, there is a canonical R-module isomorphism*

$$\mathrm{Mult}_R^k(M_{\sigma(1)}, \ldots, M_{\sigma(k)}; N) \cong \mathrm{Mult}_R^k(M_1, \ldots, M_k; N).$$

Proof. For item (1), define a map

$$\mathrm{Mult}_R^l(M_1, \ldots, M_l; \mathrm{Mult}_R^{k-l}(M_{l+1}, \ldots, M_k; N)) \to \mathrm{Mult}_R^k(M_1, \ldots, M_k; N)$$

$$\mu \mapsto \underline{\mu}$$

by putting

$$\underline{\mu}(p_1, \ldots, p_l, p_{l+1}, \ldots, p_k) := \mu(p_1, \ldots, p_l)(p_{l+1}, \ldots, p_k), \tag{1.21}$$

$p_i \in M_i$, $i = 1, \ldots, k$. It is clear that the assignment $\mu \mapsto \underline{\mu}$ is well defined and it is almost obvious that it is R-linear. Additionally, it is inverted by the assignment $\underline{\mu} \mapsto \mu$ defined by reading Formula (1.21) from the right to the left. This completes the proof of item (1).

For item (2), define a map

$$\text{Mult}_R^{k+1}(M_1, \ldots, M_k, R; N) \to \text{Mult}_R^k(M_1, \ldots, M_k; N), \quad \mu \mapsto \overline{\mu} \quad (1.22)$$

by putting

$$\overline{\mu}(p_1, \ldots, p_k) := \mu(p_1, \ldots, p_k, 1), \quad p_i \in M_i, \quad i = 1, \ldots k. \quad (1.23)$$

The assignment $\mu \mapsto \overline{\mu}$ is well defined and R-linear. We want to show that it is bijective. This essentially follows from the fact that an R-linear map defined on R itself is completely determined by its value on $1 \in R$ (as R is actually a free R-module with basis $\{1\}$). More specifically, define a map

$$\text{Mult}_R^k(M_1, \ldots, M_k; N) \to \text{Mult}_R^{k+1}(M_1, \ldots, M_k, R; N), \quad \nu \mapsto \widehat{\nu} \quad (1.24)$$

by putting

$$\widehat{\nu}(p_1, \ldots, p_k, a) := a\nu(p_1, \ldots, p_k), \quad p_i \in M_i, \quad i = 1, \ldots k. \quad (1.25)$$

This is again a well-defined R-linear map that inverts the map (1.22) defined by (1.23).

For item (3), define a map

$$\text{Mult}_R^k(M_{\sigma(1)}, \ldots, M_{\sigma(k)}; N) \to \text{Mult}_R^k(M_1, \ldots, M_k; N), \quad \mu \mapsto \mu_\sigma$$

by putting

$$\mu_\sigma(p_1, \ldots, p_k) := \mu(p_{\sigma(1)}, \ldots, p_{\sigma(k)}), \quad p_i \in M_i, \quad i = 1, \ldots k.$$

This is a well-defined R-linear map inverted by $\nu \mapsto \nu_{\sigma^{-1}}$. $\qquad\qquad \square$

Exercise 1.18. Show that the map (1.24) defined by (1.25) is well defined (i.e., $\widehat{\nu} : M_1 \times \cdots \times M_k \times R \to N$ is a multilinear map) and inverts the map (1.22) defined by (1.23) (i.e., $\widehat{\overline{\mu}} = \mu$ and $\overline{\widehat{\nu}} = \nu$ for all $\mu \in \text{Mult}_R^k(M_1, \ldots, M_k; N)$ and all $\nu \in \text{Mult}_R^{k+1}(M_1, \ldots, M_k, R; N)$).

It is clear that combining the isomorphisms (1), (2) and (3) in Proposition 1.54 we find numerous new canonical isomorphisms. Note that

- when $k = 0$, item (2) says that there is a canonical isomorphism

$$\text{Hom}_R(R, N) \xrightarrow{\cong} N,$$

and the proof of Proposition 1.54 reveals that this isomorphism is given by $\phi \mapsto \phi(1)$,

- when $k = 2$, item (1) says that there is a canonical isomorphism

$$\mathrm{Hom}_R(M_1, \mathrm{Hom}_R(M_2, N)) \xrightarrow{\cong} \mathrm{Bil}_R(M_1, M_2; N), \quad \mu \mapsto \underline{\mu},$$

given by $\underline{\mu}(p_1, p_2) = \mu(p_1)(p_2)$, and
- item (3) says that there is a canonical isomorphism

$$\mathrm{Bil}_R(M_2, M_1; N) \xrightarrow{\cong} \mathrm{Bil}_R(M_1, M_2; N), \quad \mu \mapsto \mu',$$

given by $\mu'(p_1, p_2) = \mu(p_2, p_1)$.

We are now ready to introduce the main construction in this section. Let $M_1, \ldots, M_k$ be R-modules.

Definition 1.55 (Tensor Product). A *tensor product* (over R) of $M_1, \ldots, M_k$ is a pair (T, t) consisting of an R-module T and a multilinear map $t : M_1 \times \cdots \times M_k \to T$ with the following *universal property*: for every R-module M and every multilinear map $\mu : M_1 \times \cdots \times M_k \to N$, there exists a unique R-module homomorphism $\mu_T : T \to N$ such that $\mu = \mu_T \circ t$, i.e., the diagram

$$
\begin{array}{ccc}
M_1 \times \cdots \times M_k & \xrightarrow{\ \mu\ } & N \\
{\scriptstyle t}\downarrow & \nearrow_{\mu_T} & \\
T & &
\end{array}
\tag{1.26}
$$

commutes.

In other words, a tensor product allows one to encode a multilinear map $\mu : M_1 \times \cdots \times M_k \to N$ into a plain linear map $\mu_T : T \to N$ (similarly as the direct sum encodes a family of linear maps into one single linear map). In this sense, the tensor product *represents* multilinear maps.

Theorem 1.56. *Let $M_1, \ldots, M_k$ be R-modules. Then*

(1) *there exists a tensor product (T, t) of $M_1, \ldots, M_k$;*
(2) *tensor products are unique up to unique isomorphisms, i.e., if $(T_1, t_1), (T_2, t_2)$ are two tensor products of $M_1, \ldots, M_k$, then there exists a unique R-module isomorphism $\Phi : T_1 \to T_2$ such that the diagram*

$$
\begin{array}{ccc}
& M_1 \times \cdots \times M_k & \\
{\scriptstyle t_1}\swarrow & & \searrow_{t_2} \\
T_1 & \xrightarrow{\ \Phi\ } & T_2
\end{array}
$$

commutes.

Proof. For item (1), consider the free R-module spanned by $M_1 \times \cdots \times M_k$. For simplicity, we denote it by $\widetilde{T}$ (instead of $R(M_1 \times \cdots \times M_k)$). We also denote by $\widetilde{t} : M_1 \times \cdots \times M_k \to \widetilde{T}$ (instead of χ) the inclusion. Finally, we adopt a further slight change in notation. For all $(p_1, \ldots, p_k) \in M_1 \times \cdots \times M_k$, we denote by

$$p_1 \widetilde{\otimes} \cdots \widetilde{\otimes} p_k$$

(instead of $\chi_{(p_1, \ldots, p_k)}$ or simply $(p_1, \ldots, p_k)$) the image of $(p_1, \ldots, p_k)$ under $\widetilde{t}$. It is clear that

$$\widetilde{t} : M_1 \times \cdots \times M_k \to \widetilde{T}, \quad (p_1, \ldots, p_k) \mapsto p_1 \widetilde{\otimes} \cdots \widetilde{\otimes} p_k,$$

is *not* a multilinear map. However, it can be "turned into" a multilinear map with a simple trick: in $\widetilde{T}$, consider the submodule K spanned by elements $\widetilde{\tau}$ of the form

$$\widetilde{\tau} = \Big(\cdots \widetilde{\otimes} \underbrace{(ap + bq)}_{i\text{th place}} \widetilde{\otimes} \cdots \Big) - a \Big(\cdots \widetilde{\otimes} \underbrace{p}_{i\text{th place}} \widetilde{\otimes} \cdots \Big) - b \Big(\cdots \widetilde{\otimes} \underbrace{q}_{i\text{th place}} \widetilde{\otimes} \cdots \Big)$$

$$(1.27)$$

(for all $p, q \in M_i$, all $a, b \in R$, and all $i = 1, \ldots, k$). Were these elements all zero, i.e., was $K = 0$, the map $\widetilde{t}$ would be multilinear. But $K \neq 0$, and to force it to be 0, we pass to the quotient module $T := \widetilde{T}/K$. Denote by $t : M_1 \times \cdots \times M_k \to T$ the composition of $\widetilde{t}$ followed by the projection $\pi : \widetilde{T} \to T$. We want to show that (T, t) is a tensor product of $M_1 \times \cdots \times M_k$. To do this, first check that $t : M_1 \times \cdots \times M_k \to T$ is a multilinear map. This is easy, indeed, for all $i = 1, \ldots, k$, all $p, q \in M_i$, and all $a, b \in R$,

$$t\Big(\ldots, \underbrace{ap + bq}_{i\text{th place}}, \ldots \Big)$$

$$= \pi\Big(\widetilde{t}\big(\ldots, \underbrace{ap + bq}_{i\text{th place}}, \ldots \big) \Big)$$

$$= \pi\Big(\cdots \widetilde{\otimes} \underbrace{(ap + bq)}_{i\text{th place}} \widetilde{\otimes} \cdots \Big)$$

$$= \cdots \widetilde{\otimes} \underbrace{(ap + bq)}_{i\text{th place}} \widetilde{\otimes} \cdots \bmod K$$

$$= a \Big(\cdots \widetilde{\otimes} \underbrace{p}_{i\text{th place}} \widetilde{\otimes} \cdots \bmod K \Big) + b \Big(\cdots \widetilde{\otimes} \underbrace{q}_{i\text{th place}} \widetilde{\otimes} \cdots \bmod K \Big)$$

$$= a\,\pi\big(\cdots\widetilde{\otimes}\underbrace{p}_{i\text{th place}}\widetilde{\otimes}\cdots\big) + b\,\pi\big(\cdots\widetilde{\otimes}\underbrace{q}_{i\text{th place}}\widetilde{\otimes}\cdots\big)$$

$$= a\,t\big(\ldots,\underbrace{p}_{i\text{th place}},\ldots\big) + b\,t\big(\ldots,\underbrace{q}_{i\text{th place}},\ldots\big),$$

where we used that

$$\big(\cdots\widetilde{\otimes}\underbrace{(ap+bq)}_{i\text{th place}}\widetilde{\otimes}\cdots\big) - a\big(\cdots\widetilde{\otimes}\underbrace{p}_{i\text{th place}}\widetilde{\otimes}\cdots\big) - b\big(\cdots\widetilde{\otimes}\underbrace{q}_{i\text{th place}}\widetilde{\otimes}\cdots\big) \in K,$$

hence

$$\cdots\widetilde{\otimes}\underbrace{(ap+bq)}_{i\text{th place}}\widetilde{\otimes}\cdots \bmod K$$

$$= a\big(\cdots\widetilde{\otimes}\underbrace{p}_{i\text{th place}}\widetilde{\otimes}\cdots\big) + b\big(\cdots\widetilde{\otimes}\underbrace{q}_{i\text{th place}}\widetilde{\otimes}\cdots\big) \bmod K$$

$$= a\Big(\cdots\widetilde{\otimes}\underbrace{p}_{i\text{th place}}\widetilde{\otimes}\cdots\bmod K\Big) + b\Big(\cdots\widetilde{\otimes}\underbrace{q}_{i\text{th place}}\widetilde{\otimes}\cdots\bmod K\Big).$$

Now take any other R-module N and let $\mu : M_1 \times \cdots \times M_k \to N$ be a multilinear map. By the universal property of the free module spanned by $M_1 \times \cdots \times M_k$, the map μ determines a unique R-linear map $\widetilde{\mu} : \widetilde{T} \to N$ such that $\mu = \widetilde{\mu} \circ \widetilde{t}$. As μ is multilinear, the kernel $\ker \widetilde{\mu}$ contains the submodule $K \subseteq \widetilde{T}$. Indeed, let $\widetilde{\tau} \in K$ be an elements of the form (1.27), and compute

$$\widetilde{\mu}(\widetilde{\tau})$$

$$= \widetilde{\mu}\Big(\big(\cdots\widetilde{\otimes}\underbrace{(ap+bq)}_{i\text{th place}}\widetilde{\otimes}\cdots\big)$$

$$- a\big(\cdots\widetilde{\otimes}\underbrace{p}_{i\text{th place}}\widetilde{\otimes}\cdots\big) - b\big(\cdots\widetilde{\otimes}\underbrace{q}_{i\text{th place}}\widetilde{\otimes}\cdots\big)\Big)$$

$$= \widetilde{\mu}\big(\cdots\widetilde{\otimes}\underbrace{(ap+bq)}_{i\text{th place}}\widetilde{\otimes}\cdots\big)$$

$$- a\widetilde{\mu}\big(\cdots\widetilde{\otimes}\underbrace{p}_{i\text{th place}}\widetilde{\otimes}\cdots\big) - b\widetilde{\mu}\big(\cdots\widetilde{\otimes}\underbrace{q}_{i\text{th place}}\widetilde{\otimes}\cdots\big) \quad (\widetilde{\mu} \text{ is } R\text{-linear})$$

$$= \mu(\ldots, \underbrace{ap + bq}_{i\text{th place}}, \ldots)$$

$$- a\mu(\ldots, \underbrace{p}_{i\text{th place}}, \ldots) - b\mu(\ldots, \underbrace{q}_{i\text{th place}}, \ldots) \qquad (\mu = \widetilde{\mu} \circ \widetilde{t})$$

$$= 0 \qquad\qquad\qquad (\mu \text{ is multilinear}).$$

As elements of the form (1.27) span K (by definition of K), we have $K \subseteq \ker \widetilde{\mu}$ as desired. Hence, from the Homomorphism Theorem 1.22, $\widetilde{\mu}$ descends to a (unique) linear map $\mu_T : T = \widetilde{T}/K \to N$ (such that $\widetilde{\mu} = \mu_T \circ \pi$), and we have

$$\mu_T \circ t = \mu_T \circ \pi \circ \widetilde{t} = \widetilde{\mu} \circ \widetilde{t} = \mu.$$

In order to complete the proof of item (1), it remains to check that, if $\mu'_T : T \to N$ is another linear map such that $\mu'_T \circ t = \mu$, then $\mu'_T = \mu_T$. This is indeed the case. To see this, first of all, note that, as $\widetilde{T}$ is generated by the image of $\widetilde{t}$ and the projection $\pi : \widetilde{T} \to T$ is surjective, then T is generated by $\pi(\operatorname{im} \widetilde{t})$ which consists of elements τ of the form

$$\tau = p_1 \widetilde{\otimes} \cdots \widetilde{\otimes} p_k \bmod K = t(p_1, \ldots, p_k), \quad p_i \in M_i, \quad i = 1, \ldots, k.$$

Now compute

$$\mu'_T(\tau) = \mu'_T \circ t(p_1, \ldots, p_k) = \mu(p_1, \ldots, p_k) = \mu_T \circ t(p_1, \ldots, p_k) = \mu_T(\tau).$$

So, μ_T and μ'_T agree on a set of generators, hence they agree on the whole T (do you see it? If not, try to show as an exercise that two R-module homomorphisms $f, f' : N \to P$ agreeing on a set of generators are actually the same map), i.e., $\mu_T = \mu'_T$. This concludes the proof of item (1).

For item (2), let $(T_1, t_1), (T_2, t_2)$ be two tensor products of $M_1, \ldots, M_k$. In particular, $t_2 : M_1 \times \cdots \times M_k \to T_2$ is a multilinear map and, as (T_1, t_1) satisfies the universal property of the tensor product, there exists a unique linear map $\Phi : T_1 \to T_2$ such that the diagram (1.26) commutes. We want to show that Φ is an R-module isomorphism. To do this, note that, exchanging the roles of (T_1, t_1) and (T_2, t_2) we find another linear map $\Psi : T_1 \to T_2$ such that $\Psi \circ t_2 = t_1$. It is easy to see that Ψ inverts Φ and we leave the details to the reader as Exercise 1.19. This concludes the proof. □

Exercise 1.19. Show that the homomorphisms Φ, Ψ in the end of the proof of Theorem 1.56 are mutual inverses (**Hint:** *Use the same exact argument as in the end of the proof of Theorem 1.36*).

Theorem 1.56 shows that tensor products exist and are unique up to unique isomorphisms. The tensor product (T, t) constructed in the proof is a canonical choice. For this reason, we call it *the tensor product* of $M_1, \ldots, M_k$ and denote it by $M_1 \otimes_R \cdots \otimes_R M_k$ (or simply $M_1 \otimes \cdots \otimes M_k$ if this does not lead to confusion). Given a k-tuple $(p_1, \ldots, p_k) \in M_1 \times \cdots \times M_k$, its image $t(p_1, \ldots, p_k) \in M_1 \otimes \cdots \otimes M_k$ under t will be always denoted by

$$p_1 \otimes \cdots \otimes p_k.$$

In other words,

$$p_1 \otimes \cdots \otimes p_k = p_1 \widetilde{\otimes} \cdots \widetilde{\otimes} p_k \bmod K.$$

The tensor product $M_1 \otimes \cdots \otimes M_k$ is spanned, by construction, by the image of the multilinear map $t : M_1 \times \cdots \times M_1 \to M_1 \otimes \cdots \otimes M_k$, i.e., $M_1 \otimes \cdots \otimes M_k$ is spanned by elements of the type

$$p_1 \otimes \cdots \otimes p_k, \quad p_i \in M_i, \quad i = 1, \ldots, k.$$

Such elements are sometimes called *decomposable elements* (while a generic element in $M_1 \otimes \cdots \otimes M_k$ is a linear combination of decomposable elements). If $S_1, \ldots, S_k$ are sets of generators in $M_1, \ldots, M_k$ respectively, then $M_1 \otimes \cdots \otimes M_k$ is also spanned by $t(S_1 \times \cdots \times S_k)$, i.e., by elements of the form

$$q_1 \otimes \cdots \otimes q_k, \quad q_i \in S_i \subseteq M_i, \quad i = 1, \ldots, k.$$

The universal property of tensor products says that, for any R-module N, there exists a map

$$\mathrm{Mult}_R^k(M_1, \ldots, M_k; N) \to \mathrm{Hom}_R(M_1 \otimes_R \cdots \otimes_R M_k, N), \quad \mu \mapsto \mu_T. \tag{1.28}$$

This map is injective, indeed, if $\mu, \mu' \in \mathrm{Mult}_R^k(M_1, \ldots, M_k; N)$ are two multilinear maps such that $\mu_T = \mu'_T$, then

$$\mu' = \mu'_T \circ t = \mu_T \circ t = \mu.$$

The map (1.28) is also surjective. Indeed, let $F : M_1 \otimes_R \cdots \otimes_R M_k \to N$ be a linear map. Then $\mu := F \circ t : M_1 \times \cdots \times M_k \to N$ is a multilinear map and, from the universal property again, $F = \mu_T$. Concluding, (1.28) is a canonical bijection.

Proposition 1.57. *The bijection* (1.28) *is an R-module isomorphism:*

$$\mathrm{Mult}_R^k(M_1, \ldots, M_k; N) \cong \mathrm{Hom}_R(M_1 \otimes_R \cdots \otimes_R M_k, N).$$

Proof.   Left as Exercise 1.20.   □

Exercise 1.20. Prove Proposition 1.57.

In view of its universal property, the tensor product construction inherits noteworthy properties from that of multilinear maps (see Proposition 1.54). Namely, we have the following:

Proposition 1.58. *Let R be a ring, let k be a non-negative integer, and let $M_1, \ldots, M_k$ be R-modules. Then*

(1) *for every $l \leq k$ there is a canonical R-module isomorphism*

$$\left(M_1 \otimes_R \cdots \otimes_R M_l \right) \otimes_R \left(M_{l+1} \otimes_R \cdots \otimes_R M_k \right) \cong M_1 \otimes_R \cdots \otimes_R M_k; \tag{1.29}$$

(2) *there is a canonical R-module isomorphism*

$$M_1 \otimes_R \cdots \otimes_R M_k \otimes_R R \cong M_1 \otimes_R \cdots \otimes_R M_k;$$

(3) *for every permutation $\sigma \in S_k$ there is a canonical R-module isomorphism*

$$M_{\sigma(1)} \otimes_R \cdots \otimes_R M_{\sigma(k)} \cong M_1 \otimes_R \cdots \otimes_R M_k.$$

Proof.   Begin with item (1). We prove the following refinement of the statement: there exists a unique R-module isomorphism $(M_1 \otimes \cdots \otimes M_l) \otimes (M_{l+1} \otimes \cdots \otimes M_k) \cong M_1 \otimes \cdots \otimes M_k$ identifying $(p_1 \otimes \cdots \otimes p_l) \otimes (p_{l+1} \otimes \cdots \otimes p_k)$ with $p_1 \otimes \cdots \otimes p_k$ for all $p_i \in M_i$, $i = 1, \ldots, k$. To do this, consider the map

$$\begin{aligned} \underline{t} : M_1 \times \cdots \times M_k &\to \left(M_1 \otimes \cdots \otimes M_l \right) \otimes \left(M_{l+1} \otimes \cdots \otimes M_k \right) \\ (p_1, \ldots, p_k) &\mapsto (p_1 \otimes \cdots \otimes p_l) \otimes (p_{l+1} \otimes \cdots \otimes p_k). \end{aligned}$$

It is clear that $\underline{t}$ is multilinear. In order to conclude with item (1), it is enough to show that $((M_1 \otimes \cdots \otimes M_l) \otimes (M_{l+1} \otimes \cdots \otimes M_k), \underline{t})$ is a tensor product of $M_1, \ldots, M_k$, indeed, in this case, from the uniqueness, it follows that there exists an isomorphism exactly as desired (do you see it?). So, to see that $((M_1 \otimes \cdots \otimes M_l) \otimes (M_{l+1} \otimes \cdots \otimes M_k), \underline{t})$ is a tensor product, take another multilinear map

$$\mu : M_1 \times \cdots \times M_k \to N.$$

We can identify μ with a linear map

$$\underline{\mu}_T : (M_1 \otimes \cdots \otimes M_l) \otimes (M_{l+1} \otimes \cdots \otimes M_k) \to N$$

using the following chain of R-module isomorphisms:

$$\text{Mult}^k(M_1, \ldots, M_k; N) \qquad\qquad \mu$$

$$\cong \text{Mult}^l(M_1, \ldots, M_l; \text{Mult}^{k-l}(M_{l+1}, \ldots, M_k; N)) \qquad\qquad \mu_1$$

$$\cong \text{Hom}(M_1 \otimes \cdots \otimes M_l, \text{Hom}(M_{l+1} \otimes \cdots \otimes M_k, N)) \qquad\qquad \mu_2 \qquad (1.30)$$

$$\cong \text{Bil}(M_1 \otimes \cdots \otimes M_l, M_{l+1} \otimes \cdots \otimes M_k; N) \qquad\qquad \mu_3$$

$$\cong \text{Hom}((M_1 \otimes \cdots \otimes M_l) \otimes (M_{l+1} \otimes \cdots \otimes M_k), N) \qquad\qquad \underline{\mu}_T$$

The linear map $\underline{\mu}_T$ satisfies $\mu = \underline{\mu}_T \circ \underline{t}$. To see this, we have to understand how does $\underline{\mu}_T$ act. In (1.30), we gave a name to each of the maps with which μ identifies along the chain of isomorphisms. Then, for all $(p_1, \ldots, p_k) \in M_1 \times \cdots \times M_k$, we have

$$\underline{\mu}_T \circ \underline{t}(p_1, \ldots, p_k) = \underline{\mu}_T((p_1 \otimes \cdots \otimes p_l) \otimes (p_{l+1} \otimes \cdots \otimes p_k))$$

$$= \mu_3(p_1 \otimes \cdots \otimes p_l, p_{l+1} \otimes \cdots \otimes p_k)$$

$$= \mu_2(p_1 \otimes \cdots \otimes p_l)(p_{l+1} \otimes \cdots \otimes p_k)$$

$$= \mu_1(p_1, \ldots, p_l)(p_{l+1}, \ldots, p_k)$$

$$= \mu(p_1, \ldots, p_k),$$

as desired (it might seem complicated but, in practice, we are only choosing a different symbol for the same object at each step!!). It is easy to see that $\underline{\mu}_T$ is the unique linear map such that $\mu = \underline{\mu}_T \circ \underline{t}$. This follows from the fact that the image of $\underline{t}$ generates $(M_1 \otimes \cdots \otimes M_l) \otimes (M_{l+1} \otimes \cdots \otimes M_k)$ which in turn follows from the fact that each of the two factors $M_1 \otimes \cdots \otimes M_l$ and $M_{l+1} \otimes \cdots \otimes M_k$ is generated by decomposable elements (at this point, the reader is strongly suggested to stop and think about all the details). We conclude that $((M_1 \otimes \cdots \otimes M_l) \otimes (M_{l+1} \otimes \cdots \otimes M_k), \underline{t})$ is a tensor product of $M_1, \ldots, M_k$ as desired. This concludes the proof of item (1) (in the refined version at the beginning of this proof).

Items (2) and (3) can be proved in a similar way, and we only sketch the main arguments leaving the details to the reader. One can prove the following refinement of item (2): there exists a unique R-module isomorphism $M_1 \otimes \cdots \otimes M_k \otimes R \cong M_1 \otimes \cdots \otimes M_k$ identifying $p_1 \otimes \cdots \otimes p_k \otimes 1$ with $p_1 \otimes \cdots \otimes p_k$ for all $p_i \in M_i$, $i = 1, \ldots, k$. To do this, we can consider the multilinear map

$$\bar{t}: M_1 \times \cdots \times M_k \to M_1 \otimes \cdots \otimes M_k \otimes R, \qquad (p_1, \ldots, p_k) \mapsto p_1 \otimes \cdots \otimes p_k \otimes 1$$

and show that $(M_1 \otimes \cdots \otimes M_k \otimes R, \bar{t})$ is a tensor product of $M_1, \ldots M_k$ using Proposition 1.54 in a very similar way as we did for item (1). Finally, for item (3), we can prove that there exists a unique R-module isomorphism $M_{\sigma(1)} \otimes \cdots \otimes M_{\sigma(k)} \cong M_1 \otimes \cdots \otimes M_k$ identifying $p_{\sigma(1)} \otimes \cdots \otimes p_{\sigma(k)}$ with $p_1 \otimes \cdots \otimes p_k$, where $p_i \in M_i$, $i = 1, \ldots, k$. We leave the details to the reader as Exercise 1.21. $\qquad\square$

Exercise 1.21. Complete the proof of Proposition 1.58 discussing in details items (2) and (3). (**Hint:** *Use a similar strategy as for item (1)*).

Similarly as for multilinear maps, combining the isomorphisms (1), (2) and (3) in Proposition 1.58, we find numerous new canonical isomorphisms. Note that

- when $k = 1$, item (2) says that there is a canonical isomorphism

$$M \otimes_R R \xrightarrow{\cong} M$$

  and the proof of Proposition 1.58 reveals that this isomorphism maps $p \otimes 1$ to p;
- when $k = 3$, item (1) says that there are canonical isomorphisms

$$(M_1 \otimes_R M_2) \otimes_R M_3 \xrightarrow{\cong} M_1 \otimes_R M_2 \otimes_R M_3 \xrightarrow{\cong} M_1 \otimes_R (M_2 \otimes_R M_3),$$

  identifying $(p_1 \otimes p_2) \otimes p_3$ with $p_1 \otimes p_2 \otimes p_3$ with $p_1 \otimes (p_2 \otimes p_3)$ $(p_i \in M_i, i = 1, 2, 3)$;
- when $k = 2$, item (3) says that there is a canonical isomorphism

$$M_2 \otimes_R M_1 \xrightarrow{\cong} M_1 \otimes_R M_2$$

identifying $p_2 \otimes p_1$ with $p_1 \otimes p_2$ $(p_i \in M_i, i = 1, 2)$.

Finally, we remark that the isomorphisms (1.29) can be thought of as "operations"

$$\otimes : (M_1 \otimes \cdots \otimes M_l) \times (M_{l+1} \otimes \cdots \otimes M_k) \to M_1 \otimes \cdots \otimes M_k,$$

$$(\mathcal{T}, \mathcal{S}) \mapsto \mathcal{T} \otimes \mathcal{S}$$

mapping $(p_1 \otimes \cdots \otimes p_l, p_{l+1} \otimes \cdots \otimes p_k)$ to $p_1 \otimes \cdots \otimes p_k$, and satisfying R-bilinearity, associativity, and existence of a neutral element $1 \in R$ (up to the identification $M \otimes_R R \cong M$), but beware that they are *not* commutative (see also Problem 4.52).

Remark 1.59. It is important to learn how to compute with tensor products. The main property is multilinearity:

$$\cdots \otimes (ap + bq) \otimes \cdots = a (\cdots \otimes p \otimes \cdots) + b (\cdots \otimes q \otimes \cdots).$$

for all p, q module elements and all a, b scalars. $\qquad\diamond$

When $M_1, \ldots, M_k$ are free (and finitely generated), then the tensor product $M_1 \otimes \cdots \otimes M_k$ is free (and finitely generated) as well according to the following:

Proposition 1.60. *Let $M_1, \ldots, M_k$ be free R-modules, and let $B_1 = (q_{i_1}^{(1)})_{i_1 \in I_1}, \ldots, B_k = (q_{i_k}^{(k)})_{i_k \in I_k}$ be bases in $M_1, \ldots, M_k$, respectively. Then the family*

$$B^\otimes := \left(q_{i_1}^{(1)} \otimes \cdots \otimes q_{i_k}^{(k)} \right)_{(i_1, \ldots, i_k) \in I_1 \times \cdots \times I_k}$$

is a basis of $M_1 \otimes \cdots \otimes M_k$.

Proof. As B_r generates M_r for each $r = 1, \ldots k$, then $B^\otimes$ generates $M_1 \otimes \cdots \otimes M_k$. It remains to check that $B^\otimes$ is independent. So, take a zero (finite) linear combination

$$A = \sum_{(i_1, \ldots, i_k) \in I_1 \times \cdots \times I_k} a_{i_1 \cdots i_k} q_{i_1}^{(1)} \otimes \cdots \otimes q_{i_k}^{(k)} = 0.$$

From the Multilinear Extension Theorem 1.50, for any $(j_1, \ldots, j_k) \in I_1 \times \cdots \times I_k$ there exists a unique multilinear map

$$\mu_{j_1 \cdots j_k} : M_1 \times \cdots \times M_k \to R$$

such that

$$\mu_{j_1 \cdots j_k}(q_{i_1}^{(1)}, \ldots, q_{i_k}^{(k)}) = \delta_{j_1 i_1} \cdots \delta_{j_k i_k}, \quad \text{for all } (i_1, \ldots, i_k) \in I_1 \times \cdots \times I_k.$$

From the universal property of the tensor product, we get a unique linear map

$$\mu_{j_1 \cdots j_k}^T : M_1 \otimes \cdots \otimes M_k \to R$$

such that

$$\mu_{j_1 \cdots j_k}^T(q_{i_1}^{(1)} \otimes \cdots \otimes q_{i_k}^{(k)}) = \mu_{j_1 \cdots j_k}(q_{i_1}^{(1)}, \ldots, q_{i_k}^{(k)}) = \delta_{j_1 i_1} \cdots \delta_{j_k i_k},$$

for all $(i_1, \ldots, i_k) \in I_1 \times \cdots \times I_k$. Hence, we have

$$0 = \mu_{j_1 \cdots j_k}^T(A) = \mu_{j_1 \cdots j_k}^T \left(\sum_{i_1, \ldots, i_k} a_{i_1 \cdots i_k} q_{i_1}^{(1)} \otimes \cdots \otimes q_{i_k}^{(k)} \right)$$

$$= \sum_{i_1, \ldots, i_k} a_{i_1 \cdots i_k} \mu_{j_1 \cdots j_k}^T \left(q_{i_1}^{(1)} \otimes \cdots \otimes q_{i_k}^{(k)} \right)$$

$$= \sum_{i_1, \ldots, i_k} a_{i_1 \cdots i_k} \delta_{j_1 i_1} \cdots \delta_{j_k i_k} = a_{j_1 \cdots j_k}.$$

So, $a_{j_1 \cdots j_k} = 0$ for all $(j_1, \ldots, j_k) \in I_1 \times \cdots \times I_k$ and $B^\otimes$ is independent. $\square$

Note that if the bases B_r in Proposition 1.60 are finite, then the basis $B^\otimes$ is also finite so that $M_1 \otimes \cdots \otimes M_k$ is also finitely generated.

Corollary 1.61. *Let $X_1, \ldots, X_k$ be (non-necessarily finite) sets and let R be a ring. The free module $R(X_1 \times \cdots \times X_k)$ generated by the cartesian product $X_1 \times \cdots \times X_k$ is canonically isomorphic to the tensor product of the free modules $RX_1, \ldots, RX_k$:*

$$R(X_1 \times \cdots \times X_k) \cong RX_1 \otimes_R \cdots \otimes_R RX_k.$$

Proof. The sets X_r can be seen as bases in RX_r, $r = 1, \ldots, k$. From Proposition 1.60, the family

$$X^\otimes = (x_1 \otimes \cdots \otimes x_k)_{(x_1, \ldots, x_k) \in X_1 \times \cdots \times X_r}$$

is then a basis in $RX_1 \otimes_R \cdots \otimes_R RX_k$. The set $X_1 \times \cdots \times X_k$ can be seen as a basis in $R(X_1 \times \cdots \times X_k)$. Clearly, there is a unique R-module isomorphism $R(X_1 \times \cdots \times X_k) \cong RX_1 \otimes_R \cdots \otimes_R RX_k$ identifying the basis elements $(x_1, \ldots, x_k)$ and $x_1 \otimes \cdots \otimes x_k$. $\qquad\square$

Corollary 1.62. *Let $M_1, \ldots, M_k$ be free and finitely generated R-modules. Denote by $M_r^* := \mathrm{Hom}(M_r, R)$ their dual modules, $r = 1, \ldots, k$. Then there are canonical isomorphisms*

$$M_1 \otimes \cdots \otimes M_k \cong \mathrm{Mult}^k(M_1^*, \ldots, M_k^*; R)$$

and

$$M_1^* \otimes \cdots \otimes M_k^* \cong \mathrm{Mult}^k(M_1, \ldots, M_k; R).$$

Before proving Corollary 1.62, we make some remarks that might have an independent interest. For an R-module M, the module M^{**} (dual of the dual module) is also called the *bidual*. There is a canonical linear map

$$\iota : M \to M^{**}$$

given by

$$\iota(p)(\varphi) = \varphi(p), \quad p \in M, \quad \varphi \in M^*.$$

In general, this map is neither injective nor surjective. For instance, when $R = \mathbb{Z}$ and $M = \mathbb{Z}_2$, then $M^* = 0$ (can you prove it?), hence $M^{**} = 0$ and ι is the zero map, so it is not injective. On the other hand, when M is not finitely generated, M^* is too big and M^{**} is even bigger, and ι cannot be surjective (to get an intuition of this, try to describe the dual of the free module $R\mathbb{N}$). However, when M is free and finitely generated, then ι is an isomorphism. Indeed, take a finite basis $(q_1, \ldots, q_n) \subseteq M$. Exactly as for finite-dimensional vector spaces, one can show that M^* is free and finitely generated. More precisely, there exists a unique basis $(q_1^*, \ldots, q_n^*)$ of M^* such that $q_i^*(q_j) = \delta_{ij}$ for all $i, j = 1, \ldots, n$ (try to reproduce the proof in the present case). It follows that, in this case, M^{**} is also free and finitely generated. Indeed, it is easy to check that $\iota(q_i) = q_i^{**}$ for all $i = 1, \ldots, n$ and it follows that ι is an isomorphism. We are now ready to prove Corollary 1.62.

Proof of Corollary 1.62. Note that the discussion preceding the present proof takes care of the case $k = 1$. Additionally, as $M_r^{**} \cong M_r$ canonically for all $i = 1, \ldots, k$, it is enough to prove that there is a canonical isomorphism

$$M_1 \otimes \cdots \otimes M_k \cong \mathrm{Mult}^k(M_1^*, \ldots, M_k^*; R),$$

generalizing $\iota : M \to M^{**}$ to the case $k \geq 1$. So, first of all, there is a unique linear map

$$\iota_k : M_1 \otimes \cdots \otimes M_k \to \mathrm{Mult}^k(M_1^*, \ldots, M_k^*; R),$$

such that

$$\iota_k(p_1 \otimes \cdots \otimes p_k)(\varphi_1, \ldots, \varphi_k) = \varphi_1(p_1) \cdots \varphi_k(p_k), \qquad (1.31)$$

for all $p_r \in M_r$, and all $\varphi_r \in M_r^*$, $r = 1, \ldots, k$. This follows from the universal property of the tensor product and the fact that the expression on the right-hand side of (1.31) is multilinear both in $(p_1, \ldots, p_k)$ and in $(\varphi_1, \ldots, \varphi_k)$ (we leave it to the reader to make sense of the last claim as Exercise 1.22. We also stress that ι_k exists independently of the M_i being free, finitely generated). We want to show that ι_k is the isomorphism we are looking for. We adopt a slightly different strategy than in the $k = 1$ case and prove that ι_k is injective and surjective. So, fix finite bases $B_r = (q_{i_r}^{(r)})_{i_r \in I_r}$ in M_r, $r = 1, \ldots, k$, and let

$$B^{\otimes} = (q_{i_1}^{(1)} \otimes \cdots \otimes q_{i_k}^{(k)})_{(i_1, \ldots, i_k) \in I_1 \times \cdots \times I_k}$$

be the associated basis of $M_1 \otimes \cdots \otimes M_k$. Consider

$$A = \sum_{i_1, \ldots, i_k} a_{i_1 \cdots i_k} q_{i_1}^{(1)} \otimes \cdots \otimes q_{i_k}^{(k)} \in M_1 \otimes \cdots \otimes M_k$$

and assume that $A \in \ker \iota_k$. This means that, for all $(\varphi_1, \ldots, \varphi_k) \in M_1^* \times \cdots \times M_k^*$ we have

$$0 = \iota_k(A)(\varphi_1, \ldots, \varphi_k) = \iota_k \left(\sum_{i_1, \ldots, i_k} a_{i_1 \cdots i_k} q_{i_1}^{(1)} \otimes \cdots \otimes q_{i_k}^{(k)} \right) (\varphi_1, \ldots, \varphi_k)$$

$$= \sum_{i_1, \ldots, i_k} a_{i_1 \cdots i_k} \iota_k \left(q_{i_1}^{(1)} \otimes \cdots \otimes q_{i_k}^{(k)} \right) (\varphi_1, \ldots, \varphi_k) \qquad (1.32)$$

$$= \sum_{i_1, \ldots, i_k} a_{i_1 \cdots i_k} \varphi_1(q_{i_1}^{(1)}) \cdots \varphi_k(q_{i_k}^{(k)}).$$

In particular, when $\varphi_r = (q_{j_r}^{(r)})^*$, the j_r-th element in the dual basis, $r = 1, \ldots, k$, we get

$$0 = \sum_{i_1, \ldots, i_k} a_{i_1 \cdots i_k} (q_{j_1}^{(1)})^* (q_{i_1}^{(1)}) \cdots (q_{j_k}^{(k)})^* (q_{i_k}^{(k)})$$

$$= \sum_{i_1, \ldots, i_k} a_{i_1 \cdots i_k} \delta_{i_1 j_1} \cdots \delta_{i_k j_k} = a_{j_1 \cdots j_k}. \qquad (1.33)$$

From the arbitrariness of $(j_1, \ldots j_k)$, we get $A = 0$. Hence, ι_k is injective.

For the surjectivity, take $\mu \in \mathrm{Mult}^k(M_1^*, \ldots, M_k^*; R)$. Denote

$$b_{j_1 \cdots j_k} := \mu \left((q_{j_1}^{(1)})^*, \ldots, (q_{j_k}^{(k)})^* \right) \in R, \quad (j_1, \ldots, j_k) \in I_1 \times \cdots \times I_k.$$

We want to show that $\mu = \iota_k(B)$ where

$$B := \sum_{i_1, \ldots, i_k} b_{i_1 \cdots i_k} q_{i_1}^{(1)} \otimes \cdots \otimes q_{i_k}^{(k)}.$$

To do this, recall that a multilinear map is completely determined by its values on the basis elements. Finally, a computation identical to (1.32), as continued in (1.33), shows that

$$\iota_k(B) \left((q_{j_1}^{(1)})^*, \ldots, (q_{j_k}^{(k)})^* \right) = b_{j_1 \cdots j_k} = \mu \left((q_{j_1}^{(1)})^*, \ldots, (q_{j_k}^{(k)})^* \right),$$

and therefore $\mu = \iota_k(B)$. This concludes the proof. $\qquad \square$

Exercise 1.22. Let $M_1, \ldots, M_k$ be (non-necessarily free nor finitely generated) R-modules. Prove that there is a unique R-linear map

$$\iota_k : M_1 \otimes \cdots \otimes M_k \to \mathrm{Mult}^k(M_1^*, \ldots, M_k^*; R)$$

such that

$$\iota_k(p_1 \otimes \cdots \otimes p_k)(\varphi_1, \ldots, \varphi_k) = \varphi_1(p_1) \cdots \varphi_k(p_k), \qquad (1.34)$$

for all $p_i \in M_i$, and all $\varphi_i \in M_i^*$, $i = 1, \ldots, k$ (**Hint:** *First, define a map* $\mu : M_1 \times \cdots \times M_k \to \mathrm{Mult}^k(M_1^*, \ldots, M_k^*; R)$ *by putting* $\mu(p_1, \ldots, p_k)(\varphi_1, \ldots, \varphi_k) = \varphi_1(p_1) \cdots \varphi_k(p_k)$. *Second, show that μ is well defined and multilinear. Finally, use the universal property of tensor products*).

Let R be a ring and let $M_1, \ldots, M_k, N$ be R-modules. When $M_1 = \cdots = M_k =: M$, it makes sense to talk about *symmetric* and *alternating* multilinear maps $M_1 \times \cdots \times M_k \to N$. We begin with symmetric multilinear maps.

Definition 1.63 (Symmetric Multilinear Map). A k-multilinear map

$$\mu : \underbrace{M \times \cdots \times M}_{k\text{-times}} \to N$$

is *symmetric* if $\mu(p_1, \ldots, p_k)$ doesn't change when we swap two entries, $p_1, \ldots, p_k \in M$, equivalently, when for any permutation $\sigma \in S_k$ and all $p_1, \ldots, p_k \in M$ we have

$$\mu(p_{\sigma(1)}, \ldots, p_{\sigma(k)}) = \mu(p_1, \ldots, p_k).$$

From now on, we adopt the following compact notation:

$$M^{\times k} := \underbrace{M \times \cdots \times M}_{k\text{-times}}.$$

The space of symmetric multilinear maps $M^{\times k} \to N$ is a submodule in $\mathrm{Mult}_R^k(M, \ldots, M; N)$ denoted

$$\mathrm{Sym}_R^k(M; N),$$

or simply $\mathrm{Sym}^k(M; N)$. We also put $\mathrm{Sym}^0(M; N) := N$.

Definition 1.64 (*k-th* Symmetric Power). A *k-th symmetric power* (over R) of M is a pair (S, s) consisting of an R-module S and a symmetric multilinear map $s : M^{\times k} \to S$ with the following *universal property*: for every R-module N and every symmetric multilinear map $\mu : M^{\times k} \to N$, there exists a unique R-module homomorphism $\mu_S : S \to N$ such that $\mu = \mu_S \circ s$, i.e., the diagram

$$
\begin{array}{ccc}
M^{\times k} & \xrightarrow{\ \mu\ } & N \\
{\scriptstyle s}\downarrow & \nearrow & \\
S & {\scriptstyle \mu_S} &
\end{array}
\tag{1.35}
$$

commutes.

Theorem 1.65. *Let M be an R-module, and let k be a non-negative integer. Then*

(1) *there exists a k-th symmetric power (S, s) of M;*
(2) *if (S_1, s_1), (S_2, s_2) are two k-th symmetric powers, then there exists a unique R-module isomorphism $\Phi : S_1 \to S_2$ such that the diagram*

$$
\begin{array}{ccc}
 & M^{\times k} & \\
{\scriptstyle s_1}\swarrow & & \searrow{\scriptstyle s_2} \\
S_1 & \xrightarrow{\ \Phi\ } & S_2
\end{array}
$$

commutes.

Proof. The proof is similar in spirit to that of Theorem 1.56. For $k > 0$, consider the tensor product

$$
M^{\otimes k} := \underbrace{M \otimes_R \cdots \otimes_R M}_{k\text{-times}}
$$

(it is sometimes called the *k-th tensor power of M*) and the canonical multilinear map

$$
t : M^{\times k} \to M^{\otimes k}, \quad (p_1, \dots, p_k) \mapsto p_1 \otimes \cdots \otimes p_k.
$$

For $k = 0$, we put $M^{\otimes k} = R$, and, for $k = 1$, we have $M^{\otimes 1} = M$. Clearly, t is *not* symmetric in general. However, it can be "turned into" a symmetric multilinear map with the following trick: in $M^{\otimes k}$ consider the submodule O spanned by elements τ of the form

$$
\tau = p_{\sigma(1)} \otimes \cdots \otimes p_{\sigma(k)} - p_1 \otimes \cdots \otimes p_k
$$

(for all $p_i \in M$, $i = 1, \dots, k$, and all permutations $\sigma \in S_k$). Now pass to the quotient module $S := M^{\otimes k}/O$, and denote by $s : M^{\times k} \to S$ the composition $\pi \circ t$, where $\pi : M^{\otimes k} \to S$ is the usual projection. One can show that (S, s) is a k-th symmetric power of M in a very similar way as in the proof of Theorem 1.56 and we leave the details as Exercise 1.23(1). For item (2), we can use the same exact argument as for item (2) of Theorem 1.56 (check the details as Exercise 1.23(2)). This concludes the proof. □

Exercise 1.23. Let M be an R-module and k a non-negative integer.

(1) Prove that the pair (S, s) defined in the proof of Theorem 1.65 is a k-th symmetric power of M.
(2) Prove item (2) in Theorem 1.65.

(**Hint:** *For item (1) get inspired by the proof of Theorem 1.56.*)

The proof of Theorem 1.65 reveals that there is a canonical choice of k-th symmetric power of M, namely $(M^{\otimes k}/O, s)$. We call it *the k-th symmetric power* of M and denote it by $(S_R^k M, s)$ (or simply $S^k M$). Note that $S^0 M = R$ and $S^1 M = M$. Given a k-tuple $(p_1, \ldots, p_k) \in M^{\times k}$, its image $s(p_1, \ldots, p_k)$ under s will be always denoted by

$$p_1 \vee \cdots \vee p_k.$$

In other words, $p_1 \vee \cdots \vee p_k = p_1 \otimes \cdots \otimes p_k \bmod O$. It is clear that $S^k M$ is spanned by the image of s. If S is a set of generators in M, then $S^k M$ is also spanned by $s(S^{\times k})$.

It is easy to see that the map

$$\mathrm{Sym}_R^k(M; N) \to \mathrm{Hom}_R(S_R^k M, N), \quad \mu \mapsto \mu_S$$

is an R-module isomorphism. Note that this makes sense even when $k = 0$, in which case $\mathrm{Sym}_R^k(M; N) = N$ (by definition of 0-multilinear map) and $S^k M = R$.

Finally, we remark that, for any l, m, there is a unique bilinear map

$$\vee : S^l M \times S^m M \to S^{l+m} M, \quad (\mathcal{P}, \mathcal{Q}) \mapsto \mathcal{P} \vee \mathcal{Q}$$

mapping $(p_1 \vee \cdots \vee p_l, p_1' \vee \cdots \vee p_m')$ to $p_1 \vee \cdots \vee p_l \vee p_1' \vee \cdots \vee p_m'$, called the *symmetric product*. The symmetric product is (bilinear) associative, commutative and there exists a neutral element: $1 \in R \cong S^0 M$.

Proposition 1.66. *Let M be a free and finitely generated R-module, and let $B = (q_1, \ldots, q_n)$ be an ordered basis of M. Then, for all $k \geq 0$, the family*

$$B^\vee := (q_{i_1} \vee \cdots \vee q_{i_k})_{i_1 \leq \cdots \leq i_k}$$

is a (finite) basis in $S^k M$. Additionally, there are canonical isomorphisms

$$S^k M \cong \mathrm{Sym}^k(M^*; R) \quad \text{and} \quad S^k M^* \cong \mathrm{Sym}^k(M; R).$$

Proof. We do not provide a proof. We only remark that the isomorphism $S^k M \cong \mathrm{Sym}^k(M^*; R)$ mentioned in the second part of the statement is the only isomorphism mapping $p_1 \vee \cdots \vee p_k$ to the following symmetric multilinear map:

$$(M^*)^{\times k} \to R, \quad (\varphi_1, \ldots, \varphi_k) \mapsto \sum_{\sigma \in S_k} \varphi_{\sigma(1)}(p_1) \cdots \varphi_{\sigma(k)}(p_k)$$

(and likewise for the isomorphism $S^k M^* \cong \mathrm{Sym}^k(M^*; R)$). $\qquad\square$

We conclude this chapter with a short discussion of alternating multilinear maps.

Definition 1.67 (Alternating Multilinear Map). A k-multilinear map $\mu : M^{\times k} \to N$ is *alternating* if $\mu(p_1, \ldots, p_k)$ vanishes whenever two entries coincide, $p_1, \ldots, p_k \in M$.

It is easy to see that every alternating multilinear map $\mu : M^{\times k} \to N$ is also *skew-symmetric*, i.e., $\mu(p_1, \ldots, p_k)$ changes in sign when we swap two entries or, equivalently, for any permutation $\sigma \in S_k$ and all $p_1, \ldots, p_k \in M$ we have

$$\mu(p_{\sigma(1)}, \ldots, p_{\sigma(k)}) = (-)^\sigma \mu(p_1, \ldots, p_k),$$

where $(-)^\sigma$ is the sign of σ. The converse in false in general but is true sometimes.

Remark 1.68. For any ring R, there is a ring homomorphism $\psi : \mathbb{Z} \to R$: the unique ring homomorphism mapping 1 to 1. The kernel of ψ is an ideal in $\mathbb{Z}$, hence it is of the form $\ker \psi = n\mathbb{Z}$ for some non-negative integer n, called the *characteristic* of R. If R is a field and its characteristic is not 2, then skew-symmetric multilinear maps are also alternating. For instance, for a bilinear map $\mu : M \times M \to N$, from $\mu(p, q) = -\mu(q, p)$, we have $\mu(p, p) = -\mu(p, p)$ for all $p \in M$, hence

$$0 = 2\mu(p, p) = \psi(2)\mu(p, p).$$

As $\psi(2) \neq 0$ and R is a field, then $\psi(2)$ is invertible and $\mu(p, p) = 0$. But in general, skew-symmetric multilinear maps are *not* alternating (while, as already mentioned, the converse is always true). $\qquad\diamond$

Example 1.69. The determinant is an alternating multilinear map (see Example 1.49). $\qquad\blacklozenge$

The space of alternating multilinear maps $M^{\times k} \to N$ is a submodule in $\mathrm{Mult}_R^k(M, \ldots, M; N)$ denoted

$$\mathrm{Alt}_R^k(M; N),$$

or simply $\mathrm{Alt}^k(M; N)$. We also put $\mathrm{Alt}^0(M; N) := N$.

Definition 1.70 (k-th Exterior Power). A *k-th exterior power* (over R) of M is a pair (Λ, λ) consisting of an R-module Λ and an alternating multilinear map $\lambda : M^{\times k} \to \Lambda$ with the following *universal property*: for every R-module N and every alternating multilinear map $\mu : M^{\times k} \to N$, there exists a unique R-module homomorphism $\mu_\Lambda : \Lambda \to N$ such that $\mu = \mu_\Lambda \circ \lambda$, i.e., the diagram

$$
\begin{array}{ccc}
M^{\times k} & \xrightarrow{\ \mu\ } & N \\
{\scriptstyle \lambda}\downarrow & \nearrow {\scriptstyle \mu_\Lambda} & \\
\Lambda & &
\end{array}
$$

commutes.

Theorem 1.71. *Let M be an R-module, and let k be a non-negative integer. Then*

(1) *there exists a k-th exterior power (Λ, λ) of M;*
(2) *if (Λ_1, λ_1), (Λ_2, λ_2) are two k-th exterior powers, then there exists a unique R-module isomorphism $\Phi : S_1 \to S_2$ such that the diagram*

$$
\begin{array}{ccc}
& M^{\times k} & \\
{\scriptstyle \lambda_1}\swarrow & & \searrow {\scriptstyle \lambda_2} \\
\Lambda_1 & \xrightarrow{\ \Phi\ } & \Lambda_2
\end{array}
$$

commutes.

Proof. The proof is similar in spirit to that of Theorem 1.56. We only sketch it. In $M^{\otimes k}$ consider the submodule P spanned by elements of the form

$$
p_1 \otimes \cdots \otimes \underbrace{p}_{i\text{-th place}} \otimes \cdots \otimes \underbrace{p}_{j\text{-th place}} \otimes \cdots \otimes p_k
$$

(for all $p, p_\ell \in M$, and all $i < j = 1, \ldots, k$). Denote $\Lambda := M^{\otimes k}/P$, and put $\lambda = \pi \circ t : M^{\times k} \to \Lambda$. Then (Λ, λ) is a k-th exterior power of M. Uniqueness is proved exactly as item (2) of Theorem 1.56. $\qquad\square$

As discussed in the proof of Theorem 1.71, there is a canonical choice of a k-th exterior power of M, namely $(M^{\otimes k}/P, \lambda)$. We call it *the k-th exterior power* of M and denote it by $(\wedge_R^k M, \lambda)$ (or simply $\wedge^k M$). Note that $\wedge^0 M = R$ and $\wedge^1 M = M$. Given a k-tuple $(p_1, \ldots, p_k) \in M^{\times k}$, its image under λ will be denoted by

$$
p_1 \wedge \cdots \wedge p_k.
$$

It is clear that $\wedge^k M$ is spanned by the image of λ. If S is a set of generators in M, then $\wedge^k M$ is also spanned by $\lambda(S^{\times k})$.

The map

$$\mathrm{Alt}_R^k(M; N) \to \mathrm{Hom}_R(\wedge_R^k M, N), \quad \mu \mapsto \mu_\wedge$$

is an R-module isomorphism (which makes sense for $k = 0$ as well). Finally, we remark that, for all l, m there is a unique bilinear map

$$\wedge : \wedge^l M \times \wedge^m M \to \wedge^{l+m} M, \quad (\omega, \rho) \mapsto \omega \wedge \rho$$

mapping $(p_1 \wedge \cdots \wedge p_l, p_1' \wedge \cdots \wedge p_m')$ to $p_1 \wedge \cdots \wedge p_l \wedge p_1' \wedge \cdots \wedge p_m'$, and called the *exterior product*, or the *wedge product*. The exterior product is associative and there exists a neutral element: $1 \in R \cong \wedge^0 M$. Additionally, it satisfies the following *graded commutativity* property:

$$\omega \wedge \omega' = (-)^{kk'} \omega' \wedge \omega,$$

for all $\omega \in \wedge^k M$ and all $\omega' \in \wedge^{k'} M$.

Proposition 1.72. *Let M be a free and finitely generated R-module, and let $B = (q_1, \ldots, q_n)$ be an ordered basis of M. Then, for all $k \geq 0$, the family*

$$B^\wedge := \left(q_{i_1} \wedge \cdots \wedge q_{i_k} \right)_{i_1 < \cdots < i_k}$$

is a (finite) basis in $\wedge^k M$. Additionally, there are canonical isomorphisms

$$\wedge^k M \cong \mathrm{Alt}^k(M^*; R) \quad \text{and} \quad \wedge^k M^* \cong \mathrm{Alt}^k(M; R).$$

Proof. We do not provide a proof. We only remark that the isomorphism $\wedge^k M \cong \mathrm{Alt}^k(M^*; R)$ in the second part of the statement is the only isomorphism mapping $p_1 \wedge \cdots \wedge p_k$ to the following alternating multilinear map:

$$(M^*)^{\times k} \to R, \quad (\varphi_1, \ldots, \varphi_k) \mapsto \sum_{\sigma \in S_k} (-)^\sigma \varphi_{\sigma(1)}(p_1) \cdots \varphi_{\sigma(k)}(p_k)$$

(and likewise for the isomorphism $\wedge^k M^* \cong \mathrm{Alt}^k(M; R)$). $\qquad\square$

Example 1.73 (Cross Product). Let $R = \mathbb{R}$ and $M = \mathbb{R}^3$. Let (E_1, E_2, E_3) be the canonical basis in $\mathbb{R}^3$. Then, $\wedge_{\mathbb{R}}^1 \mathbb{R}^3 = \mathbb{R}^3$ and $\wedge_{\mathbb{R}}^2 \mathbb{R}^3$ is spanned by

$$E_2 \wedge E_3, \quad E_3 \wedge E_1, \quad E_1 \wedge E_2.$$

Now take $A, B \in \mathbb{R}^3$. Then

$$A = A_1 E_1 + A_2 E_2 + A_3 E_3 \quad \text{and} \quad B = B_1 E_1 + B_2 E_2 + B_3 E_3,$$

for some $A_i, B_j \in \mathbb{R}$. A direct computation exploiting all the properties of the wedge product shows that

$$A \wedge B = \begin{vmatrix} A_2 & A_3 \\ B_2 & B_3 \end{vmatrix} E_2 \wedge E_3 - \begin{vmatrix} A_1 & A_3 \\ B_1 & B_3 \end{vmatrix} E_3 \wedge E_1 + \begin{vmatrix} A_1 & A_2 \\ B_1 & B_2 \end{vmatrix} E_1 \wedge E_2,$$

whose coefficients are the components of the cross product of A and B. This shows that the wedge product generalizes the standard cross product in $\mathbb{R}^3$. ♦

1.5 End-of-Chapter Problems

Problem 1.1. Interpret $\mathbb{Z}$ as a $\mathbb{Z}$-module. Show that the only submodules of $\mathbb{Z}$ are those of the type $n\mathbb{Z}$ for some $n \in \mathbb{Z}$.

Problem 1.2. Let R be a commutative ring with unit and let M be an R-module. Moreover, let $Q \subseteq N \subseteq M$ be submodules. Prove that the map

$$\frac{N}{Q} \to \frac{M}{Q}, \quad p \bmod Q \mapsto p \bmod Q,$$

is a well-defined injection which identifies N/Q with the submodule of M/Q consisting of equivalence classes represented by elements in N. Prove also that, if we understand the latter identification, then the map

$$\frac{M/Q}{N/Q} \mapsto \frac{M}{N}, \quad (p \bmod Q) \bmod N/Q \mapsto p \bmod N,$$

is a well-defined module isomorphism.

Problem 1.3. Show that there exists a non-injective linear map between modules which transforms every independent subset of its domain in an independent subset of its codomain (**Hint:** *Think at what does it mean "for every independent subset" and note that independent subsets might even not exists*).

Problem 1.4. Show that $\mathbb{Z}_n := \mathbb{Z}/n\mathbb{Z}$ is not a free $\mathbb{Z}$-module for all n.

Problem 1.5. Let R be a commutative ring with unit and let X be a set. Show that the R-module R^X is free if and only if X is a finite set.

Problem 1.6. Let R be a commutative ring with unit, let I, J be disjoint sets, and let $(M_i)_{i \in I}$, $(M_j)_{j \in J}$ be families of R-modules. Show that there are canonical module isomorphisms

$$\left(\bigoplus_{i \in I} M_i \right) \oplus \left(\bigoplus_{j \in J} N_j \right) \cong \bigoplus_{k \in I \cup J} M_k$$

and

$$\left(\prod_{i \in I} M_i \right) \oplus \left(\prod_{j \in J} M_j \right) \cong \prod_{k \in I \cup J} M_k.$$

Problem 1.7. Let R be a commutative ring with unit, and let $(X_i)_{i \in I}$ be a family of sets. Show that there are canonical module isomorphisms

$$\bigoplus_{i \in I} R X_i \cong R \left(\coprod_{i \in I} X_i \right)$$

and

$$\prod_{i \in I} R^{X_i} \cong R^{\coprod_{i \in I} X_i}.$$

Problem 1.8. Let R be a commutative ring with unit, and let

$$0 \longrightarrow N \xrightarrow{\alpha} M \xrightarrow{\beta} Q \longrightarrow 0$$

be a short exact sequence of R-modules. A splitting $s : Q \to M$ is also called a *right splitting*. On the other hand, a *left splitting* is a linear map $t : M \to N$ inverting α on the left: $t \circ \alpha = \mathrm{id}_N$. So, let t be a right splitting. Prove that the module M, together with the linear maps (t, β) is a direct product of (N, Q). Moreover, prove that

(1) given a right splitting s, for any $p \in M$, there exists a unique $p' \in N$ such that $\alpha(p') = p - s \circ \beta(p)$;
(2) the map $t_s : M \to N$, $p \mapsto p'$ is a right splitting;
(3) the assignment $s \mapsto t_s$ establishes a one-to-one correspondence between right and left splittings.

Finally, describe explicitly the inverse map from left to right splittings.

Problem 1.9. Let R be a commutative ring with unit, let N be an R-module, and let $X_1, \ldots, X_k$ be sets. Prove that the Multilinear Extension Theorem 1.50 establishes an R-module isomorphism

$$\mathrm{Mult}_k(RX_1, \ldots, RX_k; N) \cong N^{X_1 \times \cdots \times X_k},$$

where we are denoting by $N^{X_1 \times \cdots \times X_k} = \mathcal{F}(X_1 \times \cdots \times X_k, N)$ the module of maps $X_1 \times \cdots \times X_k \to N$.

Problem 1.10. Let R be a commutative ring with unit, let $(M_i)_{i \in I}$ be a family of R-modules, and let N be an additional R-module. Prove that there is a canonical R-module isomorphism

$$\mathrm{Hom}_R\left(\oplus_{i \in I} M_i, N\right) \cong \prod_{i \in I} \mathrm{Hom}_R\left(M_i, N\right).$$

Problem 1.11. Let R be a commutative ring with unit, let $(N_i)_{i \in I}$ be a family of R-modules, and let M be an additional R-module. Prove that there is a canonical R-module isomorphism

$$\mathrm{Hom}_R\left(M, \oplus_{i \in I} N_i\right) \cong \bigoplus_{i \in I} \mathrm{Hom}_R\left(M, N_i\right).$$

Problem 1.12. Let $\mathbb{K}$ be a field and let $M_{m,n}(\mathbb{K})$ be the space of $m \times n$ matrices over $\mathbb{K}$. Prove that, together with the bilinear map

$$\mathbb{K}^m \times \mathbb{K}^n \to M_{m,n}(\mathbb{K}),$$

$$\left((x_1, \ldots, x_m), (y_1, \ldots, y_n)\right) \mapsto \begin{pmatrix} x_1 y_1 & \cdots & x_1 y_n \\ \vdots & \ddots & \vdots \\ x_m y_1 & \cdots & x_m y_n \end{pmatrix},$$

$M_{m,n}(\mathbb{K})$ is a tensor product of $\mathbb{K}^m$ and $\mathbb{K}^n$.

Problem 1.13. Prove that the tensor product is *distributive* with respect to direct sum, i.e., let R be a commutative ring with unit, let $(M_i)_{i \in I}$ be a family of R-modules, and let N be an additional R-module; prove that there is a canonical R-module isomorphism

$$(\oplus_{i \in I} M_i) \otimes N \cong \bigoplus_{i \in I} (M_i \otimes N).$$

Problem 1.14. Let R be a commutative ring with unit, and let M, N be R-modules. Prove that, for all $k \in \mathbb{N}_0$, there are canonical R-module isomorphisms

$$(M \oplus N)^{\otimes k} \cong \bigoplus_{i+j=k} M^{\otimes i} \otimes N^{\otimes j},$$

where $M^{\otimes k} := \underbrace{M \otimes \cdots \otimes M}_{k\text{-times}}$ (likewise for $(M \oplus N)^{\otimes k}, N^{\otimes k}$), and

$$S^k(M \oplus N) \cong \bigoplus_{i+j=k} S^i M \otimes S^j N,$$

and

$$\wedge^k(M \oplus N) \cong \bigoplus_{i+j=k} \wedge^i M \otimes \wedge^j N.$$

Chapter 2

Chain and Cochain Complexes

In this chapter, we introduce and discuss our main objects of study, namely *chain* (resp. *cochain*) *complexes* and their *homology* (resp. *cohomology*). These objects appear frequently (both) in (pure and applied) mathematics, particularly topology, algebra and geometry. Here, we develop the elementary theory of (co)chain complexes, while motivations and applications are postponed to Chapters 4, 5 and 6. For more on the algebra aspects of (co)chain complexes, see, e.g., the excellent monographs on Homological Algebra (Rotman, 2009; Weibel, 1994).

2.1 (Co)Chain Complexes

Let R be a commutative ring with unit.

Definition 2.1 (Chain Complex). A *chain complex* of R-modules is a pair $(C_\bullet, d)$, where $C_\bullet = (C_i)_{i\in\mathbb{Z}}$ is a sequence of R-modules and $d = (d_i : C_i \to C_{i-1})_{i\in\mathbb{Z}}$ is a sequence of R-linear maps:

$$\cdots \xleftarrow{d_{i-1}} C_{i-1} \xleftarrow{d_i} C_i \xleftarrow{d_{i+1}} C_{i+1} \longleftarrow \cdots \qquad (2.1)$$

such that $d_i \circ d_{i+1} = 0$ for all $i \in \mathbb{Z}$, i.e., composing two successive arrows in (2.1), we get the 0 linear map. Elements in C_i are called *degree i chains*, or simply *i-chains*, while d_i is called the *i-th differential*. Degree i chains c such that $d_i c = 0$, i.e., $c \in \ker(d_i : C_i \to C_{i-1})$, are called *degree i cycles*, or *i-cycles*, while chains $a \in C_i$ such that there exists a chain $b \in C_{i+1}$ with $a = d_{i+1} b$, i.e., $a \in \operatorname{im}(d_{i+1} : C_{i+1} \to C_i)$, are called *degree i boundaries*, or *i-boundaries*. If $S \subseteq \mathbb{Z}$ is a subset, a chain complex $(C_\bullet, d)$ is *concentrated in degree S* if $C_i = 0$, the trivial module, for $i \notin S$.

If $(C_\bullet, d)$ is a chain complex of R-modules, we will often denote all the linear maps $d_i : C_i \to C_{i-1}$ by the same symbol $d : C_i \to C_{i-1}$ and write, for instance,

$$\cdots \xleftarrow{d} C_{i-1} \xleftarrow{d} C_i \xleftarrow{d} C_{i+1} \longleftarrow \cdots$$

instead of (2.1), or $d \circ d = 0$, instead of $d_i \circ d_{i+1} = 0$ for all i.

Remark 2.2. A chain complex $(C_\bullet, d)$ can also be encoded into a single R-module C together with an R-module endomorphism, also denoted $d : C \to C$, by putting $C := \bigoplus_{i \in \mathbb{Z}} C_i$. Then the endomorphism $d : C \to C$ is the unique linear map such that $dc = d_i c \in C_{i-1} \subseteq C$ for all i-chains $c \in C_i \subseteq C$. Yet in other words, $d : C \to C$ maps a sequence $(c_i)_{i \in \mathbb{Z}} \in C = \bigoplus_{i \in I} C_i$ to the sequence $(c_i' := d_{i+1} c_{i+1})_{i \in I} \in C$. Any R-module of the type $C = \bigoplus_{i \in \mathbb{Z}} C_i$ for some sequence $(C_i)_{i \in \mathbb{Z}}$ is called a *graded R-module*, and any linear map $f : C \to C$ for which there exists $k \in \mathbb{Z}$ such that $f(C_i) \subseteq C_{i+k}$ is called a *graded homomorphism of degree k*. We conclude that a chain complex can be encoded into a graded module C together with a graded endomorphism $d : C \to C$ of degree -1 such that $d \circ d = 0$. This point of view is very useful in many situations but is not adopted in this book. $\diamond$

Given a chain complex $(C_\bullet, d)$, the condition $d_i \circ d_{i+1} = 0$ is equivalent to

$$\operatorname{im} d_{i+1} \subseteq \ker d_i$$

(do you see it?). As both $\ker d_i$ and $\operatorname{im} d_{i+1}$ are submodules in C_i, we can form the quotient module

$$H_i(C, d) := \ker d_i / \operatorname{im} d_{i+1}.$$

The submodules $\ker d_i \subseteq C_i$ and $\operatorname{im} d_{i+1} \subseteq C_i$ are often denoted $Z_i(C, d)$ and $B_i(C, d)$ (B for boundaries), and we also adopt this notation in what follows.

Definition 2.3 (Homology). The *homology* of the chain complex $(C_\bullet, d)$ is the sequence of R-modules $H_\bullet(C, d) := (H_i(C, d))_{i \in \mathbb{Z}}$ with

$$H_i(C, d) := Z_i(C, d) / B_i(C, d), \quad i \in \mathbb{Z}.$$

The i-th space $H_i(C, d)$ in the sequence is called *degree i homology space*, or simply the *i-th homology*, and its elements are *degree i homology classes*. If $c \in C_i$ is an i-cycle, i.e., $dc = 0$, its class in $H_i(C, d)$ is called the *homology class* of c and it is denoted by $[c]_C$ (or simply $[c]$ if this does not lead to confusion). Two i-cycles $c, c' \in C_i$ are *homologous* if they have the same homology class: $[c] = [c']$; in other words, there exists an $(i+1)$-chain $b \in C_{i+1}$ such that $c - c' = db$. A chain complex $(C_\bullet, d)$ is *acyclic* if $H_i(C, d) = 0$ for all $i \in \mathbb{Z}$. Equivalently, a chain complex $(C_\bullet, d)$ is acyclic if $\ker d_i = \operatorname{im} d_{i+1}$ for all i, i.e., all cycles are boundaries. An acyclic chain complex is also called an *exact sequence* (of R-modules).

We now discuss a few examples.

Example 2.4. Let $C_\bullet = (C_i)_{i \in \mathbb{Z}}$ be a sequence of R-modules. We can define a chain complex $(C_\bullet, d)$ by putting $d_i = 0$ for all i:

$$\cdots \xleftarrow{0} C_{i-1} \xleftarrow{0} C_i \xleftarrow{0} C_{i+1} \longleftarrow \cdots .$$

In this case,

$$H_i(C, d) = \frac{\ker(0 : C_i \to C_{i-1})}{\operatorname{im}(0 : C_{i+1} \to C_i)} = \frac{C_i}{0} = C_i$$

for all i. ◆

Example 2.5. Every R-module homomorphism $f : M \to N$ can be seen as a chain complex concentrated, e.g., in degrees $-1, 0$ as follows. Put

$$C_i := \begin{cases} N & \text{if } i = -1 \\ M & \text{if } i = 0 \\ 0 & \text{otherwise} \end{cases}$$

and

$$d_i := \begin{cases} f & \text{if } i = 0 \\ 0 & \text{otherwise} \end{cases}.$$

Then $(C_\bullet, d)$ is a chain complex. In other words, $(C_\bullet, d)$ is the sequence

$$0 \longleftarrow \underset{-1}{N} \xleftarrow{f} \underset{0}{M} \longleftarrow 0$$

where everything else is 0, and we denoted the degrees explicitly. The homology is clearly 0 in degrees $i \neq -1, 0$. The 0-th homology is

$$H_0(C, d) = \frac{Z_0(C, d)}{B_0(C, d)} = \frac{\ker(f : M \to N)}{\operatorname{im}(0 : 0 \to M)} = \frac{\ker f}{0} = \ker f.$$

The -1-st homology is

$$H_{-1}(C, d) = \frac{Z_{-1}(C, d)}{B_{-1}(C, d)} = \frac{\ker(0 : N \to 0)}{\operatorname{im}(f : M \to N)} = \frac{N}{\operatorname{im} f}.$$

The quotient $N / \operatorname{im} f$ is also called the *cokernel* of f and it is denoted $\operatorname{coker} f$. ◆

Example 2.6. Define a chain complex $(C_\bullet, d)$ of abelian groups as follows. Put $C_i = \mathbb{Z}_8$ for all $i \in \mathbb{Z}$ and

$$d : \mathbb{Z}_8 \to \mathbb{Z}_8, \quad n \bmod 8 \mapsto 4 \cdot (n \bmod 8) = 4n \bmod 8.$$

It is clear that $d^2 = 0$ (do you see it?). So the sequence

$$\cdots \xleftarrow{d} \mathbb{Z}_8 \xleftarrow{d} \mathbb{Z}_8 \xleftarrow{d} \mathbb{Z}_8 \longleftarrow \cdots$$

is a chain complex. We want to *compute* the homology $H_\bullet(C, d)$. In general, *computing the homology* of a chain complex means describing it in the most explicit/efficient possible way. In the present case, we prove by hands that, for each $i \in \mathbb{Z}$, there is canonical isomorphism

$$\overline{\varphi} : H_i(C, d) \to \mathbb{Z}_2.$$

This will represent a good enough description for us. We begin describing the i-cycles and the i-boundaries. Note that the discussion is actually independent of i. So, let $c = n \bmod 8 \in \mathbb{Z}_8$ be a cycle. This means that $0 = dc = 4n \bmod 8$, i.e., $4n = 8k$, or equivalently $n = 2k$, for some $k \in \mathbb{Z}$. So,

$$Z_i(C, d) = \{2k \bmod 8 : k \in \mathbb{Z}\} \subseteq \mathbb{Z}_8.$$

Similarly,

$$B_i(C, d) = \{4h \bmod 8 : h \in \mathbb{Z}\} \subseteq Z_i(C, d).$$

Next, define a map $\varphi : Z_i(C, d) \to \mathbb{Z}_2$ by putting

$$\varphi(2k \bmod 8) = k \bmod 2.$$

It is easy to see that φ is a well-defined homomorphism of abelian groups, and $B_i(C, d) \subseteq \ker \varphi$ (Exercise 2.1). It follows that φ descends to a well-defined homomorphism

$$\overline{\varphi} : H_i(C, d) = \frac{Z_i(C, d)}{B_i(C, d)} \to \mathbb{Z}_2, \quad [2k \bmod 8]_C \mapsto k \bmod 2.$$

Now, prove that $\overline{\varphi}$ is an isomorphism as part of Exercise 2.1.   ◆

Exercise 2.1. Prove that the map $\varphi : Z_i(C, d) \to \mathbb{Z}_2$ in Example 2.6 is a well-defined abelian group homomorphism such that $B_i(C, d) \subseteq \ker \varphi$. Prove also that the induced homomorphism $\overline{\varphi} : H_i(C, d) \to \mathbb{Z}_2$ is an isomorphism.

Example 2.7. Every short exact sequence (1.6) can be seen as an acyclic chain complex concentrated, e.g., in degrees $-1, 0, 1$, extending it by 0 (do you see it?). ♦

Example 2.8. Let M be an R-module and let $\varphi : M \to R$ be a linear map. With these data, we can construct a chain complex $(C_\bullet, d_\varphi)$ as follows. For each $i \in \mathbb{Z}$, put

$$C_i := \begin{cases} 0 & \text{if } i < 0 \\ \wedge^i M & \text{if } i \geq 0 \end{cases},$$

and note that there exists a unique linear map

$$d_\varphi : \wedge^i M \to \wedge^{i-1} M$$

such that

$$d_\varphi(p_1 \wedge \cdots \wedge p_i) = \sum_{j=1}^{i} (-)^{j-1} \varphi(p_j) p_1 \wedge \cdots \wedge \widehat{p_j} \wedge \cdots \wedge p_i \qquad (2.2)$$

for all $p_1, \ldots, p_i \in M$, where a hat $\widehat{}$ denotes omission. To see this, it is enough to show that the right hand side of (2.2) is multilinear alternating in the arguments $(p_1, \ldots, p_i)$ and then use the universal property of the exterior power (see Exercise 2.2). The pair $(C_\bullet, d_\varphi)$ is a chain complex. Indeed, let $p_1, \ldots, p_i \in M$ and compute

$$
\begin{aligned}
& d_\varphi \circ d_\varphi(p_1 \wedge \cdots \wedge p_i) \\
&= d_\varphi \sum_j (-)^{j-1} \varphi(p_j) p_1 \wedge \cdots \wedge \widehat{p_j} \wedge \cdots \wedge p_i \\
&= \sum_{k<j} (-)^{k+j} \varphi(p_j) \varphi(p_k) p_1 \wedge \cdots \wedge \widehat{p_k} \wedge \cdots \wedge \widehat{p_j} \wedge \cdots \wedge p_i \\
&\quad + \sum_{j<k} (-)^{k+j-1} \varphi(p_j) \varphi(p_k) p_1 \wedge \cdots \wedge \widehat{p_j} \wedge \cdots \wedge \widehat{p_k} \wedge \cdots \wedge p_i = 0,
\end{aligned}
$$

where, for the last step, we just renamed the indexes. As $\wedge^i M$ is generated by elements of the form $p_1 \wedge \cdots \wedge p_i$, this is enough to conclude that $d_\varphi \circ d_\varphi = 0$ (do you see it?). We compute the homology of $(C_\bullet, d_\varphi)$ in the following section (under appropriate simplifying hypothesis). ♦

> **Exercise 2.2.** With the same notation as in Example 2.8, show that there is a unique linear map $d_\varphi : \wedge^i M \to \wedge^{i-1} M$ such that (2.2) holds (**Hint:** *Follow the indications in the Example*).

It is often convenient to interpret a chain complex as an ascending sequence (rather than a descending one). This is implemented in the following

Definition 2.9 (Cochain Complex). A *cochain complex* of R-modules is a pair $(C^\bullet, d)$ where $C^\bullet = (C^i)_{i \in \mathbb{Z}}$ is a sequence of R-modules and $d = (d^i : C^i \to C^{i+1})_{i \in \mathbb{Z}}$ is a sequence of R-linear maps:

$$\cdots \longrightarrow C^{i-1} \xrightarrow{d^{i-1}} C^i \xrightarrow{d^i} C^{i+1} \xrightarrow{d^{i+1}} \cdots \tag{2.3}$$

such that $d^{i+1} \circ d^i = 0$ for all $i \in \mathbb{Z}$. Elements in C^i are called *degree i cochains*, or simply *i-cochains*, while d^i is also called the *i-th differential*. Degree i cochains c such that $d^i c = 0$, are called *degree i cocycles*, or *i-cocycles*, while cochains $a \in C^i$ such that there exists a cochain $b \in C^{i-1}$ with $a = d_{i-1}b$ are called *degree i coboundaries*, or *i-coboundaries*.

There is no conceptual difference between chain and cochain complexes as, given a chain complex $(C_\bullet, d)$, we can construct a cochain complex $(C^\bullet, d)$ (containing exactly the same information) by putting $C^i = C_{-i}$ for all $i \in \mathbb{Z}$. So, the difference is purely conventional. However, as customary in the literature, we will keep the distinction and will adopt different notation/terminology for chain and cochain complexes. For instance, given a cochain complex $(C^\bullet, d)$, we denote

$$H^i(C, d) := \ker d^i / \operatorname{im} d^{i-1}$$

and we also put $Z^i(C, d) := \ker d^i$ and $B^i(C, d) := \operatorname{im} d^{i-1}$. Additionally, we give the following:

Definition 2.10 (Cohomology). The *cohomology* of the cochain complex $(C^\bullet, d)$ is the sequence of R-modules $H^\bullet(C, d) := (H^i(C, d))_{i \in \mathbb{Z}}$ with

$$H^i(C, d) := Z^i(C, d) / B^i(C, d), \quad i \in \mathbb{Z}.$$

The i-th space $H^i(C, d)$ in the sequence is called *degree i cohomology space*, or simply the *i-th cohomology*, and its elements are *degree i cohomology classes*. If $c \in C^i$ is an i-cocycle, its class in $H^i(C, d)$ is called the *cohomology class* of c and it is denoted by $[c]_C$ (or simply $[c]$ if this does not lead to confusion). Two i-cocycles $c, c' \in C^i$ are *cohomologous* if they have the same cohomology class: $[c] = [c']$. A cochain complex $(C^\bullet, d)$ is *acyclic* if $H^i(C, d) = 0$ for all $i \in \mathbb{Z}$. An acyclic cochain complex is also called an *exact sequence* (of R-modules).

Example 2.11. Let M be an R-module and let $q \in M$. With these data we can construct a cochain complex $(C^\bullet, d_q)$ as follows. For each $i \in \mathbb{Z}$, put

$$C^i := \begin{cases} 0 & \text{if } i < 0 \\ \wedge^i M & \text{if } i \geq 0 \end{cases},$$

and let d_q be given by

$$d_q : \wedge^i M \to \wedge^{i+1} M, \quad \omega \mapsto q \wedge \omega.$$

From $q \wedge q = 0$, and the associativity of the wedge product, it immediately follows that $d_q \circ d_q = 0$. ◆

2.2 (Co)Chain Maps

We now introduce a way to compare (co)chain complexes.

Definition 2.12 ((Co)Chain Map). A *chain map* (resp. a *cochain map*) between the chain complexes $(C_\bullet, d_C), (D_\bullet, d_D)$ (resp. the cochain complexes $(C^\bullet, d_C), (D^\bullet, d_D)$) of R-modules is a sequence $f = (f_i : C_i \to D_i)_{i \in \mathbb{Z}}$ (resp. $f = (f^i : C^i \to D^i)_{i \in \mathbb{Z}}$) of R-linear maps such that the diagram

$$\cdots \xleftarrow{d_C} C_{i-1} \xleftarrow{d_C} C_i \xleftarrow{d_C} C_{i+1} \xleftarrow{\quad} \cdots$$
$$\downarrow{f_{i-1}} \qquad \downarrow{f_i} \qquad \downarrow{f_{i+1}} \qquad \qquad (2.4)$$
$$\cdots \xleftarrow{d_D} D_{i-1} \xleftarrow{d_D} C_i \xleftarrow{d_D} D_{i+1} \xleftarrow{\quad} \cdots$$

(resp. the diagram

$$\cdots \xrightarrow{\quad} C^{i-1} \xrightarrow{d_C} C^i \xrightarrow{d_C} C^{i+1} \xrightarrow{d_C} \cdots$$
$$\downarrow{f^{i-1}} \qquad \downarrow{f^i} \qquad \downarrow{f^{i+1}} \qquad \qquad)$$
$$\cdots \xrightarrow{\quad} D^{i-1} \xrightarrow{d_D} D^i \xrightarrow{d_D} D^{i+1} \xrightarrow{d_D} \cdots$$

commutes, i.e., $f_i(d_C c) = d_D f_{i+1}(c)$ for all $c \in C_{i+1}$ (resp. $f^i(d_C c) = d_D f^{i-1}(c)$ for all $c \in C^{i-1}$), $i \in \mathbb{Z}$. In this case, we write

$$f : (C_\bullet, d_C) \to (D_\bullet, d_D) \quad (\text{resp. } f : (C^\bullet, d_C) \to (D^\bullet, d_D)).$$

Let $f : (C_\bullet, d_C) \to (D_\bullet, d_D)$ be a chain map. We often denote the linear maps $f_i : C_i \to D_i$ simply by $f : C_i \to D_i$ and write, for instance,

$$\cdots \xleftarrow{d_C} C_{i-1} \xleftarrow{d_C} C_i \xleftarrow{d_C} C_{i+1} \xleftarrow{\quad} \cdots$$
$$\downarrow{f} \qquad \downarrow{f} \qquad \downarrow{f}$$
$$\cdots \xleftarrow{d_D} D_{i-1} \xleftarrow{d_D} D_i \xleftarrow{d_D} D_{i+1} \xleftarrow{\quad} \cdots$$

instead of (2.4), or $f \circ d_C = d_D \circ f$, instead of $f_i \circ d_C = d_D \circ f_{i+1}$. Likewise for cochain maps.

Before providing examples, we discuss the main properties of (co)chain maps. We discuss the chain case and leave it to the reader to translate all the statements to the "cochain language".

Definition/Proposition 2.13. *Let* $(C_\bullet, d), (D_\bullet, d_D), (E_\bullet, d_E), (C'_\bullet, d')$ *be chain complexes.*

(1) We define the *identity chain map*

$$\mathrm{id}_C : (C_\bullet, d) \to (C_\bullet, d)$$

as the sequence $\mathrm{id}_C := (\mathrm{id}_{C_i} : C_i \to C_i)_{i \in \mathbb{Z}}$ and it is a chain map.

(2) Let

$$(C_\bullet, d) \xrightarrow{f} (D_\bullet, d_D) \xrightarrow{g} (E_\bullet, d_E)$$

be chain maps. We define the *composition*

$$g \circ f : (C_\bullet, d) \to (E_\bullet, d_E)$$

of f followed by g as the sequence $g \circ f := (g_i \circ f_i : C_i \to E_i)_{i \in \mathbb{Z}}$ and it is a chain map.

(3) If

$$\Phi : (C_\bullet, d) \to (C'_\bullet, d')$$

is an *invertible chain map*, i.e., $\Phi_i : C_i \to C'_i$ is invertible for all i, then we also call Φ a *chain isomorphism* and define its *inverse*

$$\Phi^{-1} : (C'_\bullet, d') \to (C_\bullet, d)$$

as the sequence $\Phi^{-1} := (\Phi_i^{-1} : C'_i \to C_i)_{i \in \mathbb{Z}}$ and it is a chain isomorphism. Two chain complexes that can be connected by a chain isomorphism are called *isomorphic*.

Likewise for cochain complexes and cochain maps.

Proof.   Left as Exercise 2.3. $\square$

Exercise 2.3. Prove the proposition part of Definition/Proposition 2.13.

Let $f : (C_\bullet, d_C) \to (D_\bullet, d_D)$ be a chain map. Then f maps cycles to cycles and boundaries to boundaries, i.e., for all $i \in \mathbb{Z}$,

$$f(Z_i(C, d_C)) \subseteq Z_i(D, d_D) \quad \text{and} \quad f(B_i(C, d_C)) \subseteq B_i(D, d_D),$$

indeed let $c \in Z_i(C, d_C)$ be an i-cycle. Compute

$$d_D f(c) = f(d_C c) = f(0) = 0,$$

showing that $f(c)$ is an i-cycle as well. Similarly, if c is an i-boundary, then there exists an $(i+1)$-chain b such that $c = d_C b$ and

$$f(c) = f(d_C b) = d_D f(b),$$

showing that $f(c)$ is also an i-boundary. It immediately follows from Corollary 1.24 that f *induces a linear map in homology*, i.e., the assignment

$$H_i(f) : H_i(C, d_C) \to H_i(D, d_D), \quad [c]_C \mapsto [f(c)]_D$$

is a well-defined linear map for all $i \in \mathbb{Z}$. We also use the symbol $H_\bullet(f)$ for the sequence $(H_i(f) : H_i(C, d_C) \to H_i(D, d_D))_{i \in \mathbb{Z}}$. Similarly, a cochain map $f : (C^\bullet, d_C) \to (D^\bullet, d_D)$ maps cocycles to cocycles and coboundaries to coboundaries, hence it induces a well-defined linear map

$$H^i(f) : H^i(C, d_C) \to H^i(D, d_D), \quad [c]_C \mapsto [f(c)]_D$$

in cohomology, for all i, and we put $H^\bullet(f) := (H^i(f) : H^i(C, d_C) \to H^i(D, d_D))_{i \in \mathbb{Z}}$.

Example 2.14. Let M be an R-module and let $f : M \to N$ be a linear map. First of all, note that, for any $i \in \mathbb{Z}$, there exists a unique linear map

$$\wedge^i f : \wedge^i M \to \wedge^i N$$

such that

$$\wedge^i f(p_1 \wedge \cdots \wedge p_i) = f(p_1) \wedge \cdots \wedge f(p_i) \tag{2.5}$$

for all $p_1, \ldots, p_i \in M$. This follows in the usual way from the universal property of the exterior power and the fact that the right hand side of

(2.5) is multilinear and alternating in the arguments $p_1, \ldots, p_i$. Put $\wedge^\bullet f :=$ $(\wedge^i f : \wedge^i M \to \wedge^i N)_{i \in \mathbb{Z}}$. Now, let $\varphi \in M^*$ and let $(\wedge^\bullet M, d_\varphi)$ be the associated chain complex as in Example 2.8. We claim that, when $\varphi = \psi \circ f$ for some $\psi \in N^*$, then $\wedge^\bullet f$ is a chain map:

$$\wedge^\bullet f : (\wedge^\bullet M, d_\varphi) \to (\wedge^\bullet N, d_\psi).$$

We leave it to the reader to check the details as Exercise 2.4.   ◆

Exercise 2.4. Prove all the unproven claims in Example 2.14.

Exercise 2.5. Let M be an R-module, let $q \in M$ and let $(\wedge^\bullet M, d_q)$ be the cochain complex described in Example 2.11. Prove that, for any linear map $f : M \to N$, the sequence $\wedge^\bullet f$ described in Example 2.14 is a cochain map

$$\wedge^\bullet f : (\wedge^\bullet M, d_q) \to (\wedge^\bullet N, d_{f(q)}).$$

Proposition 2.15. *Let* $(C_\bullet, d), (D_\bullet, d_D), (E_\bullet, d_E), (C'_\bullet, d')$ *be chain complexes. Then*

(1) *the identity chain map* $\mathrm{id}_C : (C_\bullet, d) \to (C_\bullet, d)$ *induces the identity in homology:*

$$H_i(\mathrm{id}_C) = \mathrm{id}_{H_i(C,d)} \quad \text{for all } i \in \mathbb{Z};$$

(2) *the composition* $g \circ f$ *of two chain maps* $(C_\bullet, d) \xrightarrow{f} (D_\bullet, d_D) \xrightarrow{g} (E_\bullet, d_E)$ *induces the compositions of the induced maps in homology:*

$$H_i(g \circ f) = H_i(g) \circ H_i(f) \quad \text{for all } i \in \mathbb{Z};$$

(3) *an isomorphism of chain complexes* $\Phi : (C_\bullet, d) \to (C'_\bullet, d')$ *induces an R-module isomorphism in homology whose inverse is the map induced by the inverse isomorphism:*

$$\text{there exists } H_i(\Phi)^{-1} \text{ and } \quad H_i(\Phi)^{-1} = H_i(\Phi^{-1}) \quad \text{for all } i \in \mathbb{Z}.$$

In particular, isomorphic chain complexes have isomorphic homologies. Likewise for cochain complexes.

Proof.   Left as Exercise 2.6.   □

> **Exercise 2.6.** Prove Proposition 2.15.

It may happen that a (co)chain map is not an isomorphism of (co)chain complexes, yet it induces an isomorphism in (co)homology. We will see various examples of this phenomenon in what follows.

Definition 2.16 (Quasi-isomorphism). A *quasi-isomorphism* of chain (resp. cochain) complexes is a chain map $f : (C_\bullet, d) \to (C'_\bullet, d')$ (resp. a cochain map $f : (C^\bullet, d) \to (C'^\bullet, d'))$ inducing an isomorphism in homology (resp. in cohomology), i.e., for all $i \in \mathbb{Z}$, the induced linear map $H_i(f) : H_i(C, d) \to H_i(C', d')$ (resp. $H^i(f) : H^i(C, d) \to H^i(C', d'))$ is an R-module isomorphism.

Example 2.17. Let $(C_\bullet, d)$ be a chain complex. Consider also the trivial complex $(0_\bullet, 0)$ where all the chains and all the differentials are zero. The latter is obviously an acyclic complex. There is a unique chain map $0 : (C_\bullet, d) \to (0_\bullet, 0)$, the zero map (do you see that it is a chain map?). Such a chain map induces the zero map in homology $0 : H_\bullet(C, d) \to 0$. It is clear that $(C_\bullet, d)$ is acyclic if and only if $0 : (C_\bullet, d) \to (0_\bullet, 0)$ is a quasi-isomorphism. Likewise for cochain complexes. ◆

The (co)homology contains an important information about a (co)chain complex. Hence, it is important to develop techniques to compute it. In the following two sections, we present two such techniques that play a particularly important role in Chapters 5 and 6.

2.3 Algebraic Homotopies

Let $f : (C_\bullet, d_C) \to (D_\bullet, d_D)$ be a chain map between chain complexes. As we already mentioned, it might happen that f is not an isomorphism of chain complexes yet $H_i(f) : H_i(C, d_C) \to H_i(D, d_D)$ is an isomorphism (for some or) for all i. In order to explore this phenomenon, we begin presenting a sufficient condition under which two chain maps $f, g : (C_\bullet, d_C) \to (D_\bullet, d_D)$ induce the same map in homology: $H_\bullet(f) = H_\bullet(g)$.

Definition 2.18 (Homotopy). A *homotopy* (more precisely an *algebraic homotopy*) between the chain maps $f, g : (C_\bullet, d_C) \to (D_\bullet, d_D)$ is a sequence $h = (h_i : C_i \to D_{i+1})_{i \in \mathbb{Z}}$ of R-linear maps such that

$$f_i - g_i = d_D \circ h_i + h_{i-1} \circ d_C, \quad \text{for all } i \in \mathbb{Z}. \tag{2.6}$$

Similarly, a *homotopy* between the cochain maps $f, g : (C^\bullet, d_C) \to (D^\bullet, d_D)$ is a sequence $h = (h^i : C^i \to D^{i-1})_{i \in \mathbb{Z}}$ of R-linear maps such that

$$f^i - g^i = d_D \circ h^i + h^{i+1} \circ d_C, \quad \text{for all } i \in \mathbb{Z}. \tag{2.7}$$

Two (co)chain maps f, g are said to be *homotopic* (or to *agree up to homotopy*) if there exists a homotopy h between them. In this case, we write $f \sim_h g$. A (co)chain map f is *null-homotopic* if it is homotopic to the zero (co)chain map, i.e., $f \sim_h 0$ for some homotopy h.

The situation in Definition 2.18 is illustrated in the following two diagrams: for chain maps,

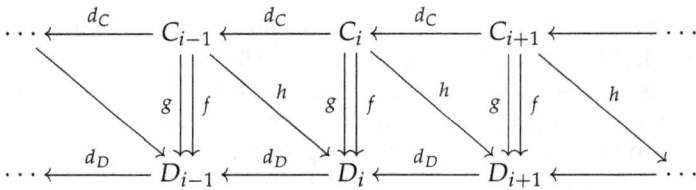

and for cochain maps,

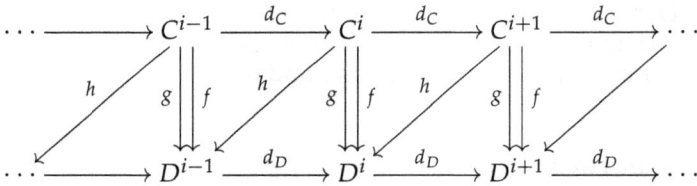

(beware that such diagrams *do not* commute). Sometimes, we simply write $f - g = d_D \circ h + h \circ d_C$ instead of (2.6) or (2.7).

Exercise 2.7. Show that "being homotopic" is an equivalence relation on the set of (co)chain maps (between two given (co)chain complexes). More precisely, if $f, g, j : (C_\bullet, d_C) \to (D_\bullet, d_D)$ are chain maps such that $f \sim_h g$ and $g \sim_k j$ for some homotopies h, k, then

- $f \sim_0 f$ (reflexivity),
- $g \sim_{-h} f$ (symmetry),
- $f \sim_{h+k} j$ (transitivity).

Likewise for cochain maps.

We discuss a few more examples at the end of this section. Now, we develop a little bit further the theory and show how homotopies may help computing (co)homologies of (co)chain complexes.

Proposition 2.19. *Homotopies respect the composition of (co)chain maps. More precisely if*

$$(C_\bullet, d_C) \underset{g}{\overset{f}{\rightrightarrows}} (D_\bullet, d_D) \underset{g'}{\overset{f'}{\rightrightarrows}} (E_\bullet, d_E)$$

are chain maps such that $f \sim_h g$ and $f' \sim_{h'} g'$ for some homotopies h, h', then there exists a homotopy H (to be specified in the proof) such that $f' \circ f \sim_H g' \circ g$. Likewise for cochain maps.

Proof.   We want to compare $f' \circ f$ and $g' \circ g$ and show that they agree up to homotopy. We use a trick: compute

$$
\begin{aligned}
f' &\circ f - g' \circ g \\
&= f' \circ f - f' \circ g + f' \circ g - g' \circ g \\
&= f' \circ (f - g) + (f' - g') \circ g & \text{(Example 1.53)} \\
&= f' \circ (d_D \circ h + h \circ d_C) + (d_E \circ h' + h' \circ d_D) \circ g & (f \sim_h g \text{ and } f' \sim_{h'} g') \\
&= f' \circ d_D \circ h + f' \circ h \circ d_C + d_E \circ h' \circ g + h' \circ d_D \circ g & \text{(Example 1.53)} \\
&= d_E \circ f' \circ h + f' \circ h \circ d_C + d_E \circ h' \circ g + h' \circ g \circ d_C & (f' \text{ and } g \text{ are chain maps}) \\
&= d_E \circ (f' \circ h + h' \circ g) + (f' \circ h + h' \circ g) \circ d_C & \text{(Example 1.53).}
\end{aligned}
$$

This shows that

$$H := f' \circ h + h' \circ g = (f'_{i+1} \circ h_i + h'_i \circ g_i : C_i \to E_{i+1})_{i \in \mathbb{Z}}$$

is the desired homotopy.   □

Proposition 2.20. *Let $f, g : (C_\bullet, d_C) \to (D_\bullet, d_D)$ be homotopic chain maps. Then f and g induce the same map in homology:*

$$H_i(f) = H_i(g), \quad \text{for all } i \in \mathbb{Z}.$$

Likewise for cochain maps.

Proof.   Let h be a homotopy such that $f \sim_h g$. Pick a cycle $c \in Z_\bullet(C, d_C)$ in (C, d_C), and let $[c]$ be its cohomology class. Compute

$$H_\bullet(f)[c] = [f(c)] = [g(c) + d_D h(c) + h(d_C c)] = [g(c)] = H_\bullet(g)[c],$$

where, in the second step, we used that $f \sim_h g$, and, in the third step, we used that $d_C c = 0$ and that $g(c) + d_D h(c)$ and $g(c)$ are homologous. It follows from the arbitrariness of c that $H_\bullet(f) = H_\bullet(g)$ as desired. The same exact proof works for cochain complexes and we invite the reader to check the details.   □

Corollary 2.21. *If $f : (C_\bullet, d_C) \to (D_\bullet, d_D)$ is a null-homotopic chain map, then $H_\bullet(f) = 0$, i.e., f induces the zero map in homology. Likewise for cochain maps.*

Corollary 2.22. *Let $(C_\bullet, d)$ be a chain complex. If there exists a sequence of maps $h = (h_i : C_i \to C_{i+1})_{i \in \mathbb{Z}}$ such that*

$$d \circ h + h \circ d = \mathrm{id}_C,$$

then $(C_\bullet, d)$ is acyclic. Likewise for cochain complexes.

Proof.   The hypothesis means that the identity chain map id_C is null-homotopic. Hence, it induces the null map in homology, i.e., for every cycle c

$$[c] = [\mathrm{id}_C(c)] = H_\bullet(\mathrm{id}_C)[c] = 0.$$

This concludes the proof.   □

Definition 2.23 (Contracting Homotopy). A sequence h as in Corollary 2.22 is called a *contracting homotopy* or simply a *contraction* for the chain complex $(C_\bullet, d)$. Likewise for cochain complexes.

The terminology "contraction" is clarified in Chapter 5. We now illustrate the theory of algebraic homotopies with various examples of (co)chain complexes equipped with a contracting homotopy showing that the (co)homology does actually vanish. We discuss more examples (and examples of a more general nature) in Chapters 5 and 6.

Example 2.24. Let M be an R-module, let $q \in M$ and let $\varphi : M \to R$ be a linear map. Consider the chain complex $(\wedge^\bullet M, d_\varphi)$ of Example 2.8 and the cochain complex $(\wedge^\bullet M, d_q)$ of Example 2.11. We claim that, if $\varphi(q) = 1$, then d_φ is a contracting homotopy for $(\wedge^\bullet M, d_q)$ and vice versa d_q is a contracting homotopy for $(\wedge^\bullet M, d_\varphi)$. We can discuss the two things at once proving that

$$d_\varphi \circ d_q + d_q \circ d_\varphi = \mathrm{id}_{\wedge^\bullet M}.$$

In order to check that $d_\varphi \circ d_q + d_q \circ d_\varphi$ and $\mathrm{id}_{\wedge^\bullet M}$ agree, it is enough to check that they agree on elements of the form $p_1 \wedge \cdots \wedge p_i$, $p_1, \ldots, p_i \in M$. So, compute

$$d_\varphi \circ d_q(p_1 \wedge \cdots \wedge p_i)$$
$$= d_\varphi(q \wedge p_1 \wedge \cdots \wedge p_i)$$
$$= \varphi(q)p_1 \wedge \cdots \wedge p_i - \sum_{j=1}^{i}(-)^{j-1}\varphi(p_j)q \wedge p_1 \wedge \cdots \wedge \widehat{p_j} \wedge \cdots \wedge p_i$$
$$= p_1 \wedge \cdots \wedge p_i - q \wedge \sum_{j=1}^{i}(-)^{j-1}\varphi(p_j)p_1 \wedge \cdots \wedge \widehat{p_j} \wedge \cdots \wedge p_i$$
$$= p_1 \wedge \cdots \wedge p_i - d_q \circ d_\varphi(p_1 \wedge \cdots \wedge p_i).$$

This shows that $d_\varphi \circ d_q + d_q \circ d_\varphi = \mathrm{id}_{\wedge^\bullet M}$ as claimed.

We conclude that, when there exists $q' \in M$ such that $a := \varphi(q') \in R$ is an invertible element (with respect to the product), then the chain complex $(\wedge^\bullet M, d_\varphi)$ is acyclic (just use $d_{a^{-1}q'}$ as a contracting homotopy). Similarly, when there exists $\varphi' \in M^*$ such that $b := \varphi'(q)$ is invertible, then $(\wedge^\bullet M, d_q)$ is acyclic (use $d_{b^{-1}\varphi'}$ as a contracting homotopy). This happens, e.g., when $R = \mathbb{K}$ is a field and both φ, q are non-zero (do you see it?). $\blacklozenge$

Example 2.25 (Polynomial de Rham Complex). Let M be an R-module. For this example, we assume that the canonical ring homomorphism $\mathbb{Z} \to R$ maps every non-zero integer to an invertible element in R. This happens, e.g., when R is a field of zero characteristic. For every integer n, we construct a different cochain complex $(C(n)^\bullet, d)$ by putting

$$C(n)^i = S^{n-i}M \otimes \wedge^i M.$$

The differential d is defined as follows. At the level i, it is the unique R-linear map

$$d^i : C(n)^i = S^{n-i}M \otimes \wedge^i M \to C(n)^{i+1} = S^{n-i-1}M \otimes \wedge^{i+1}M$$

such that

$$d^i(p_1 \vee \cdots \vee p_{n-i} \otimes \omega) := \sum_{j=1}^{n-i} p_1 \vee \cdots \vee \widehat{p_j} \vee \cdots \vee p_{n-i} \otimes p_j \wedge \omega,$$

for all $p_1, \ldots, p_{n-i} \in M$ and all $\omega \in \wedge^i M$. It is not hard to see that $d^{i+1} \circ d^i = 0$ for all i (and all n), so $(C(n)^\bullet, d)$ is a cochain complex (called the *polynomial de Rham complex*). When $n > 0$, there is a canonical contracting homotopy h for $(C(n)^\bullet, d)$. Namely,

$$h^i : C(n)^i = S^{n-i} M \otimes \wedge^i M \to C(n)^{i-1} = S^{n-i+1} M \otimes \wedge^{i-1} M$$

is the unique linear map such that

$$h^i(\mathcal{P} \otimes q_1 \wedge \cdots \wedge q_i) = \frac{1}{n} \sum_{j=1}^{i} (-)^{j-1} \mathcal{P} \vee q_j \otimes q_1 \wedge \cdots \wedge \widehat{q_j} \wedge \cdots \wedge q_i,$$

for all $\mathcal{P} \in S^{n-i} M$ and all $q_1, \ldots, q_i \in M$. Note that the factor $1/n$ in the latter formula makes sense exactly because we are assuming that n is invertible in R. A direct computation shows that

$$d \circ h + h \circ d = \mathrm{id}_{C(n)^\bullet},$$

hence $(C(n)^\bullet, d)$ is acyclic.   ◆

Exercise 2.8. Prove all unproven claims in Example 2.25.

Example 2.26. Let M be an R-module. As in Example 2.25, we assume that every non-zero integer is invertible in R. For every integer n, define a chain complex $(C(n)_\bullet, d)$ by putting

$$C(n)_i = S^{n+i} M^* \otimes \wedge^i M.$$

The differential d is defined, on $C(n)_i$, as the unique linear map such that

$$d(\varphi_1 \vee \cdots \vee \varphi_{n+i} \otimes \omega) = \sum_{j=1}^{n+i} \varphi_1 \vee \cdots \vee \widehat{\varphi_j} \vee \cdots \vee \varphi_{n+i} \otimes d_{\varphi_j} \omega$$

$$\in C(n)_{i-1} = S^{n+i-1} M^* \otimes \wedge^{i-1} M,$$

for all $\varphi_1, \ldots, \varphi_{n+i} \in M^*$, and all $\omega \in \wedge^i M$, where, for every $\varphi \in M^*$, d_φ is the differential defined in Example 2.8. Now assume that M is free and finitely generated, and let $(e_a)_{a=1,\ldots,m}$ be a finite basis of M (of cardinality m). In this special case, if $n + m > 0$, then $(C(n)_\bullet, d)$ possesses a contracting homotopy h defined as follows. Let $(e^a)_{a=1,\ldots,m}$ be the dual basis in M^*. Then h is defined, on $C(n)_i$, as the unique linear map such that

$$h(\mathcal{P} \otimes \omega) = \frac{1}{n+m} \sum_{a=1}^{m} \mathcal{P} \vee e^a \otimes e_a \wedge \omega$$

$$\in C(n)_{i+1} = S^{n+i+1}M^* \otimes \wedge^{i+1}M,$$

for all $\mathcal{P} \in S^{n+i}M^*$ and all $\omega \in \wedge^i M$. Note that the factor $1/(n+m)$ in the last formula makes sense when $n + m \neq 0$ (and non-zero integers are invertible in R). A direct computation (exploiting some little tricks) shows that $h \circ d + d \circ h - \text{id}_{C(n)_\bullet}$ vanishes on elements of the form

$$e^{a_1} \vee \cdots \vee e^{a_{n+i}} \otimes e_{b_1} \wedge \cdots \wedge e_{b_i}$$

hence it vanishes everywhere. This shows that h is a contracting homotopy and $(C(n)_\bullet, d)$ is acyclic. ♦

Exercise 2.9. Prove all unproven claims in Example 2.26.

Example 2.27 (de Rham Complex of $\mathbb{R}^3$). Let $U \subseteq \mathbb{R}^n$ be a non-empty open subset in the standard Euclidean space $\mathbb{R}^n$. We denote by $C^\infty(U, \mathbb{R}^m)$ the real vector space of *smooth* $\mathbb{R}^m$-valued maps on U, i.e., functions $U \to \mathbb{R}^m$ that are differentiable arbitrarily many times (as usual, maps taking values in a module, in this case a vector space, are added and multiplied by a scalar point-wisely). When $m = 1$, we simply denote $C^\infty(U)$ (instead of $C^\infty(U, \mathbb{R})$). Smooth maps $C^\infty(U, \mathbb{R}^n)$ (i.e., $m = n$) can also be interpreted as *vector fields* on U (see also Chapter 6).

We define a cochain complex $(C^\bullet, d)$ by putting

$$C^i = \begin{cases} \mathbb{R} & \text{if } i = -1 \\ C^\infty(\mathbb{R}^3) & \text{if } i = 0, 3 \\ C^\infty(\mathbb{R}^3, \mathbb{R}^3) & \text{if } i = 1, 2 \\ 0 & \text{otherwise} \end{cases}.$$

In order to describe the differential, we introduce the following useful notation for vector calculus on $\mathbb{R}^3$. We put an arrow "$\vec{}$" over vector (i.e., 3 component) quantities. For instance, we denote by $\vec{x} = (x_1, x_2, x_3)$ the standard coordinates on $\mathbb{R}^3$, or by $\vec{F} = (F_1, F_2, F_3)$ a vector valued map $\vec{F} : \mathbb{R}^3 \to \mathbb{R}^3$. Additionally, we denote by $\vec{\nabla}$ the standard vector operator $(\frac{\partial}{\partial x_1}, \frac{\partial}{\partial x_2}, \frac{\partial}{\partial x_3})$. Finally, we denote by a cross "$\times$" and by a dot "$\cdot$" the usual vector and scalar products in $\mathbb{R}^3$ (both can be applied to vector operators in the obvious way). The differential is now given by

$$d^{-1}c = c \quad \text{(the constant function equal to } c\text{)}$$

$$d^0 f = \text{grad } f := \vec{\nabla} f \in C^\infty(\mathbb{R}^3, \mathbb{R}^3)$$

$$d^1 \vec{F} = \text{curl } \vec{F} := \vec{\nabla} \times \vec{F} \in C^\infty(\mathbb{R}^3, \mathbb{R}^3)$$

$$d^2 \vec{G} = \text{div } \vec{G} := \vec{\nabla} \cdot \vec{G} \in C^\infty(\mathbb{R}^3)$$

$$d^i = 0 \quad \text{for } i \notin \{-1, 0, 1, 2\}$$

for all $f \in C^\infty(\mathbb{R}^3)$, all $\vec{F}, \vec{G} \in C^\infty(\mathbb{R}^3, \mathbb{R}^3)$, and all $c \in \mathbb{R}$.

In this way, we get a sequence

$$0 \longrightarrow \mathbb{R} \longrightarrow C^\infty(\mathbb{R}^3) \xrightarrow{\text{grad}} C^\infty(\mathbb{R}^3, \mathbb{R}^3) \xrightarrow{\text{curl}} C^\infty(\mathbb{R}^3, \mathbb{R}^3) \xrightarrow{\text{div}} C^\infty(\mathbb{R}^3) \longrightarrow 0.$$
$$(2.8)$$

A direct computation shows that this sequence is actually a cochain complex of $\mathbb{R}$-vector spaces (concentrated in degrees $-1, 0, 1, 2, 3$), i.e., $\text{curl} \circ \text{grad} = \text{div} \circ \text{curl} = 0$. This cochain complex possesses a canonical contracting homotopy

$$0 \longleftarrow \mathbb{R} \xleftarrow{h^0} C^\infty(\mathbb{R}^3) \xleftarrow{h^1} C^\infty(\mathbb{R}^3, \mathbb{R}^3) \xleftarrow{h^2} C^\infty(\mathbb{R}^3, \mathbb{R}^3) \xleftarrow{h^3} C^\infty(\mathbb{R}^3) \longleftarrow 0$$

given by

$$h^0 f = f(0) \in \mathbb{R}$$

$$h^1 \vec{F} = \int_0^1 \vec{F}(t\vec{x}) \cdot \vec{x}\, dt \in C^\infty(\mathbb{R}^3)$$

$$h^2 \vec{G} = \int_0^1 \vec{G}(t\vec{x}) \times \vec{x}\, t\, dt \in C^\infty(\mathbb{R}^3, \mathbb{R}^3)$$

$$h^3 g = \int_0^1 g(t\vec{x})\, \vec{x}\, t^2 dt \in C^\infty(\mathbb{R}^3, \mathbb{R}^3)$$

$$h^i = 0 \quad \text{for } i \notin \{0, 1, 2, 3\}$$

That $h = (h^i)_{i \in \mathbb{Z}}$ is a contracting homotopy can be proved by hands and we leave the details to the reader as Exercise 2.10. Here, we conclude that the cochain complex (2.8) is acyclic. This means in particular that

- the only gradient-free functions on $\mathbb{R}^3$ are the constant ones;
- the only curl-free vector fields on $\mathbb{R}^3$ are the gradients;
- the only divergence-free vector fields on $\mathbb{R}^3$ are the curls;
- every function is a divergence.

This example is greatly generalized in Chapter 6. ◆

Exercise 2.10. Prove all unproven claims in Example 2.27.

There is a special class of quasi-isomorphisms, called *homotopy equivalences*, that play an important role in Homological Algebra. In the last part of this section, we define them and discuss their basic properties. We postpone (non-trivial) examples to Chapters 5 and 6.

Definition 2.28 (Homotopy Equivalence). A chain map $F : (C_\bullet, d) \to (C'_\bullet, d')$ is a *homotopy equivalence* if there exists a chain map in the other direction $G : (C'_\bullet, d') \to (C_\bullet, d)$ such that $G \circ F$ is homotopic to the identity of $(C_\bullet, d)$ and $F \circ G$ is homotopic to the identity of $(C'_\bullet, d')$. In symbols,

$$G \circ F \sim_J \mathrm{id}_C \quad \text{and} \quad F \circ G \sim_K \mathrm{id}_{C'}$$

for some homotopies J, K. In this situation, G is clearly a homotopy equivalence as well. We also say that G is a *homotopy inverse* of F (and vice versa) or that G *inverts F up to homotopy*. If $(C_\bullet, d), (C'_\bullet, d')$ are connected by a homotopy equivalence, we say that they are *homotopy equivalent* or *isomorphic up to homotopy*. Likewise for cochain complexes.

Proposition 2.29. *Let* $F : (C_\bullet, d) \to (C_\bullet, d')$ *be a homotopy equivalence with homotopy inverse* $G : (C'_\bullet, d') \to (C_\bullet, d)$. *Then both F, G are quasi-isomorphisms inducing mutually inverse module isomorphisms in homology, i.e.,* $H_i(F) : H_i(C, d) \to H_i(C', d')$ *and* $H_i(G) : H_i(C', d') \to H_i(C, d)$ *are module isomorphisms and*

$$H_i(F)^{-1} = H_i(G) \quad \text{for all } i \in \mathbb{Z}.$$

In particular, homotopy equivalent chain complexes have isomorphic homologies. Likewise for cochain complexes.

Proof. Let J, K be homotopies such that $G \circ F \sim_J \mathrm{id}_C$ and $F \circ G \sim_K \mathrm{id}_{C'}$. From the first homotopy, we get

$$H_i(F) \circ H_i(G) = H_i(F \circ G) = H_i(\mathrm{id}_{C'}) = \mathrm{id}_{H_i(C', d')},$$

where we also used Proposition 2.15 (1)–(2) for all i. Swapping the roles of F and G we get $H_i(G) \circ H_i(F) = \mathrm{id}_{H_i(C, d)}$. This concludes the proof. $\square$

Exercise 2.11. Let $(C_\bullet, d)$ be a chain complex possessing a contracting homotopy h. Show that the only chain map $(C_\bullet, d) \to (0_\bullet, 0)$ to the zero chain complex is a homotopy equivalence. Likewise for cochain complexes.

Proposition 2.30. *Homotopy equivalence of (co)chain complexes is an equivalence relation.*

Proof. We discuss the chain complex case, and we leave to the reader the translation to the "cochain language". It is clear that the identity chain map $\mathrm{id}_C : (C_\bullet, d) \to (C_\bullet, d)$ is a homotopy equivalence, with the involved homotopies being the zero maps (do you see it?). Hence, homotopy equivalence is a reflexive relation. It is also clear that it is a symmetric relation and it remains to prove that it is transitive. So, let

$$(C_\bullet, d) \underset{G}{\overset{F}{\rightleftarrows}} (C'_\bullet, d') \underset{G'}{\overset{F'}{\rightleftarrows}} (C''_\bullet, d'')$$

be homotopy equivalences with their homotopy inverses. We want to show that $F' \circ F$ is a homotopy equivalence with homotopy inverse given by $G \circ G'$. So let h, h' be homotopies such that $G \circ F \sim_h \mathrm{id}_C$ and $G' \circ F' \sim_{h'} \mathrm{id}_{C'}$. Then, from the proof of Proposition 5.15, we have

$$G \circ G' \circ F' \sim_{G \circ h'} G \circ \mathrm{id}_{C'} = G$$

where we used that $G \sim_0 G$. Hence, again from the proof of Proposition 5.15,

$$G \circ G' \circ F' \circ F \sim_{G \circ h' \circ F} G \circ F \sim_h \mathrm{id}_C$$

where we used that $F \sim_0 F$, and, from Exercise 2.7,

$$G \circ G' \circ F' \circ F \sim_{G \circ h' \circ F + h} \mathrm{id}_C .$$

Similarly, there is a homotopy K such that $F' \circ F \circ G \circ G' \sim_K \mathrm{id}_{C''}$. This concludes the proof. $\qquad\square$

2.4 The Long Homology Exact Sequence

In this section, we present another (elementary) technique which is often useful in computing the (co)homology of a (co)chain complex. Let R be a ring and let $(C_\bullet, d)$ be a chain complex of R-modules.

Definition 2.31 (Subcomplex). A *subcomplex* in $(C_\bullet, d)$ is a family $A_\bullet = (A_i)_{i \in \mathbb{Z}}$ of submodules $A_i \subseteq C_i$ such that $d(A_i) \subseteq A_{i-1}$ for all $i \in \mathbb{Z}$. Likewise for cochain complexes.

Let $A_\bullet$ be a subcomplex in $(C_\bullet, d)$. For each $i \in \mathbb{Z}$, we can restrict the differential $d : C_i \to C_{i-1}$ to A_i in the domain and to A_{i-1} in the codomain, obtaining a (new) R-module homomorphism

$$d_A : A_i \to A_{i-i}.$$

It is clear that the pair $(A_\bullet, d_A)$ is a chain complex again. Obviously, the family $i_A := (i_{A_i} : A_i \to C_i)_{i \in \mathbb{Z}}$ of inclusions is a chain map, i.e., the diagram

$$
\begin{array}{ccccccc}
\cdots \xleftarrow{d_A} & A_{i-1} & \xleftarrow{d_A} & A_i & \xleftarrow{d_A} & A_{i+1} & \xleftarrow{} \cdots \\
& \downarrow{i_A} & & \downarrow{i_A} & & \downarrow{i_A} & \\
\cdots \xleftarrow{d} & C_{i-1} & \xleftarrow{d} & C_i & \xleftarrow{d} & C_{i+1} & \xleftarrow{} \cdots
\end{array}
$$

commutes, and we sometimes write $(A_\bullet, d_A) \subseteq (C_\bullet, d)$ (instead of $i_A : (A_\bullet, d_A) \to (C_\bullet, d)$). Likewise for cochain complexes.

Example 2.32. Let $f : (C_\bullet, d) \to (D_\bullet, d_D)$ be a chain map. The *kernel* of f is the family $\ker f := (\ker(f : C_i \to D_i))_{i \in \mathbb{Z}}$. Similarly, the *image* of f is the family $\operatorname{im} f := (\operatorname{im}(f : C_i \to D_i))_{i \in \mathbb{Z}}$. The kernel of f is a subcomplex in $(C_\bullet, d)$. Indeed, let $c \in C_n$ be an n-chain in the kernel of f, i.e., $f(c) = 0$. Then $f(dc) = d_D f(c) = 0$, i.e., dc is in the kernel of f as well. Similarly, the image of f is a subcomplex in $(D_\bullet, d_D)$ (do you see it?). Likewise for cochain maps. ◆

Now, let $(A_\bullet, d_A) \subseteq (C_\bullet, d)$ be a subcomplex in the chain complex $(C_\bullet, d)$. For each $i \in \mathbb{Z}$, we can take the quotient module C_i / A_i and get a new family $C_\bullet / A_\bullet := (C_i / A_i)_{i \in \mathbb{Z}}$ of R-modules. Additionally, from $d(A_i) \subseteq A_{i-1}$, the differential d induces unique R-linear maps $d_{C/A} : C_i / A_i \to C_{i-1} / A_{i-1}$ such that $d_{C/A}(c \bmod A_i) = dc \bmod$

A_{i-1} for all i-chains $c \in C_i$ (see Corollary 1.24). The new sequence of R-linear maps

$$\cdots \xleftarrow{d_{C/A}} C_{i-1}/A_{i-1} \xleftarrow{d_{C/A}} C_i/A_i \xleftarrow{d_{C/A}} C_{i+1}/A_{i+1} \longleftarrow \cdots$$

is a chain complex, indeed, for all $c \in C_{i+1}$,

$$d_{C/A} \circ d_{C/A}(c \bmod A_{i+1}) = d_{C/A}(dc \bmod A_i) = (d \circ d)c \bmod A_{i-1} = 0,$$

showing that $d_{C/A} \circ d_{C/A} = 0$. Additionally, the projection $\pi = (\pi : C_i \to C_i/A_i)_{i \in \mathbb{Z}}$ is a chain map, indeed, for all $c \in C_i$,

$$d_{C/A}(\pi(c)) = d_{C/A}(c \bmod A_i) = dc \bmod A_{i-1} = \pi(dc).$$

Definition 2.33 (Quotient complex). The chain complex $(C_\bullet/A_\bullet, d_{C/A})$ is called the *quotient complex* (of $(C_\bullet, d)$ over the subcomplex $(A_\bullet, d_A)$). Likewise for cochain complexes.

Example 2.34 (Relative de Rham Subcomplex). Let $(C^\bullet, d)$ be the cochain complex of real vector spaces described in Example 2.27. We consider the subcomplex $(A^\bullet, d_A) \subseteq (C^\bullet, d)$ defined as follows. First of all, denote by $W \subseteq \mathbb{R}^3$ the vector subspace

$$W := \{(x_1, x_2, 0) \in \mathbb{R}^3 : (x_1, x_2) \in \mathbb{R}^2\}.$$

It is clear that W is a 2-dimensional vector subspace spanned by $E_1 = (1, 0, 0), E_2 = (0, 1, 0)$. Hence, it identifies canonically with $\mathbb{R}^2$ via the vector space isomorphism

$$\mathbb{R}^2 \to W, \quad (x_1, x_2) \mapsto (x_1, x_2, 0).$$

In the following, we use this isomorphism to identify W and $\mathbb{R}^2$. For instance, we interpret $\mathbb{R}^2$ as a subspace in $\mathbb{R}^3$. Now put

$$A^i = \begin{cases} \{f \in C^\infty(\mathbb{R}^3) : f|_{\mathbb{R}^2} = 0\} & \text{if } i = 0 \\ \{\vec{F} \in C^\infty(\mathbb{R}^3, \mathbb{R}^3) : F_1|_{\mathbb{R}^2} = F_2|_{\mathbb{R}^2} = 0\} & \text{if } i = 1 \\ \{\vec{G} \in C^\infty(\mathbb{R}^3, \mathbb{R}^3) : G_3|_{\mathbb{R}^2} = 0\} & \text{if } i = 2 \\ C^\infty(\mathbb{R}^3) & \text{if } i = 3 \\ 0 & \text{otherwise} \end{cases}.$$

We leave it to the reader to prove that the subcomplex condition $d(A_i) \subseteq A_{i+1}$ is fulfilled. We want to describe the quotient complex $(C^\bullet/A^\bullet, d_{C/A})$.

We claim that it is isomorphic to the cochain complex

$$(B^\bullet, d_B): \quad 0 \longrightarrow \mathbb{R} \xrightarrow{d_B} C^\infty(\mathbb{R}^2) \xrightarrow{d_B} C^\infty(\mathbb{R}^2, \mathbb{R}^2) \xrightarrow{d_B} C^\infty(\mathbb{R}^2) \longrightarrow 0 ,$$
$$\quad\quad\quad\quad\quad\quad\quad\quad -1 \quad\quad\quad\quad\quad 0 \quad\quad\quad\quad\quad\quad\quad 1 \quad\quad\quad\quad\quad 2$$

$$(2.9)$$

where the differential d_B is given by

$$d_B^{-1} c = c \quad \text{(the constant function equal to } c\text{)}$$

$$d_B^0 f = \left(\frac{\partial f}{\partial x_1}, \frac{\partial f}{\partial x_2} \right) \in C^\infty(\mathbb{R}^2, \mathbb{R}^2)$$

$$d_B^1 (F_1, F_2) = \frac{\partial F_2}{\partial x_1} - \frac{\partial F_1}{\partial x_2} \in C^\infty(\mathbb{R}^2)$$

$$d_B^i = 0 \quad \text{for } i \notin \{-1, 0, 1\}$$

(do you agree that $(B^\bullet, d_B)$ is really a cochain complex?). To prove that $(C^\bullet / A^\bullet, d_{C/A}) \cong (B^\bullet, d_B)$ consider the commutative diagram

$$. \quad (2.10)$$

$$A^\bullet \quad\quad\quad\quad C^\bullet \quad\quad\quad\quad B^\bullet$$

All the columns are cochain complexes, while the horizontal arrows define cochain maps. Here the cochain map $p : (C^\bullet, d) \to (B^\bullet, d_B)$ is defined by

$$p^{-1}c = c \in \mathbb{R}$$
$$p^0 f = f|_{\mathbb{R}^2} \in C^\infty(\mathbb{R}^2)$$
$$p^1(F_1, F_2, F_3) = (F_1|_{\mathbb{R}^2}, F_2|_{\mathbb{R}^2}) \in C^\infty(\mathbb{R}^2, \mathbb{R}^2).$$
$$p^2(G_1, G_2, G_3) = G_3|_{\mathbb{R}^2} \in C^\infty(\mathbb{R}^2)$$
$$p^i = 0 \quad \text{for } i \notin \{-1, 0, 1, 2\}$$

It is easy to see that, with this definition, the diagram indeed commutes and, additionally, the rows are short exact sequence of vector spaces. Hence, from Corollary 1.23, we get vector space isomorphisms $\overline{p} : C^i/A^i \to B^i$ such that $\overline{p}(c \bmod A_i) = p(c)$. Finally, from the commutativity of (2.10), it easily follows that the diagram

$$
\begin{array}{ccccccccccc}
B^\bullet & : & 0 \longrightarrow & \mathbb{R} & \longrightarrow & C^\infty(\mathbb{R}^2) & \xrightarrow{d_B} & C^\infty(\mathbb{R}^2, \mathbb{R}^2) & \xrightarrow{d_B} & C^\infty(\mathbb{R}^2) & \xrightarrow{d_B} 0 \longrightarrow 0 \\
& & & \uparrow{\overline{p}} & & \uparrow{\overline{p}} & & \uparrow{\overline{p}} & & \uparrow{\overline{p}} & \uparrow{\overline{p}} \\
C^\bullet/A^\bullet & : & 0 \longrightarrow C^{-1}/A^{-1} \longrightarrow & C^0/A^0 & \xrightarrow{d_{C/A}} & C^1/A^1 & \xrightarrow{d_{C/A}} & C^2/A^2 & \xrightarrow{d_{A/C}} & C^3/A^3 \longrightarrow 0
\end{array}
$$

commutes as well. Hence, $\overline{p} : (C_\bullet/A_\bullet, d_{C/A}) \to (B_\bullet, d_B)$ is an isomorphism of cochain complexes as claimed. ◆

Example 2.34 suggests a slight generalization of the picture "complex, subcomplex, quotient complex". Namely, consider a sequence of chain maps:

$$0 \longrightarrow (A_\bullet, d_A) \xrightarrow{\alpha} (C_\bullet, d_C) \xrightarrow{\beta} (B_\bullet, d_B) \longrightarrow 0, \qquad (2.11)$$

where the 0s at the two extremes are the zero chain complexes $(0_\bullet, 0)$. We call any sequence of type (2.11) a *short exact sequence of chain complexes* if the sequence

$$0 \longrightarrow A_i \xrightarrow{\alpha} C_i \xrightarrow{\beta} B_i \longrightarrow 0$$

is a short exact sequence of modules for all i, i.e., $\alpha : A_i \to C_i$ is injective, $\beta_i : C_i \to B_i$ is surjective and $\ker(\beta : C_i \to B_i) = \operatorname{im}(\alpha : A_i \to C_i)$. Likewise for cochain complexes.

Example 2.35. Let $(C_\bullet, d)$ be a chain complex and let $(A_\bullet, d_A) \subseteq (C_\bullet, d)$ be a subcomplex. The sequence

$$0 \longrightarrow (A_\bullet, d_A) \xrightarrow{i_A} (C_\bullet, d) \xrightarrow{\pi} (C_\bullet/A_\bullet, d_{C/A}) \longrightarrow 0$$

is a short exact sequence of chain complexes. Every short exact sequence of chain complexes is of this type up to appropriate isomorphisms. Indeed, take a short exact sequence (2.11). As α is (degree-wise) injective, it identifies $(A_\bullet, d_A)$ with a subcomplex $(\alpha(A_\bullet), d_{\alpha(A)}) \subseteq (C_\bullet, d)$. Additionally, as $\alpha(A_\bullet) = \operatorname{im} \alpha = \ker \beta$, from Proposition 1.22, we get an isomorphism of chain complexes $(B_\bullet, d_B) \cong (C_\bullet/\alpha(A_\bullet), d_{C/\alpha(A)})$ identifying (2.11) with the short exact sequence

$$0 \longrightarrow (\alpha(A_\bullet), d_{\alpha(A)}) \xrightarrow{i_{\alpha(A)}} (C_\bullet, d) \xrightarrow{\pi} (C_\bullet/\alpha(A_\bullet), d_{C/\alpha(A)}) \longrightarrow 0$$

(the chain complex isomorphism $(B_\bullet, d_B) \cong (C_\bullet/\alpha(A_\bullet), d_{C/\alpha(A)})$ intertwines the chain maps; we leave the obvious details to the reader). Likewise for cochain complexes. ♦

Lemma 2.36. *Consider a short exact sequence of chain complexes* (2.11). *For every* $i \in \mathbb{Z}$ *the induced sequence in homology,*

$$H_i(A, d_A) \xrightarrow{H(\alpha)} H_i(C, d_C) \xrightarrow{H(\beta)} H_i(B, d_B), \tag{2.12}$$

is exact: $\ker H(\beta) = \operatorname{im} H(\alpha)$. *Likewise for cochain complexes.*

Proof. First of all, from Proposition 2.15

$$H(\beta) \circ H(\alpha) = H(\beta \circ \alpha) = H(0) = 0.$$

This shows that $\operatorname{im} H(\alpha) \subseteq \ker H(\beta)$. It remains to prove the reverse inclusion $\ker H(\beta) \subseteq \operatorname{im} H(\alpha)$. So, let $c \in Z_i(C, d_C)$ be an i-cycle in $(C_\bullet, d_C)$ and let $[c]_C \in H_i(C, d_C)$ be its homology class. Assume that

$$0 = H(\beta)[c]_C = [\beta(c)]_B.$$

This means that there exists $b \in B_{i+1}$ such that $\beta(c) = d_B b$. As β is surjective, there also exists $c' \in C_{i+1}$ such that $b = \beta(c')$. Hence,

$$\beta(c) = d_B b = d_B(\beta(c')) = \beta(d_C c'),$$

where we used that β is a chain map. In other words,

$$0 = \beta(c) - \beta(d_C c') = \beta(c - d_C c'),$$

i.e., $c - d_C c' \in \ker \beta$. But $\ker \beta = \operatorname{im} \alpha$, therefore there exists $a \in A_i$ such that

$$c - d_C c' = \alpha(a) \Rightarrow c = \alpha(a) + d_C c' \Rightarrow [c]_C = [\alpha(a)]_C = H(\alpha)[a]_A,$$

showing that $[c]_C \in \operatorname{im} H(\alpha)$ as desired (do you see that $d_A a = 0$ so that the last step makes indeed sense? Please check the details). □

Although the sequence (2.12) is exact, the map $H(\alpha) : H_{i-1}(A, d_A) \to H_{i-1}(C, d_C)$ is not injective nor the map $H(\beta) : H_i(C, d_C) \to H_i(B, d_B)$ is surjective in general. Interestingly, the failure of $H(\alpha)$ from being injective and that of $H(\beta)$ from being surjective are measured by a natural family of R-module homomorphisms $\Delta : H_i(B, d_B) \to H_{i-1}(A, d_A)$, with the property that $\operatorname{im} \Delta = \ker H(\alpha)$ and $\operatorname{im} H(\beta) = \ker \Delta$. Overall, the exact sequences

$$\vdots$$

$$H_{i+1}(A, d_A) \xrightarrow{H(\alpha)} H_{i+1}(C, d_C) \xrightarrow{H(\beta)} H_{i+1}(B, d_B)$$

$$H_i(A, d_A) \xrightarrow{H(\alpha)} H_i(C, d_C) \xrightarrow{H(\beta)} H_i(B, d_B)$$

$$H_{i-1}(A, d_A) \xrightarrow{H(\alpha)} H_{i-1}(C, d_C) \xrightarrow{H(\beta)} H_{i-1}(B, d_B)$$

$$\vdots$$

are connected by R-linear maps Δ, such that the sequence

$$
\begin{array}{c}
\cdots \\
\to H_{i+1}(A, d_A) \xrightarrow{H(\alpha)} H_{i+1}(C, d_C) \xrightarrow{H(\beta)} H_{i+1}(B, d_B) \\
\to H_i(A, d_A) \xrightarrow{H(\alpha)} H_i(C, d_C) \xrightarrow{H(\beta)} H_i(B, d_B) \\
\to H_{i-1}(A, d_A) \xrightarrow{H(\alpha)} H_{i-1}(C, d_C) \xrightarrow{H(\beta)} H_{i-1}(B, d_B) \\
\to \cdots
\end{array}
\quad (2.13)
$$

is exact. Summarizing we have the following:

Theorem 2.37 (Long Exact Homology Sequence). *Let*

$$0 \longrightarrow (A_\bullet, d_A) \xrightarrow{\alpha} (C_\bullet, d_C) \xrightarrow{\beta} (B_\bullet, d_B) \longrightarrow 0$$

be a short exact sequence of chain complexes. For every $i \in \mathbb{Z}$, *there exists a natural R-module homomorphism* $\Delta : H_i(B, d_B) \to H_{i-1}(A, d_A)$ *(to be defined in the proof) such that the sequence* (2.13) *is exact.*

Proof. Here we only define $\Delta : H_i(B, d_B) \to H_{i-1}(A, d_A)$. The rest uses similar arguments as those in the proof of Lemma 2.36 and is left as Exercise 2.12. So, let $b \in Z_i(B, d_B)$ be an i-cycle in $(B_\bullet, d_B)$ and let $[b]_B \in H_i(B, d_B)$ be its homology class. As β is surjective, there exists $c \in C_i$ such that $b = \beta(c)$. Consider the differential $d_C c \in C_{i-1}$ and note that

$$\beta(d_C c) = d_B \beta(c) = d_B b = 0,$$

where we used that β is a chain map and that b is a cycle. The latter computation shows that $d_C c \in \ker \beta = \operatorname{im} \alpha$. Hence, there exists (a unique) $a \in A_{i-1}$ such that $\alpha(a) = d_C c$. Additionally, we have

$$\alpha(d_A a) = d_C \alpha(a) = d_C d_C c = 0,$$

where we used that α is a chain map. So, $d_A a \in \ker \alpha$. But α is injective, hence $d_A a = 0$ and a is an $(i-1)$-cycle in $(A_\bullet, d_A)$. In particular, we can take its homology class $[a]_A \in H_{i-1}(A, d_A)$. We put

$$\Delta[b]_B := [a]_A.$$

We still have to show that $\Delta : H_i(B, d_B) \to H_{i-1}(A, d_A)$ is a well-defined R-linear map. In order to see that it is well defined, we have to show that $\Delta[b]_B$ does only depend on $[b]_B$, i.e., it is independent of the arbitrary choices that we made to define it. Actually, we made only two choices: we chose a representative b in the homology class $[b]_B$, and we chose an element c in the pre-image $\beta^{-1}(b)$. So, take b' homologous to b and take (another) c' such that $\beta(c') = b'$. Define $a' \in A_{i-1}$ from c' exactly as we defined a from c (i.e., a' is the unique cycle such that $\alpha(a') = d_C c'$). We want to show that a' is homologous to a so that $[a]_A = [a']_A$. We have $b' = b + db''$ for some $b'' \in B_{i+1}$. As β is surjective, there exists $c'' \in C_{i+1}$ such that $\beta(c'') = b''$. So,

$$\beta(c') = b' = b + d_B b'' = \beta(c) + d_B \beta(c'') = \beta(c + d_C c'').$$

Hence, $\beta(c' - c - d_C c'') = 0$ and $c' - c - d_C c'' \in \ker \beta$. As $\ker \beta = \operatorname{im} \alpha$, there exists $a'' \in A_i$ such that $\alpha(a'') = c' - c - d_C c''$, which in turn implies

$$\alpha(d_A a'') = d_C \alpha(a'') = d_C(c' - c - d_C c'') = d_C c' - d_C c$$
$$= \alpha(a') - \alpha(a) = \alpha(a' - a).$$

We are almost done. As α is injective, $a' - a = d_A a''$, i.e., a' and a are homologous as desired. It remains to show that Δ is R-linear. So, let b_1, b_2 be i-cycles in $(B_\bullet, d_B)$, and let $\lambda_1, \lambda_2 \in R$. We have to show that

$$\Delta(\lambda_1 [b_1]_B + \lambda_2 [b_2]_B) = \lambda_1 \Delta[b_1]_B + \lambda_2 \Delta[b_2]_B.$$

To do this, choose $c_1, c_2 \in C_i$ and $a_1, a_2 \in Z_{i-1}(A, d_A)$ such that $\beta(c_1) = b_1, \beta(c_2) = b_2$ and $\alpha(a_1) = d_C c_1, \alpha(a_2) = d_C c_2$ (as we have showed above, this is always possible). Then we have $\Delta[b_1]_B = [a_1]_A, \Delta[b_2]_B = [a_2]_A$. In order to compute $\Delta(\lambda_1 [b_1]_B + \lambda_2 [b_2]_B)$, we note that

$$\lambda_1 [b_1]_B + \lambda_2 [b_2]_B = [\lambda_1 b_1 + \lambda_2 b_2]_B.$$

Now we have to choose $c \in C_i$ and $a \in Z_{i-1}(A, d_A)$ such that $\beta(c) = \lambda_1 b_1 + \lambda_2 b_2$ and $\alpha(a) = d_C c$. It is easy to see that we can choose $c = \lambda_1 c_1 + \lambda_2 c_2$ and $a = \lambda_1 a_1 + \lambda_2 a_2$ (do you see it?). We stress that this is not the only possible choice (but any other choice will give the same result for $\Delta(\lambda_1 [b_1]_B + \lambda_2 [b_2]_B)$). Nonetheless it is a particularly convenient one for our purposes. Indeed,

$$\Delta(\lambda_1 [b_1]_B + \lambda_2 [b_2]_B) = \Delta[\lambda_1 b_1 + \lambda_2 b_2]_B = [a]_A$$
$$= [\lambda_1 a_1 + \lambda_2 a_2]_A = \lambda_1 [a_1]_A + \lambda_2 [a_2]_A$$
$$= \lambda_1 \Delta[b_1]_B + \lambda_2 \Delta[b_2]_B.$$

So, $\Delta : H_i(B, d_B) \to H_{i-1}(A, d_A)$ is a well-defined R-linear map. The rest is left to the reader. $\square$

Exercise 2.12. Complete the proof of Theorem 2.37 showing that $\ker \Delta = \operatorname{im} H(\beta)$ and $\operatorname{im} \Delta = \ker H(\alpha)$.

There is an obvious version of Theorem 2.37 for cochains giving a degree ascending (rather than descending) long exact sequence in cohomology. We state it for completeness.

Theorem 2.38 (Long Exact Cohomology Sequence). *Let*

$$0 \longrightarrow (A^\bullet, d_A) \overset{\alpha}{\longrightarrow} (C^\bullet, d_C) \overset{\beta}{\longrightarrow} (B^\bullet, d_B) \longrightarrow 0$$

be a short exact sequence of cochain complexes. For every $i \in \mathbb{Z}$, there exists a natural R-module homomorphism $\Delta : H^i(B, d_B) \rightarrow H^{i+1}(A, d_A)$ (defined in the same way as in the proof of Theorem 2.37, up to the obvious modifications) such that the following sequence:

$$\cdots$$

$$\longrightarrow H^{i+1}(A, d_A) \xrightarrow{H(\alpha)} H^{i+1}(C, d_C) \xrightarrow{H(\beta)} H^{i+1}(B, d_B) \Big) \Delta$$

$$\longrightarrow H^i(A, d_A) \xrightarrow{H(\alpha)} H^i(C, d_C) \xrightarrow{H(\beta)} H^i(B, d_B) \Big) \Delta \qquad (2.14)$$

$$\longrightarrow H^{i-1}(A, d_A) \xrightarrow{H(\alpha)} H^{i-1}(C, d_C) \xrightarrow{H(\beta)} H^{i-1}(B, d_B) \Big) \Delta$$

$$\Big) \Delta$$

$$\cdots$$

is exact.

Definition 2.39 (Connecting Homomorphism). The family

$$\Delta = (\Delta : H_i(B, d_B) \rightarrow H_{i-1}(A, d_A))_{i \in \mathbb{Z}}$$

$$\left(\text{resp. } \Delta = (\Delta : H^i(B, d_B) \rightarrow H^{i+1}(A, d_A))_{i \in \mathbb{Z}}\right)$$

in Theorem 2.37 (resp. Theorem 2.38) is called the *connecting homomorphism* and the exact sequence (2.13) (resp. (2.14)) is called the *long exact homology* (resp. *cohomology*) *sequence* determined by the short exact sequence of chain (resp. cochain) complexes (2.11)).

Corollary 2.40. *Let*

$$0 \longrightarrow (A_\bullet, d_A) \xrightarrow{\alpha} (C_\bullet, d_C) \xrightarrow{\beta} (B_\bullet, d_B) \longrightarrow 0$$

be a short exact sequence of chain complexes. Assume that two out of three among the chain complexes $(A_\bullet, d_A), (B_\bullet, d_B), (C_\bullet, d_C)$ are acyclic. Then the third one is also acyclic. Likewise for cochain complexes.

Proof. Suppose that $(A_\bullet, d_A)$ and $(B_\bullet, d_B)$ are acyclic and prove that $(C_\bullet, d_C)$ is also acyclic (the other two cases can be discussed exactly in the same way). Then the long exact homology sequence looks as follows:

$$\cdots \longrightarrow 0 \longrightarrow H_i(C, d_C) \longrightarrow 0 \longrightarrow \cdots.$$

As it is an exact sequence, the image of $0 \to H_i(C, d_C)$ (which is 0) agrees with the kernel of $H_i(C, d_C) \to 0$. This shows that $H_i(C, d_C) \to 0$ is injective. The only possibility is that $H_i(C, d_C) = 0$. $\square$

Example 2.41. Consider the short exact sequence of cochain complexes

$$0 \longrightarrow (A^\bullet, d_A) \xrightarrow{i_A} (C^\bullet, d_C) \xrightarrow{p} (B^\bullet, d_B) \longrightarrow 0$$

described in Example 2.34. We know already that $(C^\bullet, d_C)$ is acyclic (Example 2.27). Actually, $(B_\bullet, d_B)$ possesses a similar contracting homotopy (that is discussed in Chapter 4). We conclude that $(A_\bullet, d_A)$ is acyclic as well. ◆

Example 2.42 (Cohomological Integral). Here, we discuss a toy example that is relevant for the integration theory of smooth real valued functions of a real variable. Let $[a_1, a_2] \subseteq \mathbb{R}$ be a closed interval in $\mathbb{R}$ ($a_2 > a_1$) and denote by t the standard coordinate in $\mathbb{R}$. Consider the diagram

$$
\begin{array}{ccccccccc}
& & 0 & & 0 & & 0 & & \\
& & \uparrow & & \uparrow & & \uparrow & & \\
0 & \longrightarrow & C^\infty([a_1,a_2]) & \xrightarrow{\text{id}} & C^\infty([a_1,a_2]) & \longrightarrow & 0 & \longrightarrow & 0 \\
& & \uparrow{\scriptstyle \frac{d}{dt}} & & \uparrow{\scriptstyle \frac{d}{dt}} & & \uparrow & & \\
0 & \longrightarrow & C^\infty_{\text{rel}}([a_1,a_2]) & \xrightarrow{i} & C^\infty([a_1,a_2]) & \xrightarrow{p} & \mathbb{R}^2 & \longrightarrow & 0 \ , \\
& & \uparrow & & \uparrow & & \uparrow & & \\
& & 0 & & 0 & & 0 & &
\end{array}
\qquad (2.15)
$$

$$A^\bullet \qquad\qquad C^\bullet \qquad\qquad B^\bullet$$

where $C^\infty_{\text{rel}}([a_1, a_2])$ consists of those smooth functions $f : [a_1, a_2] \to \mathbb{R}$ such that $f(a_1) = f(a_2) = 0$, $i : C^\infty_{\text{rel}}([a_1, a_2]) \to C^\infty([a_1, a_2])$ is the inclusion and $p : C^\infty([a_1, a_2]) \to \mathbb{R}^2$ is given by $f \mapsto (f(a_1), f(a_2))$. The columns of Diagram (2.15) are (particularly simple) cochain complexes of real vector spaces that we denote $(A^\bullet, d_A), (C^\bullet, d_C), (B^\bullet, d_B)$, respectively (the first two are concentrated in degrees $0, 1$, the third one in concentrated in degree 0). The rows are short exact sequences of vector spaces. For the upper row, this is obvious. For the lower row, i is injective and its image is the kernel of p by definition of $C_{\mathrm{rel}}^\infty([a_1, a_2])$. Additionally p is surjective. Indeed, for $(y_1, y_2) \in \mathbb{R}^2$ we can take the smooth function

$$g(t) := y_1 + \frac{t - a_1}{a_2 - a_1}(y_2 - y_1) \tag{2.16}$$

which clearly satisfies $p(f) = (y_1, y_2)$. Finally, Diagram (2.15) obviously commutes, so it is a short exact sequence of cochain complexes. The cohomology of $(C^\bullet, d_C)$ can be computed by hands:

$$H^0(C, d_C) = \ker\left(\frac{d}{dt} : C^\infty([a_1, a_2]) \to C^\infty([a_1, a_2])\right)$$

$$= \{\text{constant functions}\}$$

$$\cong \mathbb{R},$$

and $H^1(C, d_C) = 0$, indeed the linear map $\frac{d}{dt} : C^\infty([a_1, a_2]) \to C^\infty([a_1, a_2])$ is surjective: for any $g \in C^\infty([a_1, a_2])$, there is $f \in C^\infty([a_1, a_2])$ such that $g = \frac{df}{dt}$ (just take $g(t) := \int_{a_1}^t f(s)ds$). Similarly, $H^0(A, d_A)$ consists of constant functions vanishing on both a_1, a_2, so $H^0(A, d_A) = 0$. It remains to compute $H^1(A, d_A)$. This can be done by hands. We prefer to use the long exact cohomology sequence associated to the short exact sequence of cochain complexes (2.15). Several cohomologies vanish and we remain with

$$0 \longrightarrow H^0(C, d_C) \xrightarrow{H(p)} H^0(B, d_B) \xrightarrow{\Delta} H^1(A, d_A) \longrightarrow 0$$
$$\shortparallel \qquad\qquad \shortparallel$$
$$\mathbb{R} \qquad\qquad \mathbb{R}^2$$

which is a short exact sequence of vector spaces. The map $H(p) : \mathbb{R} \to \mathbb{R}^2$ is given by $y \mapsto (y, y)$ and we can complete it to a short exact sequence of vector spaces as illustrated in the following diagram:

$$
\begin{array}{ccccccccc}
0 & \longrightarrow & H^0(C, d_C) & \xrightarrow{H(p)} & H^0(B, d_B) & \xrightarrow{\Delta} & H^1(A, d_A) & \longrightarrow & 0 \\
 & & \shortparallel & & \shortparallel & & \downarrow I & & \\
0 & \longrightarrow & \mathbb{R} & \xrightarrow{H(p)} & \mathbb{R}^2 & \xrightarrow{\Delta'} & \mathbb{R} & \longrightarrow & 0
\end{array} \tag{2.17}
$$

$$(y_1, y_2) \longmapsto y_2 - y_1$$

We conclude that $H^1(A, d_A)$ is a 1-dimensional vector space and there exists a unique vector space isomorphism

$$I : H^1(A, d_A) \to \mathbb{R}$$

such that Diagram (2.17) commutes. We claim that I is given by

$$I[f]_A = \int_{a_1}^{a_2} f(t)dt, \quad f \in A^1 = C^\infty([a_1, a_2]), \tag{2.18}$$

so giving a cohomological flavor to the usual definite integral. First, note that Formula (2.18) gives a well-defined $\mathbb{R}$-linear map $I : H^1(A, d_A) \to \mathbb{R}$. Indeed, when $f = dh/dt$ for some $h \in C^\infty_{\mathrm{rel}}([a_1, a_2])$, then

$$\int_{a_1}^{a_2} f(t)dt = \int_{a_1}^{a_2} \frac{dh}{dt}(t)dt = h(a_2) - h(a_1) = 0.$$

Now, take $(y_1, y_2) \in \mathbb{R}^2 = H^0(B, d_B)$ and compute $\Delta(y_1, y_2)$. We use the definition of the connecting homomorphism: there exists a smooth function g such that $p(g) = (g(a_1), g(a_2)) = (y_1, y_2)$, for instance (2.16). Then $\Delta(y_1, y_2) = [dg/dt]_A$. Hence,

$$I \circ \Delta(y_1, y_2) = I[dg/dt]_A = \int_{a_1}^{a_2} \frac{dg}{dt}(t)dt = g(a_2) - g(a_1)$$

$$= y_2 - y_1 = \Delta'(y_1, y_2).$$

Summarizing, the integral $\int_{a_1}^{a_2} : C^\infty([a_1, a_2]) \to \mathbb{R}$ can be characterized as the composition

$$C^\infty([a_1, a_2]) \longrightarrow H^1(A, d_A) \overset{I}{\longrightarrow} \mathbb{R},$$

where I is the unique isomorphism making the Diagram (2.17) commutative.   ◆

We conclude this section and this chapter showing that the connecting homomorphism is compatible with "transforming short exact sequences of (co)chain complexes". We begin explaining what does it mean "transforming short exact sequences".

Definition 2.43 (Morphism of Short Exact Sequences of (Co)Chain Complexes). A *morphism of short exact sequence of chain complexes* is a commutative diagram

$$0 \longrightarrow (A_\bullet, d_A) \overset{\alpha}{\longrightarrow} (C_\bullet, d_C) \overset{\beta}{\longrightarrow} (B_\bullet, d_B) \longrightarrow 0 \qquad (2.19)$$
$$\Big\downarrow F_A \qquad\qquad \Big\downarrow F_C \qquad\qquad \Big\downarrow F_B$$
$$0 \longrightarrow (A'_\bullet, d_{A'}) \overset{\alpha'}{\longrightarrow} (C'_\bullet, d_{C'}) \overset{\beta'}{\longrightarrow} (B'_\bullet, d_{B'}) \longrightarrow 0$$

such that the rows are short exact sequences of chain complexes, and the columns are chain maps. Like-wise for cochain complexes.

Exercise 2.13. In this exercise, we ask the reader to guess various definitions (and prove that they are well posed):

(1) Define the identity morphism of short exact sequences of (co)chain complexes.
(2) Define the composition of morphisms of short exact sequence of (co)chain complexes and prove that it is a morphism again.
(3) Define isomorphisms of short exact sequences of (co)chain complexes and their inverses. Prove that the inverse of an isomorphism is an isomorphism again.

Proposition 2.44 (Naturality of the Connecting Homomorphism). *For any morphism (2.19) of short exact sequences of chain complexes, the diagram*

$$\cdots \longrightarrow H_i(C, d_C) \overset{H(\beta)}{\longrightarrow} H_i(B, d_B) \overset{\Delta}{\longrightarrow} H_{i-1}(A, d_A) \overset{H(\alpha)}{\longrightarrow} H_{i-1}(C, d_C) \overset{H(\beta)}{\longrightarrow} \cdots$$
$$\Big\downarrow H(F_C) \qquad \Big\downarrow H(F_B) \qquad \Big\downarrow H(F_A) \qquad \Big\downarrow H(F_C)$$
$$\cdots \longrightarrow H_i(C', d_{C'}) \overset{H(\beta')}{\longrightarrow} H_i(B', d_{B'}) \overset{\Delta}{\longrightarrow} H_{i-1}(A', d_{A'}) \overset{H(\alpha')}{\longrightarrow} H_{i-1}(C', d_{C'}) \overset{H(\beta)}{\longrightarrow} \cdots$$
$$(2.20)$$

commutes. Likewise for cochain complexes.

Proof. We know more or less already that the squares in (2.20) not involving the connecting homomorphism commute. Indeed, it immediately follows from the properties of the induced map in homology (Proposition 2.15) that the induced diagram in homology from a commuting diagram of chain complexes is commutative as well. For instance, from $F_B \circ \beta = \beta' \circ F_C$, we get

$$H_i(F_C) \circ H_i(\beta) = H_i(F_C \circ \beta) = H_i(\beta' \circ F_C) = H_i(\beta') \circ H_i(F_C)$$

for all $i \in \mathbb{Z}$. It remains to show that $H(F_A) \circ \Delta = \Delta \circ H(F_B)$. So, let $b \in Z_i(B, d_B)$ be an i-cycle in $(B_\bullet, d_B)$ and let $[b]_B \in H_i(B, d_B)$ be its homology class. Choose $a \in Z_{i-1}(A, d_A)$ and $c \in C_i$ such that $\alpha(a) = d_C c$ and $b = \beta(c)$. This is always possible and $\Delta[b]_B = [a]_A$. Now put $a' = F_A(a)$, $c' = F_C(c)$ and $b' = F_B(b)$. Then we have

$$\alpha'(a') = \alpha'(F_A(a)) = F_C(\alpha(a)) = F_C(d_C c) = d_{C'} F_C(c) = d_{C'} c'$$

and

$$\beta'(c') = \beta'(F_C(c)) = F_B(\beta(c)) = F_B(b) = b'.$$

This shows that $\Delta[b']_{B'} = [a']_{A'}$, hence

$$\begin{aligned}
\Delta \circ H(F_B)[b]_B = \Delta[F_B(b)]_{B'} &= \Delta[b']_{B'} \\
&= [a']_{A'} = [F_A(a)]_{A'} = H(F_A)[a]_A \\
&= H(F_A) \circ \Delta[b]_B
\end{aligned}$$

and the claim follows from the arbitrariness of b. $\square$

2.5   End-of-Chapter Problems

Problem 2.1. Compute the homologies of the following chain complex of abelian groups:

$$\cdots \xleftarrow{3} \mathbb{Z}_9 \xleftarrow{6} \mathbb{Z}_9 \xleftarrow{3} \mathbb{Z}_9 \xleftarrow{6} \mathbb{Z}_9 \xleftarrow{} \cdots.$$

Problem 2.2. Compute the cohomologies of the following cochain complex of real vector spaces:

$$0 \longrightarrow \mathbb{R}^2 \xrightarrow{\left(\begin{smallmatrix} 1 & 0 \\ 0 & 0 \\ 0 & -1 \end{smallmatrix}\right)} \mathbb{R}^3 \xrightarrow{\left(\begin{smallmatrix} 0 & 1 & 0 \\ 0 & -1 & 0 \\ 0 & 0 & 0 \\ 0 & 1 & 0 \end{smallmatrix}\right)} \mathbb{R}^4 \xrightarrow{\left(\begin{smallmatrix} -1 & -1 & 0 & 0 \\ 1 & -1 & 0 & -2 \\ 0 & -2 & 0 & -2 \end{smallmatrix}\right)} \mathbb{R}^3 \xrightarrow{\left(\begin{smallmatrix} 1 & 1 & -1 \\ -1 & -1 & 1 \end{smallmatrix}\right)} \mathbb{R}^2 \longrightarrow 0.$$

Problem 2.3. Let

$$(V_\bullet, d): \quad 0 \longleftarrow V_0 \overset{d}{\longleftarrow} V_1 \longleftarrow \cdots \overset{d}{\longleftarrow} V_n \longleftarrow 0$$

be a chain complex of finite-dimensional vector spaces, concentrated in finitely many degrees $0, 1, \ldots, n$. The *Euler characteristic* of $(V_\bullet, d)$ is, by definition, the integer

$$\varepsilon(V, d) := \sum_{i=0}^{n} (-)^i \dim H_i(V, d).$$

Prove that $\varepsilon(V, d)$ is also given by the following formula:

$$\varepsilon(V, d) = \sum_{i=0}^{n} (-)^i \dim V_i.$$

Problem 2.4. Let R be a commutative ring with unit and let $(C_\bullet, d_C), (D_\bullet, d_D)$ be chain complexes. For any $i \in \mathbb{Z}$, put

$$\mathrm{Hom}(C, D)_i := \bigoplus_{k-j=i} \mathrm{Hom}_R(C_j, D_k),$$

Consider the sequence

$$\cdots \overset{d_{\mathrm{Hom}}}{\longleftarrow} \mathrm{Hom}(C, D)_{i-1} \overset{d_{\mathrm{Hom}}}{\longleftarrow} \mathrm{Hom}(C, D)_i \overset{d_{\mathrm{Hom}}}{\longleftarrow} \mathrm{Hom}(C, D)_{i+1} \longleftarrow \cdots$$

$$(2.21)$$

where, for any $i \in \mathbb{Z}$,

$$d_i : \mathrm{Hom}(C, D)_i \to \mathrm{Hom}(C, D)_{i-1}$$

is the $\mathbb{R}$-linear map uniquely defined by

$$d_i f = d_D \circ f - (-)^i f \circ d_C, \quad f \in \mathrm{Hom}_R(C_j, D_k), \quad k - j = i.$$

Prove that (2.21) is a well-defined chain complex whose 0-cycles are chain maps $f : (C_\bullet, d_C) \to (D_\bullet, d_D)$. Prove also that two 0-cycles f, g in (2.21) are homologous if and only if they are homotopic chain maps $f, g : (C_\bullet, d_C) \to (D_\bullet, d_D)$ (the homotopy being exactly the 1-chain $h \in \mathrm{Hom}(C, D)_1$ such that $f - g = d_D \circ h + h \circ d_C$).

Problem 2.5. Let R be a commutative ring with unit and let $(C_\bullet, d_C), (D_\bullet, d_D)$ be chain complexes. For any $i \in \mathbb{Z}$, put

$$(C \otimes D)_i := \bigoplus_{j+k=i} C_j \otimes D_k,$$

Consider the sequence

$$\cdots \xleftarrow{d_\otimes} (C \otimes D)_{i-1} \xleftarrow{d_\otimes} (C \otimes D)_i \xleftarrow{d_\otimes} (C \otimes D)_{i+1} \longleftarrow \cdots \tag{2.22}$$

where, for any $i \in \mathbb{Z}$,

$$d_i : (C \otimes D)_i \to (C \otimes D)_{i-1}$$

is the $\mathbb{R}$-linear map uniquely defined by

$$d_i(c \otimes d) = d_C c \otimes d + (-)^j c \otimes d_D d, \quad c \in C_j, \ d \in D_k, \quad j + k = i.$$

Prove that (2.22) is a well-defined chain complex. The chain complex (2.22) is called the *tensor product* of $(C_\bullet, d_C)$ and $(D_\bullet, d_D)$.

Problem 2.6. Let $(C_\bullet, d)$ be a chain complex of vector spaces. Prove that there exists a homotopy equivalence $F : (C_\bullet, d) \to (H_\bullet(C, d), 0)$. Conclude that every acyclic chain complex of vector spaces possesses a contracting homotopy.

Problem 2.7. Let

$$0 \longrightarrow (A_\bullet, d_A) \xrightarrow{\alpha} (C_\bullet, d) \xrightarrow{\beta} (B_\bullet, d_B) \longrightarrow 0 \tag{2.23}$$

be a short exact sequence of chain complexes. Show that, if (2.23) *splits*, i.e., there exists a cochain map $s : (B_\bullet, d_B) \to (C_\bullet, d)$ such that $\beta \circ s = \mathrm{id}_B$, then the connecting homomorphism $\Delta : H_i(B, d_B) \to H_{i-1}(A, d_A)$ vanishes for all $i \in \mathbb{Z}$. Conclude that, in this case, $H_i(C, d) \cong H_i(A, d_A) \oplus H_i(B, d_B)$ for all $i \in \mathbb{Z}$.

Chapter 3

Categories and Functors

In this short chapter, we briefly introduce *categories* and *functors*. This language puts under the same umbrella several (similar) situations in mathematics. As a byproduct, it also allows a compact formulation of various statements. Roughly, a category is a collection of objects together with arrows that we use to compare two objects. The arrows come with a composition law with appropriate properties. A functor is a correspondence of categories that maps objects to objects and arrows to arrows preserving the composition law of arrows. See below for a precise statement. Beware however that, in the following discussion, we skip most of the foundational aspects. For more on categories and functors, see, e.g., MacLane (1978).

3.1 Categories

Not every collection of objects in mathematics can be safely called a *set*. If we insisted in doing so, we would incur in paradoxes (i.e., statements that are equivalent to their negations, hence if they are true, they are also false and vice versa, in simple words, self-contradictions) like the famous *Russell Paradox*.

Example 3.1 (Russell Paradox). Assume that for any property there is a *set* consisting exactly of all objects with that property. Now consider the set

$$\mathcal{R} := \{x \text{ is a set such that } x \notin x\}$$

of all sets that do not contain themselves as elements. It is then clear that, by the very definition of $\mathcal{R}$, the set $\mathcal{R}$ belongs to $\mathcal{R}$ itself if and only if $\mathcal{R}$ does not belong to $\mathcal{R}$ and we have a paradox. ♦

The Russell paradox stems from the (unsafe) assumption that every property defines a set, in other words that every collection of objects (defined via a property) is a set. The easiest way to avoid Russell (and related) paradoxes is giving up on insisting that every collection is a set. If we do so, we need a new terminology for those collections of objects that cannot be sets.

Definition 3.2 (Class). A *class* is a collection of objects that can be defined via a property (that its elements share) without producing paradoxes. A *proper class* is a class which is not a set.

For instance, the collection $\mathcal{R}$ in Example 3.1 is a proper class. Every set is a class, while other popular examples of classes which are not sets are the class of all sets and the class of all groups. For classes, we use a similar notation as for sets (including $\in$ for "belongs to" and $\{x : x$ satisfies $\mathcal{P}\}$ for "the class of objects x satisfying the property $\mathcal{P}$").

Our next aim is defining *categories*. The main motivation here is the following principle of a *meta-mathematical* nature: *given a class* Ob *of mathematical objects* (groups, modules, topological spaces, (co)chain complexes, etc.), *there is also a class of arrows suitable for comparing objects of* Ob. Any two such arrows can be composed to produce a new one. This is the case, for instance, for groups and group homomorphisms, for modules and linear maps, for topological spaces and continuous maps, for (co)chain complexes and (co)chain maps, etc.

Definition 3.3 (Category). A *category* $\mathcal{C}$ is a pair $(\mathrm{Ob}_{\mathcal{C}}, \mathrm{Hom}_{\mathcal{C}})$, where $\mathrm{Ob}_{\mathcal{C}}$ is a class while $\mathrm{Hom}_{\mathcal{C}}$ is a family of sets

$$\mathrm{Hom}_{\mathcal{C}} = \{\mathrm{Hom}_{\mathcal{C}}(X,Y)\}_{X,Y \in \mathrm{Ob}_{\mathcal{C}}}$$

parameterized by pairs of elements in $\mathrm{Ob}_{\mathcal{C}}$. Additionally, for any $X, Y, Z \in \mathrm{Ob}_{\mathcal{C}}$, there is a *composition law*

$$\circ : \mathrm{Hom}_{\mathcal{C}}(Y,Z) \times \mathrm{Hom}_{\mathcal{C}}(X,Y) \to \mathrm{Hom}_{\mathcal{C}}(X,Z), \quad (f,g) \mapsto f \circ g$$

such that

(1) $\circ$ is *associative*, i.e., for all $X, Y, Z, W \in \mathrm{Ob}_{\mathcal{C}}$ and all $f \in \mathrm{Hom}_{\mathcal{C}}(Z,W)$, $g \in \mathrm{Hom}_{\mathcal{C}}(Y,Z)$ and $h \in \mathrm{Hom}_{\mathcal{C}}(X,Y)$, we have

$$(f \circ g) \circ h = f \circ (g \circ h);$$

(2) $\circ$ admits *units*, i.e., for all $X \in \mathrm{Ob}_{\mathcal{C}}$ there exists a, necessarily unique, element $\mathrm{id}_X \in \mathrm{Hom}_{\mathcal{C}}(X,X)$ such that for all $Y, Z \in \mathrm{Ob}_{\mathcal{C}}$, all $f \in \mathrm{Hom}_{\mathcal{C}}(X,Y)$ and all $g \in \mathrm{Hom}_{\mathcal{C}}(Z,X)$, we have

$$f \circ \mathrm{id}_X = f \quad \text{and} \quad \mathrm{id}_X \circ g = g.$$

The elements of $\mathrm{Ob}_{\mathcal{C}}$ are called *objects* of $\mathcal{C}$, while the elements of $\mathrm{Hom}_{\mathcal{C}}(X, Y)$ are called *morphisms*, or *arrows*, between X and Y. Given two objects $X, Y \in \mathrm{Ob}_{\mathcal{C}}$, a morphism $f \in \mathrm{Hom}_{\mathcal{C}}(X, Y)$ will be also denoted by

$$f : X \to Y \quad \text{or} \quad X \xrightarrow{f} Y.$$

Then X is called the *source* and Y is called the *target* of f. The morphism id_X is called the *identity morphism* of X or the *unit*. An *isomorphism* between two objects X, X' is a morphism $\Phi : X \to X'$ such that there exists a, necessarily unique, morphism $\Phi^{-1} : X' \to X$, called the *inverse* of Φ, such that $\Phi^{-1} \circ \Phi = \mathrm{id}_X$ and $\Phi \circ \Phi^{-1} = \mathrm{id}_{X'}$. A *small category* is a category whose class of objects is a set.

Clearly, it makes sense to talk about *commutative diagrams* in any category. We now present a long list of examples that should help the reader getting an intuition of what a category really is.

Example 3.4 (The Category of Sets). Sets (as objects) and maps (as morphisms) form a category called the *category of sets* and denoted **Set**. The composition law of morphisms in **Set** is the usual composition of maps and the units are the identity maps. The isomorphisms in **Set** are the invertible maps. ◆

Example 3.5 (The Category of Groups). Groups and group homomorphisms form a category called the *category of groups* and denoted **Gr**. The composition law of morphisms in **Gr** is the usual composition of maps and the units are the identity homomorphisms. The isomorphisms in **Gr** are the group isomorphisms. ◆

Next, fix a ring R.

Example 3.6 (The Category of R-Modules). R-modules and R-module homomorphisms form a category called the *category of R-modules* and denoted $\mathbf{Mod}_R$. The composition law of morphisms in $\mathbf{Mod}_R$ is the usual composition of (linear) maps and the units are the identity homomorphisms. The isomorphisms in $\mathbf{Mod}_R$ are the R-module isomorphisms. The category $\mathbf{Mod}_{\mathbb{Z}}$ is also called the *category of abelian groups* and denoted **Ab**. If $R = \mathbb{K}$ is a field, we often write $\mathbf{Vect}_{\mathbb{K}}$ instead of $\mathbf{Mod}_{\mathbb{K}}$ and call it the *category of $\mathbb{K}$-vector spaces*. ◆

Example 3.7 (The Category of (Co)Chain Complexes). (Co)Chain complexes of R-modules and (co)chain maps between them form a category called the *category of (co)chain complexes* and denoted $(\mathbf{Co})\mathbf{Ch}_R$. The composition law of morphisms in $(\mathbf{Co})\mathbf{Ch}_R$ is the composition of (co)chain maps and the units are the identity (co)chain maps. The isomorphisms in $(\mathbf{Co})\mathbf{Ch}_R$ are the (co)chain isomorphisms. ◆

Example 3.8 (The Category of Short Exact Sequences of (Co)Chain Complexes). Exercise 2.13 shows that short exact sequences of (co)chain complexes and their morphisms form a category. ◆

Example 3.9 (The Category of Topological Spaces). Topological spaces and continuous maps (with the composition of maps) form a category called the *category of topological spaces* and denoted **Top**. The isomorphisms is **Top** are the homeomorphisms. ◆

Exercise 3.1. Show that, in every category, the composition $\Phi \circ \Psi$ of two isomorphisms is an isomorphism with inverse $\Psi^{-1} \circ \Phi^{-1}$.

Not all categories consist of a class of sets together with a family of maps.

Example 3.10 (Monoids as Categories). A monoid $\mathbb{M}$ (in particular, a group) can be seen as a (small) category $(\mathrm{Ob}_\mathbb{M}, \mathrm{Hom}_\mathbb{M})$, where $\mathrm{Ob}_\mathbb{M} := \{*\}$ is a one element class (the only element in $\mathrm{Ob}_\mathbb{M}$ is conventionally denoted $*$) and $\mathrm{Hom}_\mathbb{M} = \mathrm{Hom}_\mathbb{M}(*, *) := \mathbb{M}$. The composition law of morphisms in $\mathbb{M}$ is just the monoid product and the unit is just the unit of the monoid. Isomorphisms in $\mathbb{M}$ are invertible elements in the monoid. ◆

Example 3.11 (Preordered Sets as Categories). Remember that a preordered set is a set $\mathbb{P}$ with a *preorder*, i.e., a reflexive and transitive relation $\leq$. A preordered set $\mathbb{P}$ can be seen as a (small) category $(\mathrm{Ob}_\mathbb{P}, \mathrm{Hom}_\mathbb{P})$, where $\mathrm{Ob}_\mathbb{P} := \mathbb{P}$ and, for all $a, b \in \mathbb{P}$, the set of arrows $\mathrm{Hom}_\mathbb{P}(a, b)$ is either a singleton (when $a \leq b$) or empty (when $a \not\leq b$). The composition law of morphisms in $\mathbb{P}$ is the only possible one (write it explicitly!). ◆

Exercise 3.2. Show that every category with just one object $*$ is a monoid. Show also that every small category with at most one arrow between any two objects is a preordered set.

Example 3.12 (The Category of Matrices). Fix a field $\mathbb{K}$. Matrices on $\mathbb{K}$ of arbitrary order can be seen as a (small) category $\mathbf{Mat}(\mathbb{K}) = (\mathrm{Ob}_{\mathbf{Mat}(\mathbb{K})}, \mathrm{Hom}_{\mathbf{Mat}(\mathbb{K})})$ as follows. Put $\mathrm{Ob}_{\mathbf{Mat}(\mathbb{K})} = \mathbb{N}_0$, and for all

$n, m \in \mathbb{N}_0$, put $\mathrm{Hom}_{\mathbf{Mat}(\mathbb{K})}(n, m) = M_{m,n}(\mathbb{K})$. The composition law is the matrix product. The units are the identity matrices and the isomorphisms are the invertible matrices. ♦

Example 3.13 (The Homotopy Category of (Co)Chain Complexes). Fix a ring R again. We consider the category $\mathbf{Ch}_R$ of chain complexes of R-modules. For simplicity, we denote it simply by $\mathbf{Ch}$. Define a new category $\mathbf{hCh}$ (or $\mathbf{hCh}_R$ if we want to insist on the fact that we work on the ring R) as follows. The objects in $\mathbf{hCh}$ are chain complexes of R-module, i.e., $\mathrm{Ob}_{\mathbf{hCh}} = \mathrm{Ob}_{\mathbf{Ch}}$. In order to define morphisms, recall that "being homotopic" is an equivalence relation on the set $\mathrm{Hom}_{\mathbf{Ch}}\left((C_\bullet, d_C), (D_\bullet, d_D)\right)$ of chain maps between the chain complexes $(C_\bullet, d_C), (D_\bullet, d_D)$ (Exercise 2.7). Denote by $\sim$ this equivalence relation and, for any two chain complexes $(C_\bullet, d_C), (D_\bullet, d_D)$, put

$$\mathrm{Hom}_{\mathbf{hCh}}\left((C_\bullet, d_C), (D_\bullet, d_D)\right) = \mathrm{Hom}_{\mathbf{Ch}}\left((C_\bullet, d_C), (D_\bullet, d_D)\right)\Big/ \sim,$$

the set of homotopy classes of chain maps. Given a chain map $f : (C_\bullet, d_C) \to (D_\bullet, d_D)$, we denote $[f]_\sim \in \mathrm{Hom}_{\mathbf{hCh}}((C_\bullet, d_C), (D_\bullet, d_D))$ its homotopy class. The composition law of morphisms in $\mathbf{hCh}$ is defined as follows. Let

$$(C_\bullet, d_C) \xrightarrow{f} (D_\bullet, d_D) \xrightarrow{g} (E_\bullet, d_E)$$

be chain maps. We put

$$[g]_\sim \circ [f]_\sim := [g \circ f]_\sim.$$

As homotopies respect the composition of chain maps (Proposition 5.15), this is well defined (do you see it?). The composition law of morphisms in $\mathbf{hCh}$ defined in this way is clearly associative. The units are the homotopy classes of the identity chain maps. The isomorphisms in $\mathbf{hCh}$ are the (homotopy classes of) homotopy equivalences of chain complexes (do you see it?). The category $\mathbf{hCh}$ is called the *homotopy category of chain complexes* of R-modules and it is extremely useful when one wants to study chain complexes only up to homotopy equivalence. The homotopy category of cochain complexes is defined is a similar (obvious) way. This example shows that the structure of a category can change significantly changing the morphisms without changing the objects. For more on homotopy categories, see, e.g., Rotman (1998). ♦

3.2 Functors

Roughly, functors are *maps of categories*: they map objects to objects and morphisms to morphisms preserving the category structure. Let $\mathcal{C}, \mathcal{D}$ be categories.

Definition 3.14 (Functor). A *functor* $\mathbb{F} : \mathcal{C} \to \mathcal{D}$ between $\mathcal{C}$ and $\mathcal{D}$ is the assignment

(1) of an object $\mathbb{F}(X) \in \mathrm{Ob}_{\mathcal{D}}$ for every object $X \in \mathrm{Ob}_{\mathcal{C}}$ and
(2) of an arrow $\mathbb{F}(f) : \mathbb{F}(X) \to \mathbb{F}(Y) \in \mathrm{Hom}_{\mathcal{D}}$ for every arrow $f : X \to Y \in \mathrm{Hom}_{\mathcal{C}}$, where $X, Y \in \mathrm{Ob}_{\mathcal{C}}$,

in such a way that

- $\mathbb{F}(\mathrm{id}_X) = \mathrm{id}_{\mathbb{F}(X)}$ for all $X \in \mathrm{Ob}_{\mathcal{C}}$;
- $\mathbb{F}(f \circ g) = \mathbb{F}(f) \circ \mathbb{F}(g)$ for all pairs (f, g) of composable arrows in $\mathcal{C}$.

More precisely, an $\mathbb{F}$ as in Definition 3.15 is a *covariant functor*. There is also a notion of a *contravariant functor* which is often useful. A contravariant functor is the same as a (covariant) functor except that it inverts the arrows (and their compositions).

Definition 3.15 (Contravariant Functor). A *contravariant functor* $\mathbb{G} : \mathcal{C} \to \mathcal{D}$ between the categories $\mathcal{C}$ and $\mathcal{D}$ is the assignment

(1) of an object $\mathbb{G}(X) \in \mathrm{Ob}_{\mathcal{D}}$ for every object $X \in \mathrm{Ob}_{\mathcal{C}}$ and
(2) of an arrow $\mathbb{G}(f) : \mathbb{G}(Y) \to \mathbb{G}(X) \in \mathrm{Hom}_{\mathcal{D}}$ for every arrow $f : X \to Y \in \mathrm{Hom}_{\mathcal{C}}$, where $X, Y \in \mathrm{Ob}_{\mathcal{C}}$,

in such a way that

- $\mathbb{G}(\mathrm{id}_X) = \mathrm{id}_{\mathbb{G}(X)}$ for all $X \in \mathrm{Ob}_{\mathcal{C}}$;
- $\mathbb{G}(f \circ g) = \mathbb{G}(g) \circ \mathbb{G}(f)$ for all pairs (f, g) of composable arrows in $\mathcal{C}$.

Several natural constructions in mathematics are functors (either covariant or contravariant). Here is a short list. Many more examples will pop up in the sequel of the book. Fix again a ring R.

Example 3.16 (The Free Module Construction is a Functor). The free module construction can be seen as a (covariant) functor **Free** : **Set** $\to$ **Mod**$_R$ as follows. An object X in **Set** is just a set and we put $\mathbf{Free}(X) := RX$ (the free module spanned by X) which is duly an object in **Mod**$_R$. Next, a morphism $f : X \to Y$ in **Set** is just a map of sets. Note that, from the universal property of free modules, there exists a unique linear map $Rf : RX \to RY$ such that $Rf(x) = f(x) \in Y \subseteq RY$ for all $x \in X \subseteq RX$ (here we interpret

X, resp. Y, as a subset in RX, resp. RY, as usual). Put $\mathbf{Free}(f) := Rf$. We leave it as Exercise 3.3 to check that the assignment $\mathbf{Free} : \mathbf{Set} \to \mathbf{Mod}_R$ defined in this way is indeed a functor. ♦

> **Exercise 3.3.** Show that the assignment $\mathbf{Free} : \mathbf{Set} \to \mathbf{Mod}_R$ defined in Example 3.16 is a covariant functor.

Example 3.17 (The Dual Module Construction is a Functor). Let M be an R-module. Recall that its dual module is $M^* = \mathrm{Hom}(M, R)$ (with the module structure on linear maps). If M, N are two R-modules and $f : M \to N$ is a linear map, we can define a linear map $f^* : N^* \to M^*$, called the *transpose* of f, by putting

$$f^*(\varphi) := \varphi \circ f, \quad \varphi \in N^*.$$

It is clear that $f^*(\varphi) : M \to R$ is a linear map for all $\varphi \in N^*$ (the composition of linear maps is a linear map). Additionally, the map $f^* : N^* \to M^*$ defined in this way is indeed linear (Exercise 3.4, see also Example 1.53). This construction defines a contravariant functor $* : \mathbf{Mod}_R \to \mathbf{Mod}_R$. Namely, for every object in $\mathbf{Mod}_R$, i.e., every R-module M, put $*(M) := M^*$, and, for every arrow (in $\mathbf{Mod}_R$), i.e., every linear map $f : M \to N$, put $*(f) = f^* : N^* \to M^*$. We leave it to the reader to check the details as Exercise 3.4. ♦

> **Exercise 3.4.** Prove that the transpose map $f^* : N^* \to M^*$ defined in Example 3.17 is a linear map. Prove also that the assignment $* : \mathbf{Mod}_R \to \mathbf{Mod}_R$ is a contravariant functor.

Example 3.18 ((Co)Homology is a Functor). For all $n \in \mathbb{Z}$, the n-th homology of chain complexes is a covariant functor $H_n : \mathbf{Ch}_R \to \mathbf{Mod}_R$. Namely, an object in $\mathbf{Ch}_R$ is a chain complex $(C_\bullet, d)$. Its n-th homology $H_n(C, d)$ is an R-module, i.e., an object in $\mathbf{Mod}_R$. A morphism $f : (C_\bullet, d_C) \to (D_\bullet, d_D)$ in $\mathbf{Ch}_R$ is a chain map and the induced map in n-th homology $H_n(f) : H_n(C, d_C) \to H_n(D, d_D)$ is an R-module homomorphism, i.e., a morphism in $\mathbf{Mod}_R$. Finally, the assignment $H_n : \mathbf{Ch}_R \to \mathbf{Mod}_R$ defined in this way preserves the identity morphisms and the composition of morphisms (Proposition 2.15). Similarly, the n-th cohomology of cochain complexes is a covariant functor $H^n : \mathbf{CoCh}_R \to \mathbf{Mod}_R$. ♦

Exercise 3.5. Let $\mathbb{M}, \mathbb{N}$ be monoids. Show that a covariant functor between the corresponding categories is "essentially the same" as a monoid homomorphism $f : \mathbb{M} \to \mathbb{N}$.

Exercise 3.6. Let $\mathbb{P}, \mathbb{Q}$ be preordered sets. Show that a covariant (resp. contravariant) functor between the corresponding categories is "essentially the same" as an increasing (resp. decreasing) map $f : \mathbb{P} \to \mathbb{Q}$ (remember that a map $f : \mathbb{P} \to \mathbb{Q}$ is increasing if, whenever $a, b \in \mathbb{P}$ are such that $a \leq b$, then $f(a) \leq f(b)$, and it is decreasing if, whenever $a \leq b$, then $f(b) \leq f(a)$).

Lemma 3.19. *Functors transform isomorphisms to isomorphisms.*

Proof.   Let $\mathcal{C}, \mathcal{D}$ be categories and let $\mathbb{F} : \mathcal{C} \to \mathcal{D}$ be a functor. We want to show that, for every isomorphism $\Phi : X \to Y$ in $\mathcal{C}$, its image $\mathbb{F}(\Phi) : \mathbb{F}(X) \to \mathbb{F}(Y)$ under $\mathbb{F}$ is an isomorphism (in $\mathcal{D}$). So, let $\Phi^{-1} : Y \to X$ be the inverse isomorphism. Then

$$\mathbb{F}(\Phi) \circ \mathbb{F}(\Phi^{-1}) = \mathbb{F}(\Phi \circ \Phi^{-1}) = \mathbb{F}(\mathrm{id}_Y) = \mathrm{id}_{\mathbb{F}(Y)}.$$

Swapping the roles of Φ and Φ^{-1} we see that $\mathbb{F}(\Phi^{-1}) \circ \mathbb{F}(\Phi) = \mathrm{id}_{\mathbb{F}(X)}$. This shows that $\mathbb{F}(\Phi)$ is an isomorphism and that $\mathbb{F}(\Phi^{-1})$ is its inverse. $\square$

Functors can be composed obtaining new functors. Namely, let $\mathcal{C}, \mathcal{D}, \mathcal{E}$ be categories and let

$$\mathcal{C} \xrightarrow{\ \mathbb{F}\ } \mathcal{D} \xrightarrow{\ \mathbf{G}\ } \mathcal{E}$$

be functors. We define a new functor $\mathbf{G} \circ \mathbb{F} : \mathcal{C} \to \mathcal{D}$ by putting

$$\mathbf{G} \circ \mathbb{F}(X) := \mathbf{G}(\mathbb{F}(X)) \in \mathrm{Ob}_{\mathcal{E}}, \quad \text{for all } X \in \mathrm{Ob}_{\mathcal{C}},$$

and

$$\mathbf{G} \circ \mathbb{F}(f) := \mathbf{G}(\mathbb{F}(f)) \in \mathrm{Hom}_{\mathcal{E}}, \quad \text{for all } f \in \mathrm{Hom}_{\mathcal{C}}.$$

Note that

- if $\mathbb{F}, \mathbf{G}$ are covariant functors, then $\mathbf{G} \circ \mathbb{F}$ is a covariant functor;
- if $\mathbb{F}, \mathbf{G}$ are contravariant functors, then $\mathbf{G} \circ \mathbb{F}$ is a covariant functor;
- if $\mathbb{F}, \mathbf{G}$ are a covariant and a contravariant functor (not necessarily in this order), then $\mathbf{G} \circ \mathbb{F}$ is a contravariant functor.

We leave it to the reader to check all the details.

Example 3.20 (Biduality Functor). Composing the duality functor $* :$ $\mathbf{Mod}_R \to \mathbf{Mod}_R$ with itself we get a covariant functor $** : \mathbf{Mod}_R \to$ $\mathbf{Mod}_R$ mapping an R-module M to its bidual M^{**} and an R-module homomorphism $f : M \to N$ to the transpose $f^{**} : M^{**} \to N^{**}$ of its transpose. ◆

More examples of composition of functors pop up in the sequel of this book.

3.3 Natural Transformations

We conclude this chapter defining *natural transformations* of functors, which allow to compare two functors between the same two categories. So, let $\mathcal{C}, \mathcal{D}$ be categories and let $\mathbb{F}, \mathbb{G} : \mathcal{C} \to \mathcal{D}$ be functors.

Definition 3.21 (Natural Transformation). A *natural transformation* $\tau :$ $\mathbb{F} \to \mathbb{G}$ between the functors $\mathbb{F}$ and $\mathbb{G}$ is the assignment of a morphism $\tau_X : \mathbb{F}(X) \to \mathbb{G}(X) \in \mathrm{Hom}_{\mathcal{D}}$ for every object $X \in \mathrm{Ob}_{\mathcal{C}}$ in such a way that, for every morphism $f : X \to Y \in \mathrm{Hom}_{\mathcal{C}}$, the diagram

$$
\begin{array}{ccc}
\mathbb{F}(X) & \xrightarrow{\ \mathbb{F}(f)\ } & \mathbb{F}(Y) \\
{\scriptstyle \tau_X}\downarrow & & \downarrow{\scriptstyle \tau_Y} \\
\mathbb{G}(X) & \xrightarrow{\ \mathbb{G}(f)\ } & \mathbb{G}(Y)
\end{array}
$$

of arrows in $\mathcal{D}$ commutes. A natural transformation $\tau : \mathbb{F} \to \mathbb{G}$ is a *natural isomorphism* of functors if the arrow τ_X is an isomorphism in $\mathcal{D}$ for every object X in $\mathcal{C}$.

Several natural constructions in mathematics are actually natural transformations of functors. Here we present just two examples. More examples actually pop up in the sequel but we do not (always) highlight them (see also the End-of-Chapter Problems section of this chapter). We invite the reader to look themselves at natural transformations throughout this book.

Example 3.22 (Biduality Map). Let R be a ring. Remember that, for any R-module there is a *natural* linear map $\iota : M \to M^{**}$ defined by putting $\iota(p)(\varphi) = \varphi(p)$ for all $p \in M$ and $\varphi \in M^*$ (see the discussion immediately after the statement of Corollary 1.62). This construction can be seen as a natural transformation ι between the following two functors. The source functor of ι is the *identity functor* id : $\mathbf{Mod}_R \to \mathbf{Mod}_R$ mapping every object and every arrow to itself (do you see that it is indeed a functor?). The target functor of ι is the biduality functor $** : \mathbf{Mod}_R \to \mathbf{Mod}_R$. Finally, for every object in $\mathbf{Mod}_R$, i.e., every R-module M, we put $\iota_M := \iota : M \to M^{**}$. We leave it to the reader to check that this assignment defines indeed a natural transformation $\iota : \mathrm{id} \to **$ as Exercise 3.7.   ◆

> **Exercise 3.7.** Prove that the assignment $\iota : \mathrm{id} \to **$ defined in Example 3.22 is a natural transformation of functors.

We also provide an example of a natural isomorphism.

Example 3.23. Let R be a ring. We begin noticing that the function module construction can be seen as a contravariant functor $\mathbf{Fun} : \mathbf{Set} \to \mathbf{Mod}_R$ as follows. For any set $X \in \mathrm{Ob}_{\mathbf{Set}}$ we put $\mathbf{Fun}(X) := R^X \in \mathrm{Ob}_{\mathbf{Mod}_R}$ (the module of functions $f : X \to R$). Next, for any map of sets $f : X \to Y$, we define a map

$$R^f : R^Y \to R^X$$

between the corresponding function modules (but in the reverse order) by putting

$$R^f(a) = a \circ f : X \to R$$

for all functions $a : Y \to R$. The function $R^f(a)$ is also called the *pull-back* of a along f and it is sometimes denoted by $f^*(a)$. It is easy to see that R^f is a linear map, hence it is an arrow in $\mathbf{Mod}_R$. Put $\mathbf{Fun}(f) := R^f$. The assignment $\mathbf{Fun} : \mathbf{Set} \to \mathbf{Mod}_R$ is a contravariant functor. We want to show that there is a natural isomorphism $\iota : \mathbf{Fun} \to * \circ \mathbf{Free}$. To see this, recall from Example 1.51 that, for any set X, there is a natural module isomorphism

$$\iota : R^X = \mathbf{Fun}(X) \to (RX)^* = * \circ \mathbf{Free}(X),$$

defined by putting

$$\iota(f)\left(\sum_i a_i x_i\right) := \sum_i a_i f(x_i),$$

for all $f \in R^X$ and all linear combinations $\sum_i a_i x_i \in RX$, $a_i \in R$, $x_i \in X$. We put $\iota_X := \iota : R^X \to (RX)^*$. The rest is left to the reader.   ◆

> **Exercise 3.8.** Prove all unproven claims in Example 3.23.

Exercise 3.9. Let $\tau : \mathbb{F} \to \mathbb{G}$ be a natural isomorphism between the functors $\mathbb{F}, \mathbb{G} : \mathcal{C} \to \mathcal{D}$. Prove that the assignment $\tau^{-1} : \mathbb{G} \to \mathbb{F}$ defined by putting $(\tau^{-1})_X := (\tau_X)^{-1} : \mathbb{G}(X) \to \mathbb{F}(X) \in \mathrm{Hom}_\mathcal{D}$ for all $X \in \mathrm{Ob}_\mathcal{C}$ is a natural isomorphism between $\mathbb{G}$ and $\mathbb{F}$ (such natural isomorphism is called the *inverse* of the natural isomorphism τ).

3.4 End-of-Chapter Problems

Problem 3.1. Let $\mathcal{C}$ be a category. An *initial* (resp., *terminal*) object in $\mathcal{C}$ is an object $X \in \mathrm{Ob}_\mathcal{C}$ such that, for any other object $Y \in \mathrm{Ob}_\mathcal{C}$, there exists a unique morphism $X \to Y$ (resp., $Y \to X$). Show that an initial (resp., terminal) object (if it exists) is unique up to a unique isomorphism, i.e., if $X, X' \in \mathrm{Ob}_\mathcal{C}$ are initial (terminal) objects, then there exists a unique isomorphism $X \to X'$.

Problem 3.2. Let R be a commutative ring with unit, and let X be a set. Define a category $\mathcal{C}$ as follows. An object in $\mathcal{C}$ is a pair (N, ϕ) where N is an R-module, and $\phi : X \to N$ is a map. A morphism in $\mathcal{C}$ between the objects $(N_1, \phi_1), (N_2, \phi_2)$ is an R-linear map $f : N_1 \to N_2$ such that the diagram

$$
\begin{array}{ccc}
 & X & \\
\phi_1 \swarrow & & \searrow \phi_2 \\
N_1 & \xrightarrow{\;\;f\;\;} & N_2
\end{array}
$$

commutes. The composition of morphisms in $\mathcal{C}$ is just the composition of maps. Show that $\mathcal{C}$ is a well-defined category. Show also that a free module spanned by X is an initial object in $\mathcal{C}$, see Problem 3.1 (conclude that the free module spanned by X is unique up to a unique isomorphism).

Problem 3.3. Let R be a commutative ring with unit, and let $(M_i)_{i \in I}$ be a family of R-modules parameterized by an index set I. Define a category $\mathcal{C}$ as follows. An object in $\mathcal{C}$ is a pair (N, λ) where N is an R-module and $\lambda = (\lambda_i : M_i \to N)_{i \in I}$ is a family of R-linear maps.

A morphism in $\mathcal{C}$ between the objects $(N_1, \lambda_1), (N_2, \lambda_2)$ is an R-linear map $f : N_1 \to N_2$ such that the diagram

$$
\begin{array}{ccc}
 & M_i & \\
\lambda_{1i} \swarrow & & \searrow \lambda_{2i} \\
N_1 \xrightarrow{\quad f \quad} & & N_2
\end{array}
$$

commutes for all $i \in I$. The composition of morphisms in $\mathcal{C}$ is just the composition of maps. Show that $\mathcal{C}$ is a well-defined category. Show also that a direct sum of the modules $(M_i)_{i \in I}$ is an initial object in $\mathcal{C}$, see Problem 3.1 (conclude that the direct sum of $(M_i)_{i \in I}$ is unique up to a unique isomorphism).

Problem 3.4. Show that the direct product of a family of modules is a terminal object in an appropriate category and conclude its uniqueness from this feature.

Problem 3.5. Show that the tensor product, the symmetric power and the exterior power of modules are all initial objects in appropriate categories and conclude their uniqueness from this feature.

Problem 3.6. Characterize groups among monoids purely in terms of properties of the corresponding category (see Example 3.10).

Problem 3.7. Characterize posets and sets with an equivalence relation among preordered sets purely in terms of properties of the corresponding category (see Example 3.11).

Problem 3.8. Let $\mathcal{C}$ be a category, and let $X \in \mathrm{Ob}_\mathcal{C}$ be an object. Prove that the assignment

$$
\mathrm{Hom}_\mathcal{C}(X, -) : \mathcal{C} \to \mathbf{Set}
$$

(resp.,

$$\text{Hom}_{\mathcal{C}}(-, X) : \mathcal{C} \to \textbf{Set})$$

mapping

(1) an object $Y \in \text{Ob}_{\mathcal{C}}$ to $\text{Hom}_{\mathcal{C}}(Y, X)$ (resp., $\text{Hom}_{\mathcal{C}}(X, Y)$) and
(2) a morphism $f : Y_1 \to Y_2 \in \text{Hom}_{\mathcal{C}}$ to the map:

$$\text{Hom}_{\mathcal{C}}(X, f) : \text{Hom}_{\mathcal{C}}(X, Y_1) \to \text{Hom}_{\mathcal{C}}(X, Y_2), \quad \phi \mapsto f \circ \phi$$

(resp.,

$$\text{Hom}_{\mathcal{C}}(f, X) : \text{Hom}_{\mathcal{C}}(Y_2, X) \to \text{Hom}_{\mathcal{C}}(Y_1, X), \quad \phi \mapsto \phi \circ f)$$

is a well-defined covariant (resp., contravariant) functor.

Problem 3.9. An *equivalence of categories* is a functor $\mathbb{F} : \mathcal{C} \to \mathcal{D}$ such that

- $\mathbb{F}$ is *fully faithful*, i.e., for all $X, Y \in \text{Ob}_{\mathcal{C}}$, the map $\mathbb{F} : \text{Hom}_{\mathcal{C}}(X, Y) \to \text{Hom}_{\mathcal{D}}(\mathbb{F}(X), \mathbb{F}(Y))$ is a bijection;
- $\mathbb{F}$ is *essentially surjective*, i.e., for any $Z \in \text{Ob}_{\mathcal{D}}$, there exist an $X \in \text{Ob}_{\mathcal{C}}$ and an isomorphism $\mathbb{F}(X) \to Z$.

Now, let $\mathbb{K}$ be a field. Prove that the assignment $L : \textbf{Mat}(\mathbb{K}) \to \textbf{Vect}_{\mathbb{K}}$ mapping

(1) a non-negative integer $n \in \mathbb{N}_0 = \text{Ob}_{\textbf{Mat}(\mathbb{K})}$ to the vector space $\mathbb{K}^n$ and
(2) an $m \times n$ matrix $M \in M_{m,n}(\mathbb{K}) = \text{Hom}_{\textbf{Mat}(\mathbb{K})}$ to the linear map

$$L_M : \mathbb{K}^n \to \mathbb{K}^m, \quad x \mapsto Mx,$$

is an equivalence of categories.

Problem 3.10. Let R be a commutative ring with unit, and let M be an R-module. Prove that the assignment

$$- \otimes_R M : \textbf{Mod}_R \to \textbf{Set}$$

mapping

(1) an R-module N to the R-module $N \otimes_R M$ (seen as a set) and
(2) an R-module homomorphism $f : N_1 \to N_2$ to the R-linear map

$$f \otimes_R M : N_1 \otimes_R M \to N_2 \otimes_R M$$

(seen as a map of sets) uniquely defined by

$$(f \otimes_R M)(q_1 \otimes p) := f(q_1) \otimes p, \quad q_1 \in N_1, \quad p \in M,$$

is a well-defined (covariant) functor.

Problem 3.11. Let R be a commutative ring with unit, and let M be an R-module. For any R-module N, consider the R-linear map

$$\iota_N : N^* \otimes_R M \to \mathrm{Hom}_R(N, M)$$

uniquely defined by

$$\iota_N(\varphi \otimes p)(q) := \varphi(q)p, \quad \varphi \in N^*, \quad p \in M, \quad q \in M.$$

Prove that the assignment $\iota : N \mapsto \iota_N$ is a well-defined natural transformation between the composition $(-)^* \otimes_R M$ of the duality (contravariant) functor $* : \mathbf{Mod}_R \to \mathbf{Mod}_R$ followed by the functor $- \otimes_R M : \mathbf{Mod}_R \to \mathbf{Set}$ of Problem 3.10 and the covariant functor $\mathrm{Hom}_R(M, -) : \mathbf{Mod}_R \to \mathbf{Set}$ of Problem 3.8.

Problem 3.12. Let $\mathcal{C}, \mathcal{D}$ be categories, let $\mathbb{F}, \mathbb{G}, \mathbb{H} : \mathcal{C} \to \mathcal{D}$ be functors, and let $\tau : \mathbb{F} \to \mathbb{G}$ and $\sigma : \mathbb{G} \to \mathbb{H}$ be natural transformations. Prove that the assignment $X \mapsto (\sigma \circ_{\mathrm{vert}} \tau)_X := \sigma_X \circ \tau_X$, $X \in \mathrm{Ob}_{\mathcal{C}}$, is a well-defined natural transformation between the functors $\mathbb{F}$ and $\mathbb{H}$ (called the *vertical composition of τ followed by σ*)

$$\sigma \circ_{\mathrm{vert}} \tau : \mathbb{F} \to \mathbb{H}.$$

Problem 3.13. Let $\mathcal{C}, \mathcal{D}, \mathcal{E}$ be categories, let $\mathbb{F}, \mathbb{F}' : \mathcal{C} \to \mathcal{D}$ and $\mathbb{G}, \mathbb{G}' : \mathcal{D} \to \mathcal{E}$ be functors, and let $\tau : \mathbb{F} \to \mathbb{F}'$ and $\sigma : \mathbb{G} \to \mathbb{G}'$ be natural transformations. Prove that for every $X \in \mathrm{Ob}_{\mathcal{C}}$ the following morphisms in $\mathcal{E}$ do actually agree: $\sigma_{\mathbb{F}'(X)} \circ G(\tau_X)$ and $G'(\tau_X) \circ \sigma_{\mathbb{F}(X)}$. Prove also that the assignment $X \mapsto (\sigma \circ_{\text{hor}} \tau)_X := \sigma_{\mathbb{F}'(X)} \circ G(\tau_X) = G'(\tau_X) \circ \sigma_{\mathbb{F}(X)}$ is a well-defined natural transformation between the functors $G \circ \mathbb{F}$ and $G' \circ \mathbb{F}'$ (called the *horizontal composition of τ followed by σ*):

$$\sigma \circ_{\text{hor}} \tau : G \circ \mathbb{F} \to G' \circ \mathbb{F}'.$$

Problem 3.14. Let $\mathcal{C}$ be a category, let $X \in \text{Ob}_{\mathcal{C}}$ and let $\mathbb{F} : \mathcal{C} \to \textbf{Set}$ be a contravariant functor. For any $x \in \mathbb{F}(X)$ and any $Y \in \text{Ob}_{\mathcal{C}}$, consider the map

$$\sigma(x)_Y : \text{Hom}_{\mathcal{C}}(Y, X) \to \mathbb{F}(Y), \quad f \mapsto \mathbb{F}(f)(x).$$

Prove that the assignment $Y \mapsto \sigma(x)_Y$ is a(n x-dependent) natural transformation between the contravariant functor $\text{Hom}_{\mathcal{C}}(-, X)$ of Problem 3.8 and $\mathbb{F}$. Prove also the following *Yoneda Lemma*, i.e., the assignment $x \mapsto \sigma(x)$, is a bijection between the set $\mathbb{F}(X)$ and the class of natural transformations $\tau : \text{Hom}_{\mathcal{C}}(-, X) \to \mathbb{F}$ whose inverse is given by the assignment $\tau \mapsto \sigma^{-1}(\tau)$ mapping a natural transformation $\tau : \text{Hom}_{\mathcal{C}}(-, X) \to \mathbb{F}$, to the element $\sigma^{-1}(\tau) \in \mathbb{F}(X)$ given by

$$\sigma^{-1}(\tau) := \tau_X(\text{id}_X).$$

Problem 3.15. Let R be a commutative ring with unit. Prove that the assignment

$$* : \textbf{Ch}_R \to \textbf{CoCh}_R$$

mapping

(1) a chain complex $(C_\bullet, d)$ to the *dual cochain complex* $(*C^\bullet, d^*)$ where $*C^i = C_i^*$, for all $i \in \mathbb{Z}$ and d^* is the transpose of d and
(2) a cochain map $f : (C_\bullet, d_C) \to (D_\bullet, d_D)$ to the *transpose cochain map* $f^* : (*D^\bullet, d_D^*) \to (*C^\bullet, d_C^*)$

is a well-defined contravariant functor.

Additionally, for any chain complex $(C_\bullet, d_C)$ of R-modules, and any $i \in \mathbb{Z}$, consider the linear map

$$\tau(C_\bullet, d) : H^i(*C, d^*) \to H_i(C, d)^*$$

mapping the cohomology class $[\varphi]_{*C}$ of a cocycle φ in the dual complex $(*C^\bullet, d^*)$ to the linear form

$$\tau(C_\bullet, d)(\varphi) : H_i(C, d) \to R, \quad [c]_C \mapsto \varphi(c),$$

where c is a cycle in $(C_\bullet, d)$ (of the same degree as φ). Prove that the assignment $(C_\bullet, d_C) \mapsto \tau(C_\bullet, d)$ is a well-defined natural transformation τ between the functors $H^i \circ *$ and $* \circ H_i$. Prove also that, when $R = \mathbb{K}$ is a field, τ is actually a natural isomorphism (in other words *the cohomology of the dual complex is the dual of the homology* in the field case).

Chapter 4

Applications in Algebra

In this chapter, we show that (co)chain complexes arise naturally in algebra from various algebraic structures. We discuss groups, associative algebras and Lie algebras. All these structures play an important role both in algebra and in geometry. Our focus is on how associating (co)chain complexes to these structures (usually in a functorial way) rather than on computing the associated (co)homology. We also provide an interpretation of low degree (co)homologies. All the (co)chain complexes in this chapter have natural generalizations to the case when the algebraic structure acts (group modules, algebra bimodules and Lie algebra representations) that we shortly discuss in the End-of-Chapter Problem Section. We recommend the book Weibel (1994) for much more on algebraic structures and chain complexes.

4.1 Simplicial Objects

It is possible to construct a (co)chain complex from certain data, called *semi-(co)simplicial modules* (and, more generally, *semi-(co)simplicial sets*). There are several important (co)chain complexes that arise in this way. We present two examples in this chapter and one more example in Chapter 5. We begin with the definition of semi-simplicial set.

Definition 4.1 (Semi-Simplicial Set). A *semi-simplicial set* is a pair $(X^\bullet, d)$ where

(1) $X^\bullet = (X^n)_{n \in \mathbb{N}_0}$ is a family of sets (indexed by non-negative integers) and

(2) $d = (d_i^n : X^n \to X^{n-1})_{0 \leq i \leq n \in \mathbb{N}}$ is a family of maps called the *face maps* (or simply the *faces*),

satisfying the following *semi-simplicial identities*:

$$d_i^{n-1} \circ d_j^n = d_{j-1}^{n-1} \circ d_i^n, \quad \text{for all } 0 \leq i < j \leq n. \tag{4.1}$$

When it is clear what is the value of n, we denote simply by $d_i : X^n \to X^{n-1}$ the i-th face map on X^n (instead of $d_i^n : X^n \to X^{n-1}$). In this short notation, the semi-simplicial identities become $d_i \circ d_j = d_{j-1} \circ d_i$ for all $i < j$. Sometimes, a semi-simplicial set $(X^\bullet, d)$ is schematically indicated

$$\cdots \Longrightarrow\!\!\!\!\!\Longrightarrow X^2 \Longrightarrow\!\!\!\!\!\Longrightarrow X^1 \Longrightarrow X^0,$$

where the arrows stand for the maps d_i. Inverting all the arrows in the definition of a semi-simplicial set, we get a (dual) definition of *semi-cosimplicial set*. More precisely, a *semi-cosimplicial set* is a pair $(Y_\bullet, d)$, where $Y_\bullet = (Y_n)_{n \in \mathbb{N}_0}$ is a family of sets and $d = (d_i^n : Y_{n-1} \to Y_n)_{0 \leq i \leq n \in \mathbb{N}}$ is a family of maps (sometimes called the *coface maps* or simply the *cofaces*), and also denoted simply $d_i : Y_{n-1} \to Y_n$, satisfying the *semi-cosimplicial identities*:

$$d_j^n \circ d_i^{n-1} = d_i^n \circ d_{j-1}^{n-1}, \quad \text{for all } 0 \leq i < j \leq n.$$

A semi-cosimplicial set $(Y_\bullet, d)$ is also indicated

$$\cdots \Longleftarrow\!\!\!\!\!\Longleftarrow Y_2 \Longleftarrow\!\!\!\!\!\Longleftarrow Y_1 \Longleftarrow Y_0.$$

Remark 4.2. The prefix "semi" in "semi-(co)simplicial set" refers to the fact that there exists a (more fundamental) notion of *(co)simplicial set* where faces and semi-(co)simplicial identities are complemented by certain *degeneracy maps* (going in the other direction) satisfying appropriate *(co)simplicial identities*. We do not need these notions in these notes, but the interested reader may consult (Weibel, 1994, Chapter 8) (see also May, 1967). ◇

Example 4.3 (Nerve of a Group). Let G be a (non-necessarily abelian) group. Consider the family $N^\bullet(G) = (G^{\times n})_{n \in \mathbb{N}_0}$ of sets, where we put $G^{\times 0} := \{*\}$, a one point set, whose only element is conventionally denoted $*$. The family $N^\bullet(G)$ can be given the structure of a semi-simplicial set

$$\cdots \rightrightarrows G^{\times 2} \rightrightarrows G \rightrightarrows \{*\}$$

with faces $d = (d_i^n : G^{\times n} \to G^{\times (n-1)})_{0 \le i \le n \in \mathbb{N}}$ given by

$$d_i^n(g_1,\ldots,g_n) = \begin{cases} (g_2,\ldots,g_n) & \text{if } i = 0 \\ (g_1,\ldots,g_{i-1},g_ig_{i+1},g_{i+2},\ldots,g_n) & \text{if } i = 1,\ldots,n-1, \\ (g_1,\ldots,g_{n-1}) & \text{if } i = n \end{cases}$$

$(g_1,\ldots,g_n) \in G^{\times n}$. We leave it to the reader to check the semi-simplicial identities as Exercise 4.1. The semi-simplicial set $(N^\bullet(G),d)$ is called the *nerve* of the group G. ◆

> **Exercise 4.1.** Prove the semi-simplicial identities for the faces of the nerve of a group G (see Example 4.3).

Example 4.4 (Standard Simplex). The following example motivates the terminology "semi-(co)simplicial set" and "(co)face map". Let $n \in \mathbb{N}_0$ be a non-negative integer. The *standard n-dimensional simplex* (or, for short, *n-simplex*) is the subset Δ_n in $\mathbb{R}^{n+1}$ defined by

$$\Delta_n := \left\{ (x_0,\ldots,x_n) \in \mathbb{R}^{n+1} : \sum_{i=0}^{n} x_i = 1 \text{ and } x_j \ge 0 \text{ for all } j = 0,\ldots,n \right\}$$
$$\subseteq \mathbb{R}^{n+1}.$$

So, Δ_0 is a point, Δ_1 is a segment, Δ_2 is an equilateral triangle, Δ_3 is a regular tetrahedron and so on (see Figure 4.1).

The family of sets $\Delta_\bullet = (\Delta_n)_{n \in \mathbb{N}_0}$ can be given the structure of a semi-cosimplicial set

$$\cdots \leftleftarrows \Delta_2 \leftleftarrows \Delta_1 \leftarrow \Delta_0,$$

with cofaces $d = (d_i^n : \Delta_{n-1} \to \Delta_n)_{0 \le i \le n \in \mathbb{N}}$ given by

$$d_i^n(x_0,\ldots,x_{n-1}) = (x_0,\ldots,x_{i-1},0,x_i,\ldots,x_{n-1}) \in \Delta_n,$$

for all $(x_0,\ldots,x_{n-1}) \in \Delta_{n-1}$. Note that d_i^n identifies the $(n-1)$-simplex with the face in the n-simplex opposite to the i-th vertex

$$E_i = (0,\ldots,\underbrace{1}_{i\text{-th place}},\ldots,0), \quad i = 0,\ldots,n$$

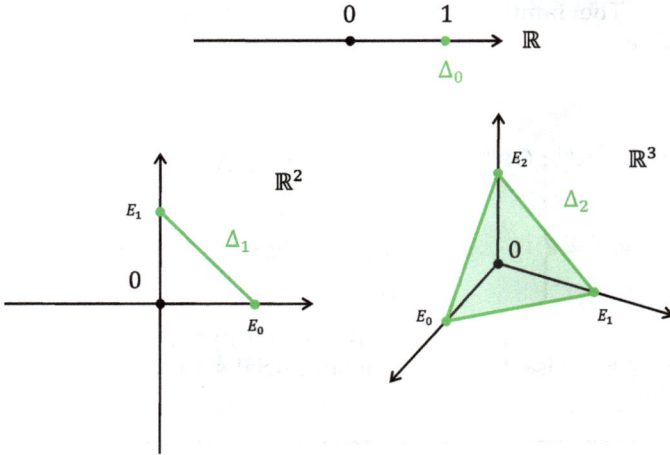

Figure 4.1.   The first three standard simplexes.

(this should explain the term "faces" for the structure maps of a semi-simplicial set). We leave it to the reader to check the semi-cosimplicial identities as Exercise 4.2. The semi-cosimplicial set $(\Delta_\bullet, d)$ is called the *standard simplex*.

We note for later use that, for all n, the standard n-simplex Δ_n is a topological space when equipped with the subspace topology induced from the standard topology in $\mathbb{R}^{n+1}$. Additionally, the coface maps $d_i : \Delta_{n-1} \to \Delta_n$ of the standard simplex $(\Delta_\bullet, d)$ are continuous maps (with respect to this topology). Indeed, they are the restrictions to a subspace both in the domain and the codomain of the linear, hence continuous (even smooth) maps

$$\mathbb{R}^n \to \mathbb{R}^{n+1}, \quad (x_0, \ldots, x_{n-1}) \mapsto (x_0, \ldots, x_{i-1}, 0, x_i, \ldots, x_{n-1}). \quad \blacklozenge$$

Exercise 4.2. Prove the semi-cosimplicial identities for the cofaces of the standard simplex (see Example 4.4).

We can compare two semi-(co)simplicial sets using appropriate "maps" that we now define.

Definition 4.5 (Semi-(Co)Simplicial Map). A *semi-simplicial map* between the semi-simplicial sets $(X^\bullet, {}_Xd), (Y^\bullet, {}_Yd)$ is a family $f = (f^n : X^n \to Y^n)_{n \in \mathbb{N}_0}$ of maps *preserving the faces* in the sense that

$$f^{n-1} \circ {}_Xd_i^n = {}_Yd_i^n \circ f^n, \quad \text{for all } 0 \le i \le n \in \mathbb{N}_0. \tag{4.2}$$

In this case, we write

$$f : (X^\bullet, {}_X d) \to (Y^\bullet, {}_Y d).$$

Likewise for semi-cosimplicial sets.

Similarly as we do for the face maps (and for cochain maps), given a semi-simplicial map $f = (f^n)_{n \in \mathbb{N}_0}$ we often drop the "n" from f^n when it is clear on which set we are acting. For instance, we simply write $f \circ {}_X d_i = {}_Y d_i \circ f$ for the identity (4.2).

Now, let R be a ring.

Definition 4.6 (Semi-(Co)Simplicial Module). A *semi-simplicial R-module* is a semi-simplicial set $(M^\bullet, d)$ such that every set M^n in the family $M^\bullet = (M^n)_{n \in \mathbb{N}_0}$ is an R-module and all the faces are R-linear. *Semi-cosimplicial modules* are defined in a similar way. A *semi-simplicial homomorphism* between the semi-simplicial modules $(M^\bullet, {}_M d), (N^\bullet, {}_N d)$ is a semi-simplicial map $f : (M^\bullet, {}_M d) \to (N^\bullet, {}_N d)$ which is additionally component-wise linear. Likewise for semi-cosimplicial modules.

It is clear that the *identity semi-simplicial map* $\mathrm{id} = (\mathrm{id} : M^n \to M^n)_{n \in \mathbb{N}_0}$ is a semi-simplicial map. The composition of semi-simplicial maps is defined component-wise and it is a semi-simplicial map as well. A component-wise invertible semi-simplicial map is a *semi-simplicial isomorphism*. Like-wise for semi-cosimplicial sets and semi-(co)simplicial modules. We conclude that semi-simplicial (resp. semi-cosimplicial) sets and semi-simplicial (resp. semi-cosimplicial) maps form a category, denoted **ssSet** (resp. **sCosSet**). Similarly, semi-simplicial (resp. semi-cosimplicial) modules over a fixed ring R and semi-simplicial (resp. semi-cosimplicial) homomorphisms form a category, denoted **ssMod**$_R$ (resp. **sCosMod**$_R$). The details are left es an exercise.

Exercise 4.3. Prove that semi-(co)simplicial sets and semi-(co)simplicial maps form a category.

The category **s(Co)sMod**$_{\mathbb{Z}}$ is denoted simply **s(Co)sAb**, and when $R = \mathbb{K}$ is a field, we write **s(Co)sVect**$_{\mathbb{K}}$ (instead of **s(Co)sMod**$_{\mathbb{K}}$).

We now present the main construction in this section. Namely, we show that there is a functor

$$\textbf{ssMod}_R \to \textbf{Ch}_R,$$

from semi-simplicial modules to chain complexes (and like-wise for semi-cosimplicial modules and cochain complexes). This construction is important because several (co)chain complexes in algebra and geometry arise in this way.

Theorem 4.7 ((Co)Chain Complexes from Semi-(Co)Simplicial Modules). *Let R be a ring and let $(M^\bullet, d)$ be a semi-simplicial R-module. For all $n \in \mathbb{Z}$, define*

$$C_n(M) = \begin{cases} 0 & \text{if } n < 0 \\ M^n & \text{if } n \geq 0 \end{cases}$$

and

$$D_n = \sum_{i=0}^{n} (-)^i d_i^n : C_n(M) \to C_{n-1}(M).$$

Then $(C_\bullet(M), D)$ is a chain complex. Likewise for semi-cosimplicial modules (in which case the analogous construction gives a cochain complex).

If $f : (M^\bullet, {}_M d) \to (N^\bullet, {}_N d)$ is a semi-simplicial map, then it is also a chain map

$$f : (C_\bullet(M), D_M) \to (C_\bullet(N), D_N)$$

between the associated chain complexes. The assignment

$$\mathbf{ssMod}_R \to \mathbf{Ch}_R$$

mapping a semi-simplicial module $(M^\bullet, d)$ to the chain complex $(C_\bullet(M), D)$ and the semi-simplicial map $f : (M^\bullet, {}_M d) \to (N^\bullet, {}_N d)$ to the chain map $f : (C_\bullet(M), D_M) \to (C_\bullet(N), D_N)$ is a functor. Likewise for semi-cosimplicial maps.

Proof. For the first part of the statement, we have to show that $0 = D \circ D : C_n(M) \to C_{n-2}(M)$ for all n. So, compute

$$D \circ D = \sum_{i=0}^{n-1} (-)^i d_i \circ \sum_{j=0}^{n} (-)^j d_j = \sum_{j=0}^{n} \sum_{i=0}^{n-1} (-)^{i+j} d_i \circ d_j, \qquad (4.3)$$

where we used that the composition of linear maps is a bilinear operation. We now split the double sum in the right hand side of (4.3) into two parts.

In this way, we are able to use the semi-simplicial identities:

$D \circ D$

$$= \sum_{j=0}^{n} \sum_{i=0}^{n-1} (-)^{i+j} d_i \circ d_j$$

$$= \sum_{j=1}^{n} \sum_{i=0}^{j-1} (-)^{i+j} d_i \circ d_j + \sum_{j=0}^{n-1} \sum_{i=j}^{n-1} (-)^{i+j} d_i \circ d_j$$

$$= \sum_{j=1}^{n} \sum_{i=0}^{j-1} (-)^{i+j} d_{j-1} \circ d_i + \sum_{j=0}^{n-1} \sum_{i=j}^{n-1} (-)^{i+j} d_i \circ d_j \quad \text{(semi-simplicial identities)}.$$

Now, rename the indexes in the first sum as follows: $\bar{j} = i$ and $\bar{i} = j - 1$. Then we have $\bar{i} \geq \bar{j}$ and $i + j = \bar{i} + \bar{j} - 1$ so that

$$D \circ D = \sum_{\bar{i}=0}^{n-1} \sum_{\bar{j}=0}^{\bar{i}} (-)^{\bar{i}+\bar{j}+1} d_{\bar{i}} \circ d_{\bar{j}} + \sum_{i \geq j} (-)^{i+j} d_i \circ d_j$$

$$= - \sum_{\bar{j}=0}^{n-1} \sum_{\bar{i}=\bar{j}}^{n-1} (-)^{\bar{i}+\bar{j}} d_{\bar{i}} \circ d_{\bar{j}} + \sum_{j=0}^{n-1} \sum_{i=j}^{n-1} (-)^{i+j} d_i \circ d_j = 0.$$

The second part of the statement is straightforward: for any n,

$$f \circ D_M = f \circ \sum_{i=0}^{n} (-)^i {}_M d_i = \sum_{i=0}^{n} (-)^i f \circ {}_M d_i$$

$$= \sum_{i=0}^{n} (-)^i {}_N d_i \circ f = \left(\sum_{i=0}^{n} (-)^i {}_N d_i \right) \circ f$$

$$= D_N \circ f,$$

where we used the bilinearity of composing linear maps. The final part of the statement is obvious (do you see it?). $\qquad\square$

Fix a ring R. Actually, there are functors

$$\mathbf{ssFree} : \mathbf{ssSet} \to \mathbf{ssMod}_R \quad \text{and} \quad \mathbf{ssFun} : \mathbf{ssSet} \to \mathbf{sCosMod}_R$$

defined as follows. First of all, every semi-simplicial set (together with the choice of a ring R) gives rise to both a semi-simplicial and a semi-cosimplicial module. Let $(X^\bullet, d)$ be a semi-simplicial set. We define a semi-simplicial module $(RX^\bullet, Rd)$, where $RX^\bullet := (RX^n)_{n \in \mathbb{N}_0}$ and

$Rd := (Rd_i : RX^n \to RX^{n-1})_{0 \le i \le n \in \mathbb{N}}$ (as usual RX^n denotes the free R-module generated by X^n and $Rd_i : RX^n \to RX^{n-1}$ the unique linear map such that $Rd_i(x) = d_i(x) \in X^{n-1} \subseteq RX^{n-1}$ for all $x \in X^n \subseteq RX^n$). In other words, $(RX^\bullet, Rd)$ is defined by applying the functor **Free** : **Set** $\to$ **Mod**$_R$ to all the sets and all the structure maps of $(X^\bullet, d)$. From the functorial properties of **Free**, it easily follows that the Rd_i satisfy the semi-simplicial identities. Next, every semi-simplicial map $f : (X^\bullet, {}_X d) \to (Y^\bullet, {}_Y d)$ gives rise to a semi-simplicial homomorphism $Rf : (RX^\bullet, R_X d) \to (RY^\bullet, R_Y d)$. Namely, for $f = (f^n : X^n \to Y^n)_{n \in \mathbb{N}_0}$, we define $Rf := (Rf^n : RX^n \to RY^n)_{n \in \mathbb{N}_0}$ (in other words, we apply the functor **Free** : **Set** $\to$ **Mod**$_R$ to all the maps in the family f). Again from the functorial properties of **Free**, the family Rf is a semi-simplicial homomorphism. Finally, the assignment **ssFree** : **ssSet** $\to$ **ssMod**$_R$ defined by putting **ssFree**$(X^\bullet, d) := (RX^\bullet, Rd)$ for every semi-simplicial set, and **ssFree**$(f) = Rf$ for every semi-simplicial map $f : (X^\bullet, {}_X d) \to (Y^\bullet, {}_Y d)$ is a functor. We leave the details to the reader as the following

---

Exercise 4.4. Use the functorial properties of **Free** to show that the linear maps $Rd = (Rd_i : RX^n \to RX^{n-1})_{0 \le i \le n \in \mathbb{N}}$ defined from a semi-simplicial set $(X^\bullet, d)$ satisfy the semi-simplicial identities. Prove also that the family $Rf = (Rf^n : RX^n \to RY^n)_{n \in \mathbb{N}_0}$ defined from a semi-simplicial map $f : (X^\bullet, {}_X d) \to (Y^\bullet, {}_Y d)$ is a semi-simplicial homomorphism. Finally, show that the assignment **ssFree** : **ssSet** $\to$ **ssMod**$_R$ defined above is a functor.

Composing the functor **ssFree** : **ssSet** $\to$ **ssMod**$_R$ with the functor **ssMod**$_R \to$ **Ch**$_R$ we get a functor **ssSet** $\to$ **Ch**$_R$. In other words, every semi-simplicial set gives rise to a chain complex and every semi-simplicial map gives rise to a chain map in a functorial way.

We conclude this section defining a functor **ssFun** : **ssSet** $\to$ **sCosMod**$_R$. Let $(X^\bullet, d)$ be a semi-simplicial set. By applying the function module functor **Fun** : **Set** $\to$ **Mod**$_R$ to all sets and all structure maps in $(X^\bullet, d)$, we get a semi-cosimplicial module $(R_\bullet^X, R^d)$, do you see it? Similarly, let $f : (X^\bullet, {}_X d) \to (Y^\bullet, {}_Y d)$ be a semi-simplicial map. Applying the functor **Fun** to all the maps in the family f, we get a semi-cosimplicial homomorphism $R^f : (R_\bullet^Y, R^{Y_d}) \to (R_\bullet^X, R^{X_d})$. The assignment **ssFun** : **ssSet** $\to$ **sCosMod**$_R$ defined by putting **ssFun**$(X^\bullet, d) := (R_\bullet^X, R^d)$ for every semi-simplicial set $(X^\bullet, d)$, and **ssFun**$(f) := R^f$ for every semi-simplicial map $f : (X^\bullet, {}_X d) \to (Y^\bullet, {}_Y d)$ is a contravariant functor.

> **Exercise 4.5.** Use the functorial properties of **Fun** to show that the linear maps $R^d := (R^{d_i} : R^{X^{n-1}} \to R^{X^n})_{0 \le i \le n \in \mathbb{N}}$ defined from a semi-simplicial set $(X^\bullet, d)$ satisfy the semi-cosimplicial identities. Prove also that the family $R^f = (R^{f^n} : R^{Y^n} \to R^{X^n})_{n \in \mathbb{N}_0}$ defined from a semi-simplicial map $f : (X^\bullet, {_X}d) \to (Y^\bullet, {_Y}d)$ is a semi-cosimplicial homomorphism. Finally, show that the assignment **ssFun** : **ssSet** $\to$ **sCosMod**$_R$ defined above is a contravariant functor.

Composing the functor **ssFun** : **ssSet** $\to$ **sCosMod**$_R$ with the functor **sCosMod**$_R$ $\to$ **CoCh**$_R$ we get a contravariant functor **ssSet** $\to$ **CoCh**$_R$. In other words, every semi-simplicial set gives rise to a cochain complex and every semi-simplicial map gives rise to a cochain map in a functorial way.

Remark 4.8. There is a duality functor $* :$ **sCosMod**$_R$ $\to$ **sCosMod**$_R$ defined by applying the contravariant functor $* :$ **Mod**$_R$ $\to$ **Mod**$_R$ to all modules and all maps. It should be clear that the functors **ssFun** and $* \circ$ **ssFree** are naturally isomorphic (i.e., there exists a natural isomorphism of functors $\iota :$ **ssFun** $\to * \circ$ **ssFree**). Do you see it? $\diamond$

4.2   Group (Co)Homology

Let G be a (non-necessarily abelian) group. In this section, we show that there is a (co)chain complex naturally associated with G. Actually, there is a (co)chain complex for any choice of a ring R (of coefficients). Indeed, let $(N^\bullet(G), d)$ be the nerve of G (Example 4.3). It is a semi-simplicial set. Hence, for any ring R, we can consider the semi-simplicial module $(RN^\bullet(G), Rd)$:

$$\cdots \, \substack{\Longrightarrow\\[-0.6em]\Longrightarrow\\[-0.6em]\Longrightarrow} \, RG^{\times 2} \, \substack{\Longrightarrow\\[-0.6em]\Longrightarrow} \, RG \, \substack{\longrightarrow\\[-0.6em]\longrightarrow} \, R\{*\}$$

(in other words, we act on the nerve with the functor **ssFree** : **ssSet** $\to$ **ssMod**$_R$). In its turn, $(RN^\bullet(G), Rd)$ determines a chain complex, denoted $(C_\bullet(G, R), D)$, via Theorem 4.7. Let's make this complex explicit. According to the definition, $C_0(G, R) = RN^0(G)$ is the free module spanned by $N^0(G) = \{*\}$, hence it is the free module with 1 generator, i.e., $C_0(G, R) = R$. In higher degree $n > 0$, $C_n(G, R) = RN^n(G) = RG^{\times n}$ is the free module spanned by $G^{\times n}$. We now describe the differential. Begin with $D : RG \to R$. As RG is a free module, $D : RG \to R$ is completely determined by its action on the basis elements, i.e., elements $g \in G \subseteq RG$.

According to Theorem 4.7, we have

$$Dg = d_0g - d_1g = * - * = 0.$$

Using Theorem 4.7 again, we see that, in higher degree, the differential $D : RG^{\times n} \to RG^{\times (n-1)}$ acts as follows:

$$
\begin{aligned}
D(g_1, \ldots, g_n) \\
= (d_0 - d_1 + \cdots + (-)^n d_n)(g_1, \ldots, g_n) \\
= (g_2, \ldots, g_n) + \sum_{i=1}^{n-1} (-)^i (g_1, \ldots, g_i g_{i+1}, \ldots, g_n) + (-)^n (g_1, \ldots, g_{n-1})
\end{aligned}
\tag{4.4}
$$

on basis elements $(g_1, \ldots, g_n) \in G^{\times n} \subseteq RG^{\times n}$ (beware that the one on the right hand side is just a formal linear combination of $(n-1)$-tuples). Summarizing, the chain complex $(C_\bullet(G, R), D)$ reads

$$0 \longleftarrow R \xleftarrow{\ 0\ } RG \xleftarrow{\ D\ } RG^{\times 2} \longleftarrow \cdots \xleftarrow{\ D\ } RG^{\times (n-1)} \xleftarrow{\ D\ } RG^{\times n} \longleftarrow \cdots,$$

where D is given by (4.4).

We actually consider only the case $R = \mathbb{Z}$ (except for Example 4.17). In this case, we simply write $(C_\bullet(G), D)$ (instead of $(C_\bullet(G, \mathbb{Z}), D)$).

Definition 4.9 (Group Chain Complex). The chain complex $(C_\bullet(G), D)$ is called the *group chain complex* of G (with integer coefficients) and the homology $H_\bullet(G) := H_\bullet(C(G), D)$ is called the *group homology* of G. Cycles in $(C_\bullet(G), D)$ are denoted $Z_\bullet(G)$ and boundaries are denoted $B_\bullet(G)$.

Our first aim is showing that isomorphic groups have isomorphic group homologies. We adopt the following strategy: we show that the nerve construction is a functor

$$N : \mathbf{Gr} \to \mathbf{ssSet}$$

from the category of groups to the category of semi-simplicial sets. Then, for all n, from its very definition, the n-th group homology becomes a functor itself: namely, the composition

$$\mathbf{Gr} \xrightarrow{\ N\ } \mathbf{ssSet} \xrightarrow{\ \text{ssFree}\ } \mathbf{ssAb} \xrightarrow{\ \text{Thm. 4.7}\ } \mathbf{Ch}_{\mathbb{Z}} \xrightarrow{\ H_n\ } \mathbf{Ab}$$

(do you see it?). As the composition of functors is a functor and functors map isomorphisms to isomorphisms, the claim follows.

In order to show that the nerve construction is a functor, we have to define how does it act on morphisms. So, let $f : G \to H$ be a group homomorphism. We define $N(f) : N(G) \to N(H)$ to be the family of maps $N(f) = (f^n : N^n(G) \to N^n(H))_{n \in \mathbb{N}_0}$ defined by

$$f^n : N^n(G) = G^{\times n} \to N^n(H) = H^{\times n},$$

$$(g_1, \ldots, g_n) \mapsto f^n(g_1, \ldots, g_n) := (f(g_1), \ldots, f(g_n)).$$

It is easy to see that $N(f)$ is a semi-simplicial map. Additionally, the assignment $N : \mathbf{Gr} \to \mathbf{ssSet}$ defined in this way is a functor. We leave the details as

---

Exercise 4.6. Prove that, for any group homomorphism $f : G \to H$, the family of maps $N(f) : N(G) \to N(H)$ defined above is a semi-simplicial map. Prove also that the assignment $N : \mathbf{Gr} \to \mathbf{ssSet}$ defined in this way is a functor.

Next, we define *group cohomology*. For any ring R, we can consider the semi-cosimplicial module $(R_\bullet^{N(G)}, R^d)$

$$\cdots \rightrightarrows R^{G^{\times 2}} \leftleftarrows R^G \longleftarrow R^{\{*\}}$$

(in other words, we act on the nerve with the functor $\mathbf{ssFun} : \mathbf{ssSet} \to \mathbf{sCosMod}_R$). In its turn, $(R_\bullet^{N(G)}, R^d)$ determines a cochain complex, denoted $(C^\bullet(G, R), D)$ via Theorem 4.7 again. We have $C^0(G, R) = R^{N^0(G)} = R^{\{*\}} = R$. In higher degree $n > 0$, $C^n(G, R) = R^{N_n(G)} = R^{G^{\times n}}$. As for the differential, $D : R \to R^G$ is the zero map. Indeed, for all $a \in R$, the differential Da is the function $Da : G \to R$ given by

$$Da(g) = a(d_0 g) - a(d_1 g) = a(*) - a(*) = a - a = 0,$$

for all $g \in G$, where we also interpreted a as the constant function on the one element set $N^0(G) = \{*\}$. In higher degree, the differential $D : R^{G^{\times n}} \to R^{G^{\times(n+1)}}$ acts as follows:

$$
\begin{aligned}
Dc&(g_1, \ldots, g_{n+1}) \\
&= \left(c \circ d_0 - c \circ d_1 + \cdots + (-)^{n+1} c \circ d_{n+1} \right)(g_1, \ldots, g_{n+1}) \\
&= c(g_2, \ldots, g_{n+1}) + \sum_{i=1}^{n} (-)^i c(g_1, \ldots, g_i g_{i+1}, \ldots, g_{n+1}) \\
&\quad + (-)^{n+1} c(g_1, \ldots, g_n),
\end{aligned}
\tag{4.5}
$$

for all $c \in R^{G^{\times n}}$, and $g_1, \ldots, g_{n+1} \in G$. Summarizing, the cochain complex $(C^\bullet(G, R), D)$ reads

$$0 \longrightarrow R \xrightarrow{\ 0\ } R^G \xrightarrow{\ D\ } R^{G^{\times 2}} \xrightarrow{\ D\ } \cdots \longrightarrow R^{G^{\times n}} \xrightarrow{\ D\ } R^{G^{\times(n+1)}} \xrightarrow{\ D\ } \cdots,$$

where D is given by (4.5).

Now, go back to the case $R = \mathbb{Z}$ and consider the sequence of functors

$$\mathbf{Gr} \xrightarrow{\ N\ } \mathbf{ssSet} \xrightarrow{\ \mathbf{ssFun}\ } \mathbf{sCosAb} \xrightarrow{\ \text{Thm. 4.7}\ } \mathbf{CoChz}. \tag{4.6}$$

Their composition is a functor $\mathbf{Gr} \to \mathbf{CoChz}$.

Definition 4.10 (Group Cochain Complex). The image $(C^\bullet(G), D)$ of a group G under the composition of functors (4.6) is called the *group cochain complex* of G (with integer coefficients) and its cohomology $H^\bullet(G) := H^\bullet(C(G), D)$ is called the *group cohomology* of G. Cocycles in $(C^\bullet(G), D)$ are denoted $Z^\bullet(G)$ and coboundaries are denoted $B^\bullet(G)$.

The n-th group cohomology is the functor $H^n : \mathbf{Gr} \to \mathbf{Ab}$ obtained composing the group cochain complex functor $\mathbf{Gr} \to \mathbf{CoChz}$ with the n-th cohomology functor $H^n : \mathbf{CoChz} \to \mathbf{Ab}$. We conclude that isomorphic groups have isomorphic group cohomologies.

In the sequel, we try to convince the reader that the group (co)homology of G contains relevant information about G by explicitly describing the group (co)homology in low degree. Specifically, given a group G, we provide descriptions for the first homology $H_1(G)$ of G, and the first and the second cohomologies $H^1(G), H^2(G)$ of G. We begin with the low degree part of the group chain complex:

$$0 \longleftarrow \mathbb{Z} \xleftarrow{\ 0\ } \mathbb{Z}G \xleftarrow{\ D\ } \mathbb{Z}G^{\times 2} \longleftarrow \cdots. \tag{4.7}$$

As the differential $D : \mathbb{Z}G \to \mathbb{Z}$ is the 0 map, $H_0(G) = \mathbb{Z}$. Moreover, $\ker(D : \mathbb{Z}G \to \mathbb{Z}) = \mathbb{Z}G$ and

$$H_1(G) = \operatorname{coker}(D : \mathbb{Z}G^{\times 2} \to \mathbb{Z}G) = \frac{\mathbb{Z}G}{\operatorname{im}(D : \mathbb{Z}G^{\times 2} \to \mathbb{Z}G)}.$$

We show in the following that $H_1(G)$ is canonically isomorphic to the *abelianization* of G. Recall that, given two elements $g_1, g_2 \in G$, their *commutator* is the element

$$[g_1, g_2] := g_1^{-1} g_2^{-1} g_1 g_2.$$

The commutators span a normal subgroup $G' \subseteq G$ called the *commutator subgroup* or *derived subgroup*. The quotient $G^{ab} := G/G'$ is an abelian group, hence a $\mathbb{Z}$-module, called the *abelianization* of G.

Theorem 4.11. *There is a natural abelian group isomorphism*

$$H_1(G) \cong G^{ab}$$

which identifies the homology class $[g]$ of a basis element $g \in G \subseteq \mathbb{Z}G$ with the lateral $gG' \in G^{ab} = G/G'$.

Proof. Before starting, we fix our notation. The group G is not abelian in general, and we adopt the multiplicative notation for the composition law in it. However, the group G^{ab} is always abelian and we adopt the additive notation for the composition law in it. According to these conventions, as the quotient map $\pi : G \to G^{ab}$, $g \mapsto \pi(g) = gG'$ is a group homomorphism, we have

$$\pi(1_G) = 0, \quad \pi(g_1 g_2) = \pi(g_1) + \pi(g_2) \quad \text{and} \quad \pi(g^{-1}) = -\pi(g)$$

for all $g, g_1, g_2 \in G$.

We now come to the proof. We begin defining a linear map $\Phi : H_1(G) \to G^{ab}$. We do this in two stages. First of all, from the universal property of the free module, the quotient map $\pi : G \to G^{ab}$ uniquely extends to a linear map $\varphi : \mathbb{Z}G \to G^{ab}$. We want to show that

$$\mathrm{im}(D : \mathbb{Z}G^{\times 2} \to \mathbb{Z}G) \subseteq \ker(\varphi : \mathbb{Z}G \to G^{ab}).$$

i.e., $\varphi \circ D = 0$. As $\varphi \circ D : \mathbb{Z}G^{\times 2} \to G^{ab}$ is a linear map, it is completely determined by its action on basis elements $(g_1, g_2) \in G^{\times 2} \subseteq \mathbb{Z}G^{\times 2}$. So, let $(g_1, g_2) \in G^{\times 2}$, and compute

$$\varphi \circ D(g_1, g_2) = \varphi\,(g_2 - g_1 g_2 + g_1) \qquad \text{(Formula (4.4))}$$
$$= \varphi(g_2) - \varphi(g_1 g_2) + \varphi(g_1) \qquad \text{(φ is a linear map)}$$
$$= \pi(g_2) - \pi(g_1 g_2) + \pi(g_2) \qquad \text{(definition of φ)}$$
$$= 0. \qquad \text{(π is a group homomorphism)}$$

Now, from the Homomorphism Theorem (Proposition 1.22), the linear map $\varphi : \mathbb{Z}G \to G^{ab}$ descends to a linear map

$$\Phi : H_1(G) \to G^{ab}.$$

Next, we define a linear map $\Psi : G^{ab} \to H_1(G)$. We do this in two stages again. Consider the composition

$$G \to \mathbb{Z}G \to H_1(G),$$

where the first arrow is the inclusion and the second one is the projection. Denote it $\psi : G \to H_1(G)$. We want to show that ψ is a group homomorphism. So, let $g_1, g_2 \in G$ and compute

$$\psi(g_1 g_2) - \psi(g_1) - \psi(g_2) = [g_1 g_2] - [g_1] - [g_2], \qquad (4.8)$$

where, as usual, we used square brackets "$[-]$" to denote homology classes. But the right hand side of (4.8) is

$$[g_1 g_2] - [g_1] - [g_1] = [g_1 g_2 - g_1 - g_2] = -[D(g_1, g_2)] = 0.$$

We conclude that

$$\psi(g_1 g_2) - \psi(g_1) - \psi(g_2) = 0 \quad \Rightarrow \quad \psi(g_1 g_2) = \psi(g_1) + \psi(g_2),$$

for all $g_1, g_2 \in G$, i.e., ψ is a group homomorphism as claimed. Next, we show that $G' \subseteq \ker \psi$. It is enough to show that ψ annihilates all the commutators. But this immediately follows from the fact that every group homomorphism maps commutators to commutators (do you see it?), and ψ takes values in an abelian group (where all commutators vanish). From the homomorphism theorem for groups, ψ descends to a group homomorphism, hence a linear map,

$$\Psi : G^{ab} \to H_1(G).$$

We leave it to the reader to check that Φ, Ψ are mutually inverse abelian group homomorphisms as Exercise 4.7. $\qquad\square$

Exercise 4.7. Complete the proof of Theorem 4.11 showing that the linear maps $\Phi : H_1(G) \to G^{ab}$ and $\Psi : G^{ab} \to H_1(G)$ are mutually inverse.

Remark 4.12. The abelianization construction is actually functorial, namely every group homomorphism $g : G \to H$ preserves the commutator subgroups, i.e., $f(G') \subseteq H'$, hence it induces a group homomorphism $f^{ab} : G^{ab} \to H^{ab}$. The assignment $ab : \mathbf{Gr} \to \mathbf{Mod}_{\mathbb{Z}}$ defined by putting $ab(G) := G^{ab}$ for every group, and $ab(f) := f^{ab}$ for every group homomorphism $f : G \to H$, is a functor. The isomorphism provided by Theorem 4.11 is actually a natural isomorphism of functors (see Problem 4.3). $\qquad\diamond$

We now turn to cohomology and we concentrate on the low degree part of the group cochain complex:

$$0 \longrightarrow \mathbb{Z} \xrightarrow{\ 0\ } \mathbb{Z}^G \xrightarrow{\ D\ } \mathbb{Z}^{G^{\times 2}} \xrightarrow{\ D\ } \cdots. \tag{4.9}$$

As $D : \mathbb{Z} \to \mathbb{Z}^G$ is the 0 map, we have $H^0(G) = \mathbb{Z}$, $\mathrm{im}(D : \mathbb{Z} \to \mathbb{Z}^G) = 0$ and

$$H^1(G) = \ker\left(D : \mathbb{Z}^G \to \mathbb{Z}^{G^{\times 2}}\right).$$

We show in the following that $H^1(G)$ coincides with the abelian group of group homomorphisms $\mathrm{Hom}(G, \mathbb{Z}) := \mathrm{Hom}_{\mathbf{Gr}}(G, \mathbb{Z})$. The abelian group structure in $\mathrm{Hom}(G, \mathbb{Z})$ is given by the point-wise sum.

Theorem 4.13. *The first group cohomology $H^1(G)$ consists of group homomorphisms $G \to \mathbb{Z}$:*

$$H^1(G) = \mathrm{Hom}(G, \mathbb{Z}).$$

Proof. Take a function $f : G \to \mathbb{Z}$ in $\mathbb{Z}^G = C^0(G)$. According to Formula (4.5), its differential is the function $Df : G^{\times 2} \to \mathbb{Z}$ in $\mathbb{Z}^{G^{\times 2}} = C^2(G)$ given by

$$Df(g_1, g_2) = f(g_2) - f(g_1 g_2) + f(g_1).$$

So, Df vanishes if and only if

$$f(g_1 g_2) = f(g_1) + f(g_2)$$

for all $(g_1, g_2) \in G^{\times 2}$, i.e., $f : G \to \mathbb{Z}$ is a group homomorphism. $\qquad\square$

Group homomorphisms $G \to \mathbb{Z}$ are sometimes called *multiplicative functions* on G.

Remark 4.14. There is a natural abelian group isomorphism $\mathrm{Hom}(G, \mathbb{Z}) \cong \mathrm{Hom}(G^{ab}, \mathbb{Z}) = (G^{ab})^*$ defined as follows. Let $f : G \to \mathbb{Z}$ be a group homomorphism. As $\mathbb{Z}$ is an abelian group f vanishes on commutators (do you see it?). So, f descends to a group homomorphism $f^{ab} : G^{ab} \to \mathbb{Z}$, hence a $\mathbb{Z}$-linear map, given by $f^{ab}(gG') = f(g)$. Conversely, given a linear map $f^{ab} : G^{ab} \to \mathbb{Z}$, composing with the projection $\pi : G \to G^{ab}$ we get a group homomorphism $f = f^{ab} \circ \pi : G \to \mathbb{Z}$. We conclude that the assignment $\mathrm{Hom}(G, \mathbb{Z}) \to \mathrm{Hom}(G^{ab}, \mathbb{Z})$, $f \mapsto f^{ab}$ is an abelian group isomorphism. Therefore we also have that the first cohomology $H^1(G)$ is canonically isomorphic to the dual of the first homology: $H^1(G) \cong H_1(G)^*$. $\qquad\diamond$

Corollary 4.15. *If G is an abelian group, then $H_1(G) \cong G$ and $H^1(G) \cong G^*$.*

Next, we discuss the second group cohomology $H^2(G)$. To do this, we first need to define *group extensions*. Let G, K be groups. A *group extension* of the group G by the group K is (another group H together with) a *short exact sequence of groups*

$$1 \longrightarrow K \xrightarrow{\alpha} H \xrightarrow{\beta} G \longrightarrow 1. \qquad (4.10)$$

This means that α is a group monomorphism, β is a group epimorphism and, additionally, $\operatorname{im}\alpha = \ker\beta$. Two group extensions $1 \to K \to H \to G \to 1$ and $1 \to K \to H' \to G \to 1$ of G by K are *equivalent* if there exists a group isomorphism $\Phi : H \to H'$ such that the diagram

$$
\begin{array}{ccccccccc}
1 & \longrightarrow & K & \longrightarrow & H & \longrightarrow & G & \longrightarrow & 1 \\
& & \| & & \downarrow{\scriptstyle\Phi} & & \| & & \\
1 & \longrightarrow & K & \longrightarrow & H' & \longrightarrow & G & \longrightarrow & 1
\end{array}
$$

commutes. Here the vertical "=" denote the identity maps. It should be clear that "equivalence" is indeed an equivalence relation on the collection of group extensions of G by K. The *classification problem* of group extensions consists in describing the set of equivalence classes of group extensions and, if possible, identifying a distinguished representative in each class. A group extension (4.10) is called *central* if $\operatorname{im}\alpha$ is in the center of H, i.e., for all $k \in K$ and all $h \in H$ we have $\alpha(k)h = h\alpha(k)$. Note that every group extension equivalent to a central extension is also central. Group extensions are standard tools in group theory. Here we relate them to group cohomology (see, e.g., MacLane, 1975, Chapter IV for much more on group extensions and group cohomology).

Theorem 4.16. *Let G be a group, then central extensions of G by $\mathbb{Z}$ are classified by the second cohomology module $H^2(G)$, i.e., there exists a natural bijection between $H^2(G)$ and equivalence classes of central extensions of G by $\mathbb{Z}$.*

Proof. We begin showing that every 2-cocycle $c \in Z^2(G)$ determines a central extension of G by $\mathbb{Z}$. Recall that a cochain $c \in C^2(G)$ is a function $c : G^{\times 2} \to \mathbb{Z}$. According to (4.5), its differential is the function $Dc : G^{\times 3} \to \mathbb{Z}$ given by

$$Dc(g_1, g_2, g_3) = c(g_2, g_3) - c(g_1 g_2, g_3) + c(g_1, g_2 g_3) - c(g_1, g_2),$$

for all $(g_1, g_2, g_3) \in G^{\times 3}$. So, $Dc = 0$ if and only if

$$c(g_2, g_3) - c(g_1 g_2, g_3) + c(g_1, g_2 g_3) - c(g_1, g_2) = 0 \qquad (4.11)$$

for all $g_1, g_2, g_3 \in G$. From the *cocycle condition* (4.11), it follows that

$$c(1_G, g) = c(1_G, 1_G) = c(g, 1_G) \qquad (4.12)$$

for all $g \in G$. To see this, just use $g_1 = g_2 = 1$, $g_3 = g$, and $g_1 = g$, $g_2 = g_3 = 1$ in (4.11). Our strategy is using c to define a group product $\star_c$ in the set $\mathbb{Z} \times G$. For every $(m_1, g_1), (m_2, g_2) \in \mathbb{Z} \times G$ put

$$(m_1, g_1) \star_c (m_2, g_2) := (m_1 + m_2 + c(g_1, g_2), g_1 g_2).$$

If c is a cocycle, then $(\mathbb{Z} \times G, \star_c)$ is a group. Indeed,

- **(associativity)** left as Exercise 4.8(1);
- **(unit)** the element $(-c(1_G, 1_G), 1_G)$ is a unit with respect to the product $\star_c$ (Exercise 4.8);
- **(inversion)** the element $(-m - c(g, g^{-1}) - c(1_G, 1_G), g^{-1})$ is an inverse of (m, g) with respect to $\star_c$ (see Exercise 4.8 again).

So, $(\mathbb{Z} \times G, \star_c)$ is a group that we denote H_c. Consider the sequence

$$
\begin{array}{ccccccccc}
1 & \longrightarrow & \mathbb{Z} & \xrightarrow{\ \alpha_c\ } & H_c & \xrightarrow{\ \beta\ } & G & \longrightarrow & 1 \\
 & & m & \longmapsto & (m - c(1_G, 1_G), 1_G) & & & & \\
 & & & & (m, g) & \longmapsto & g & &
\end{array}
\qquad (4.13)
$$

A straightforward computation shows that both α_c, β are group homomorphisms. It is also clear that α_c is injective, β is surjective and $\operatorname{im} \alpha_c \subseteq \ker \beta$ (do you see it?). Finally, an element $(m, g) \in H_c = \mathbb{Z} \times G$ is in the kernel of β if and only if $g = 1_G$. In this case,

$$(m, g) = (m, 1_G) = (m + c(1_G, 1_G) - c(1_G, 1_G), 1_G)$$

$$= \alpha_c (m + c(1_G, 1_G)) \in \operatorname{im} \alpha_c.$$

We conclude that $\operatorname{im} \alpha_c = \ker \beta$ so that (4.13) is a group extension of G by $\mathbb{Z}$. It is actually a central extension, indeed, for all $m \in \mathbb{Z}$ and all $(n, g) \in H_c = \mathbb{Z} \times G$, we have

$$\alpha_c(m) \star_c (n, g) = (m - c(1_G, 1_G), 1_G) \star_c (n, g)$$

$$= (m - c(1_G, 1_G) + n + c(1_G, g), g)$$

$$= (m + n, g) = (n, g) \star_c \alpha_c(m),$$

where we used that, for all $g \in G$, $c(1_G, g) = c(1_G, 1_G)$. So, we constructed a central extension of G by $\mathbb{Z}$ for every 2-cocycle in the group cochain complex.

Next we show that, if $c, c' \in Z^2(G)$ are cohomologous 2-cocycles in $(C^\bullet(G), D)$, then the associated central extensions

$$1 \longrightarrow \mathbb{Z} \xrightarrow{\alpha_c} H_c \xrightarrow{\beta} G \longrightarrow 1 \quad \text{and} \quad 1 \longrightarrow \mathbb{Z} \xrightarrow{\alpha_{c'}} H_{c'} \xrightarrow{\beta} G \longrightarrow 1$$

are equivalent. So, let $a \in C^1(G) = \mathbb{Z}^G$ be such that $c - c' = Da$. This means that $a : G \to \mathbb{Z}$ is a function such that

$$c(g_1, g_2) - c'(g_1, g_2) = a(g_2) - a(g_1 g_2) + a(g_1) \tag{4.14}$$

for all $g_1, g_2 \in G$. In particular,

$$c(1_G, 1_G) - c'(1_G, 1_G) = a(1_G). \tag{4.15}$$

Consider the map

$$\Phi_a : H_c \to H_{c'}, \quad (m, g) \mapsto \Phi_a(m, g) := (m + a(g), g).$$

We claim that Φ_a is a group isomorphism and that the diagram

$$\begin{array}{ccccccccc}
1 & \longrightarrow & \mathbb{Z} & \longrightarrow & H_c & \longrightarrow & G & \longrightarrow & 1 \\
& & \| & & \downarrow{\scriptstyle \Phi_a} & & \| & & \\
1 & \longrightarrow & \mathbb{Z} & \longrightarrow & H_{c'} & \longrightarrow & G & \longrightarrow & 1
\end{array} \tag{4.16}$$

commutes. First of all, for any two elements $(m_1, g_1), (m_2, g_2) \in H_c$ we have

$$\begin{aligned}
\Phi_a\big((m_1, g_1) \star_c (m_2, g_2)\big) \\
= \Phi_a\big(m_1 + m_2 + c(g_1, g_2), g_1 g_2\big) \\
= \big(m_1 + m_2 + c(g_1, g_2) + a(g_1 g_2), g_1 g_2\big)
\end{aligned} \tag{4.17}$$

and

$$\begin{aligned}
\Phi_a(m_1, g_1) \star_{c'} \Phi_a(m_2, g_2) \\
= \big(m_1 + a(g_1), g_1\big) \star_{c'} \big(m_2 + a(g_2), g_2\big) \\
= \big(m_1 + m_2 + a(g_1) + a(g_2) + c'(g_1, g_2), g_1 g_2\big).
\end{aligned} \tag{4.18}$$

It immediately follows from (4.14) that

$$\Phi_a\big((m_1, g_1) \star_c (m_2, g_2)\big) = \Phi_a(m_1, g_1) \star_{c'} \Phi_a(m_2, g_2),$$

i.e., Φ_a is a group homomorphism. It is also invertible, the inverse $\Phi_a^{-1} : H_{c'} \to H_c$ being given by $\Phi_a^{-1}(n, h) = \Phi_{-a}(n, h) := (n - a(h), h)$ (do you see it?). So, it is a group isomorphism as claimed. We leave it to the reader to check that the diagram (4.16) commutes as part of Exercise 4.8.

In the following, we denote by $\sim$ the equivalence of central extensions and by $[H]_\sim$ the equivalence class of a central extension $0 \to \mathbb{Z} \to H \to G \to 0$. As cohomologous 2-cocycles determine equivalent central extensions, we have a well-defined map

$$
\begin{aligned}
H^2(G) &\to \{\text{central extensions of } G \text{ by } \mathbb{Z}\}/\sim \\
[c] &\mapsto [H_c]_\sim
\end{aligned}
\tag{4.19}
$$

It remains to prove that this map is bijective. We begin with some general remarks on the extension $1 \to \mathbb{Z} \to H_c \to G \to 1$ determined by a cocycle $c \in Z^2(G)$. First of all, we denote by $h_c^{-1} \in H_c$ the inverse of an element $h \in H_c$. Second, a straightforward computation shows that, for every $(m, g) \in H_c = \mathbb{Z} \times G$, we have

$$
(m, g) = \alpha_c(m) \star_c (0, g)
\tag{4.20}
$$

(check Identity (4.20) as an exercise). This remark is useful in what follows. Now, in order to prove the injectivity of (4.19), take two cohomology classes $[c], [c'] \in H^2(G)$ and assume that $[H_c]_\sim = [H_{c'}]_\sim$. This means that the extensions $0 \to \mathbb{Z} \to H_c \to G \to 0$ and $0 \to \mathbb{Z} \to H_{c'} \to G \to 0$ are equivalent. Let $\Phi : H_c \to H_{c'}$ be a group isomorphism such that the diagram

$$
\begin{array}{ccccccccc}
1 & \longrightarrow & \mathbb{Z} & \longrightarrow & H_c & \longrightarrow & G & \longrightarrow & 1 \\
& & \| & & \downarrow{\scriptstyle \Phi} & & \| & & \\
1 & \longrightarrow & \mathbb{Z} & \longrightarrow & H_{c'} & \longrightarrow & G & \longrightarrow & 1
\end{array}
$$

commutes. Take $g \in G$ and consider the element $A(g) := \Phi(0, g) \star_{c'} (0, g)_{c'}^{-1} \in H_{c'}$. We have

$$
\beta\left(A(g)\right) = \beta\left(\Phi(0, g) \star_{c'} (0, g)_{c'}^{-1}\right) = \beta\left(\Phi(0, g)\right)\beta\left((0, g)_{c'}^{-1}\right)
$$
$$
= \beta(0, g)\beta(0, g)^{-1} = 1_G,
$$

where we used that β is a group homomorphism, together with $\beta \circ \Phi = \beta$. This computation shows that $A(g) \in \ker \beta = \operatorname{im} \alpha_{c'}$, hence there exists a (unique) $a(g) \in \mathbb{Z}$ such that $A(g) = \alpha_{c'}(a(g))$. This in turn implies that

$$\Phi(0, g) = A(g) \star_{c'} (0, g) = \alpha_{c'}(a(g)) \star_{c'} (0, g). \qquad (4.21)$$

We are now ready to better describe Φ. Namely, for every $(m, g) \in H_c$, we have

$$\begin{aligned}
\Phi(m, g) &= \Phi(\alpha_c(m) \star_c (0, g)) && \text{(Identity (4.20))} \\
&= \Phi(\alpha_c(m)) \star_{c'} \Phi(0, g) && \text{(Φ is a group homomorphism)} \\
&= \alpha_{c'}(m) \star_{c'} \alpha_{c'}(a(g)) \star_{c'} (0, g) && \text{($\Phi \circ \alpha_c = \alpha_{c'}$ and (4.21))} \\
&= \alpha_{c'}(m + a(g)) \star_{c'} (0, g) && \text{($\alpha_{c'}$ is a group homomorphism)} \\
&= (m + a(g), g) && \text{(Identity (4.20) for $c = c'$).}
\end{aligned}$$

Summarizing, we have proved that there exists a function $a : G \to \mathbb{Z}$ such that

$$\Phi(m, g) = \Phi_a(m, g) = (m + a(g), g)$$

for all $(m, g) \in H_c$. Finally, the same computations as in (4.17) and (4.18), but in reverse order, reveal that (4.14) holds for all $g_1, g_2 \in G$, i.e., $c - c' = Da$. In other words, c and c' are cohomologous cocycles, i.e., $[c] = [c']$ and we conclude that the map (4.19) is injective as desired.

In order to prove the surjectivity of (4.19), let

$$0 \longrightarrow \mathbb{Z} \overset{\alpha}{\longrightarrow} H \overset{\beta}{\longrightarrow} G \longrightarrow 0$$

be a central extension of G by $\mathbb{Z}$. We have to show that the latter is equivalent to an extension of the type $0 \to \mathbb{Z} \to H_c \to G \to 0$ for some $c \in Z^2(G)$. To do this, choose any right inverse s of $\beta : H \to G$, i.e., any map $s : G \to H$ such that $\beta \circ s = \mathrm{id}_G$ (it exists by the Axiom of Choice). We claim that, for every $h \in H$, there exists a unique $m \in \mathbb{Z}$ and a unique $g \in G$ such that $h = \alpha(m)s(g)$. For the existence, put $g := \beta(h) \in G$ and consider $h\, s(g)^{-1} \in H$. Then we have

$$\beta(h\, s(g)^{-1}) = \beta(h)\beta(s(g))^{-1} = gg^{-1} = 1_G,$$

where we used that β is a group homomorphism. This shows that $h\, s(g)^{-1} \in \ker \beta = \mathrm{im}\, \alpha$. Hence, there exists a unique $m \in \mathbb{Z}$ such that $h\, s(g)^{-1} = \alpha(m)$, i.e., $h = \alpha(m)s(g)$ as desired. For the uniqueness, assume

$h = \alpha(m')s(g')$ for some (other) $(m', g') \in \mathbb{Z} \times G$. We have

$$g = \beta(h) = \beta(\alpha(m')s(g')) = \beta(\alpha(m'))\beta(s(g'))) = 1_G g' = g',$$

where we used that $\beta \circ \alpha = 1_G$ (the trivial homomorphism), that β is a group homomorphism and that $\beta \circ s = \text{id}_G$. As $g = g'$, we also have

$$\alpha(m) = hs(g)^{-1} = hs(g')^{-1} = \alpha(m')$$

and, from the injectivity of α, we conclude that $m = m'$ as well. We are now ready to define a bijection $\Phi : H \to \mathbb{Z} \times G$. For any $h \in H$, we put $\Phi(h) = (m, g)$ where $(m, g) \in \mathbb{Z} \times G$ is the pair uniquely defined by $\alpha(m)s(g) = h$. The map Φ is clearly inverted by $\Phi^{-1} : \mathbb{Z} \times G \to H$, $(m, g) \mapsto \alpha(m)s(g)$.

Now, we can use the bijection Φ to transport the group structure from H to $\mathbb{Z} \times G$. In other words, we define a product $\star$ in $\mathbb{Z} \times G$ by putting

$$(m_1, g_1) \star (m_2, g_2) := \Phi(\Phi^{-1}(m_1, g_1)\Phi^{-1}(m_2, g_2)),$$

for all $(m_1, g_1), (m_2, g_2) \in \mathbb{Z} \times G$. It is clear that $(\mathbb{Z} \times G, \star)$ is a group and $\Phi : H \to \mathbb{Z} \times G$ is a group isomorphism. We can also define group homomorphisms $\bar{\alpha} = \Phi \circ \alpha : \mathbb{Z} \to \mathbb{Z} \times G$ and $\bar{\beta} = \beta \circ \Phi^{-1} : \mathbb{Z} \times G$. Then

$$0 \longrightarrow \mathbb{Z} \xrightarrow{\bar{\alpha}} \mathbb{Z} \times G \xrightarrow{\bar{\beta}} G \longrightarrow 0$$

is a central extension of G by $\mathbb{Z}$, and the diagram

$$
\begin{array}{ccccccccc}
1 & \longrightarrow & \mathbb{Z} & \xrightarrow{\alpha} & H & \xrightarrow{\beta} & G & \longrightarrow & 1 \\
& & \| & & \downarrow{\Phi} & & \| & & \\
1 & \longrightarrow & \mathbb{Z} & \xrightarrow{\bar{\alpha}} & \mathbb{Z} \times G & \xrightarrow{\bar{\beta}} & G & \longrightarrow & 1
\end{array}
$$

is an equivalence of central extensions. It immediately follows from the explicit expression of Φ^{-1} that $\bar{\beta}$ is the projection onto the second factor (do you see it? If not, check the details). It remains to prove that $\star = \star_c$, and $\bar{\alpha} = \alpha_c$ for some cocycle $c \in Z^2(G)$. The latter two facts are a consequence of the group and the central extension axioms as we now show.

Take $g_1, g_2 \in G$ and consider the element $C(g_1, g_2) := s(g_1)s(g_2)s(g_1 g_2)^{-1}$. We have

$$\beta(C(g_1, g_2)) = \beta\left(s(g_1)s(g_2)s(g_1 g_2)^{-1}\right) = \beta(s(g_1))\beta(s(g_2))\beta(s(g_1 g_2))^{-1}$$

$$= g_1 g_2 (g_1 g_2)^{-1} = 1_G,$$

where we used that β is a group homomorphism and that $\beta \circ s = \mathrm{id}_G$. This computation shows that $C(g_1, g_2) \in \ker \beta = \operatorname{im} \alpha$, hence there exists a (unique) $c(g_1, g_2) \in \mathbb{Z}$ such that $C(g_1, g_2) = \alpha(c(g_1, g_2))$, which in turn implies

$$s(g_1)s(g_2) = \alpha(c(g_1, g_2))s(g_1 g_2). \tag{4.22}$$

We are now ready to better describe the product $\star$. Namely, for every $(m_1, g_1), (m_2, g_2) \in \mathbb{Z} \times G$, we have

$$
\begin{aligned}
(m_1, g_1) &\star (m_2, g_2) \\
&= \Phi(\Phi^{-1}(m_1, g_1)\Phi^{-1}(m_2, g_2)) && \text{(definition of } \star) \\
&= \Phi(\alpha(m_1)s(g_1)\alpha(m_2)s(g_2)) && \text{(explicit expression of } \Phi^{-1}) \\
&= \Phi(\alpha(m_1)\alpha(m_2)s(g_1)s(g_2)) && \text{(im } \alpha \text{ is in the center of } H) \\
&= \Phi(\alpha(m_1)\alpha(m_2)\alpha(c(g_1, g_2))s(g_1 g_2)) && \text{(Identity (4.22))} \\
&= \Phi(\alpha(m_1 + m_2 + c(g_1, g_2))s(g_1 g_2)) && (\alpha \text{ is a group homomorphism)} \\
&= (m_1 + m_2 + c(g_1, g_2), g_1 g_2) && \text{(definition of } \Phi).
\end{aligned}
\tag{4.23}
$$

Summarizing, we have proved that there exists a 2-cochain $c : G^{\times 2} \to \mathbb{Z}$ in $(C^\bullet(G), D)$ such that, for all $(m_1, g_1), (m_2, g_2) \in \mathbb{Z} \times G$, we have

$$(m_1, g_1) \star (m_2, g_2) = (m_1, g_1) \star_c (m_2, g_2) = (m_1 + m_2 + c(g_1, g_2), g_1 g_2).$$

As $\star$ is a group product, it is associative, and Exercise 4.8(1) reveals that c is a 2-cocycle. Finally, a similar computation as that in (4.23) shows that $\bar{\alpha} = \alpha_c$, and this concludes the proof. $\qquad\square$

Exercise 4.8. Fill all the gaps in the proof of Theorem 4.16: first of all prove the following:

(1) Let $c \in Z^2(G)$ be a 2-cochain in the group cochain complex of the group G, and let $\star_c$ be the product in $\mathbb{Z} \times G$ defined in the proof of Theorem 4.16. Prove that $\star_c$ is associative if and only if c is a cocycle. In this case, prove also that

  - $(-c(1_G, 1_G), 1_G)$ is a unit with respect to $\star_c$;
  - $(-m - c(g, g^{-1}) - c(1_G, 1_G), g^{-1})$ is an inverse of (m, g) with respect to $\star_c$.

(2) Let $c, c' \in Z^2(G)$ be cohomologous 2-cocycles, and let $a \in C^1(G)$ be such that $c - c' = Da$. Prove that Diagram (4.16) commutes.

Finally, prove that $\bar{\alpha} = \alpha_c$, where $\bar{\alpha}, c$ are those defined at the end of the proof.

Example 4.17. We conclude this section discussing the (co)homology of finite groups with coefficients in rational numbers $\mathbb{Q}$. So, let $G = \{1_G = g_1, g_2, \ldots, g_k\}$ be a finite group of order k. We want to prove that

$$H_n(G, \mathbb{Q}) = \begin{cases} \mathbb{Q} & \text{if } n = 0 \\ 0 & \text{otherwise} \end{cases}.$$

The group chain complex is

$$0 \longleftarrow \mathbb{Q} \xleftarrow{0} \mathbb{Q}G \xleftarrow{D} \mathbb{Q}G^{\times 2} \xleftarrow{D} \mathbb{Q}G^{\times 3} \xleftarrow{D} \cdots,$$

and we know already that $H_0(G, \mathbb{Q}) = \mathbb{Q}$. As the first differential $D : \mathbb{Q}G \to \mathbb{Q}$ vanishes, it remains to show that the "truncated chain complex"

$$0 \longleftarrow \mathbb{Q}G \xleftarrow{D} \mathbb{Q}G^{\times 2} \xleftarrow{D} \mathbb{Q}G^{\times 3} \xleftarrow{D} \cdots \tag{4.24}$$

is acyclic. Actually, the chain complex (4.24) possesses a canonical contracting homotopy h acting as follows:

$$h(g_{i_1}, \ldots, g_{i_n}) := \frac{1}{k} \sum_{j=1}^{k} (g_j, g_{i_1}, \ldots, g_{i_n}) \tag{4.25}$$

on basis elements $(g_{i_1}, \ldots, g_{i_n}) \in G^{\times n} \subseteq \mathbb{Q}G^{\times n}$. We stress that, in order to define h, we used both that the ring of coefficients is $\mathbb{Q}$ (in the overall factor $1/k$) and that G is finite (the sum in (4.25) is finite). We leave it to the reader to check that $D \circ h + h \circ D = \mathrm{id}$ in Exercise 4.9.

One can show in a similar way that

$$H^n(G, \mathbb{Q}) = \begin{cases} \mathbb{Q} & \text{if } n = 0 \\ 0 & \text{otherwise} \end{cases}.$$

As for the group chain complex, it is enough to look at the truncated cochain complex

$$0 \longrightarrow \mathbb{Q}^G \xrightarrow{D} \mathbb{Q}^{G^{\times 2}} \xrightarrow{D} \mathbb{Q}^{G^{\times 3}} \xrightarrow{D} \cdots, \tag{4.26}$$

which possesses a contracting homotopy h^* defined by

$$h^*(c)(g_{i_1}, \ldots, g_{i_{n-1}}) = \frac{1}{k} \sum_{j=1}^{k} c(g_j, g_{i_1}, \ldots, g_{i_{n-1}})$$

for all $c \in \mathbb{Q}^{G^{\times n}}$ and all $g_{i_1}, \ldots, g_{i_{n-1}} \in G$. ◆

Exercise 4.9. Prove that the maps h and h^* in Example 4.17 are well-defined contracting homotopies for the truncated complexes (4.24) and (4.26).

4.3   Hochschild (Co)Homology

In this section, we introduce the notion of *associative algebra* and show that any such an algebraic structure determines both a chain complex (in a functorial way) and a cochain complex. Such complexes contain information on the associative algebra. Let $\mathbb{K}$ be a field.

Definition 4.18 (Associative Algebra). An *associative $\mathbb{K}$-algebra* (or simply an *algebra*) is a $\mathbb{K}$-vector space $(A, +, \cdot)$ equipped with an additional composition law $\star : A \times A \to A$ (*product*) such that

- $\star$ is $\mathbb{K}$-bilinear;
- $\star$ is associative.

In particular, $(A, +, \star)$ is a ring. A *commutative algebra* is an algebra whose associative product is commutative. If A, B are $\mathbb{K}$-algebras, an algebra homomorphism between them is a $\mathbb{K}$-linear map $f : A \to B$ preserving the associative product, i.e., for every $\alpha, \beta \in A$

$$f(\alpha \star \beta) = f(\alpha) \star f(\beta).$$

An algebra *isomorphism* is an invertible algebra homomorphism. $\mathbb{K}$-algebras and their homomorphisms form a category denoted $\mathbf{Alg}_\mathbb{K}$ (Exercise 4.10).

> **Exercise 4.10.** Show that $\mathbb{K}$-algebras and $\mathbb{K}$-algebra homomorphisms form a category whose isomorphisms are algebra isomorphisms.

As for rings and vector spaces, the symbols $\star, \cdot$ are usually omitted in products, and we write $a\alpha$, and $\alpha\beta$ instead of $a \cdot \alpha$ and $\alpha \star \beta$, for $a \in \mathbb{K}$, and α, β algebra elements. As a first trivial example, note that $\mathbb{K}$ itself is a (commutative) algebra in the obvious way.

Example 4.19 (Real Algebra of Complex Numbers). Complex numbers (with their real vector space and their ring structures) form a commutative $\mathbb{R}$-algebra.   ◆

Example 4.20 (Endomorphism Algebra). The space $M_n(\mathbb{K})$ of $n \times n$ matrices over $\mathbb{K}$ is both a vector space and a ring (whose associative product is the matrix product). The ring and the vector space structures are compatible in the sense that, with all its operations, $M_n(\mathbb{K})$ is a $\mathbb{K}$-algebra. More generally, given a $\mathbb{K}$-vector space V, the space $\mathrm{End}_\mathbb{K} V$ of $\mathbb{K}$-linear endomorphisms $f : V \to V$ is a (generically non-commutative) $\mathbb{K}$-algebra.   ◆

Example 4.21 (Group Algebra). Let G be a (non-necessarily abelian) group. The vector space $\mathbb{K}G$ spanned by G can be given the structure of an associative algebra as follows. By the Multilinear Extension Theorem (Theorem 1.50), the group product $G \times G \to G$ extends to a unique bilinear map

$$\star : \mathbb{K}G \times \mathbb{K}G \to \mathbb{K}G.$$

In other words, for any two formal linear combinations

$$\alpha = \sum_i a_i g_i, \quad \beta = \sum_j b_j h_j, \quad a_i, b_j \in \mathbb{K}, \quad g_i, h_j \in G,$$

we put

$$\alpha \star \beta = \sum_{i,j} a_i b_j \, (g_i h_j) \in \mathbb{K}G.$$

The pair $(\mathbb{K}G, \star)$ is an associative algebra called the *group algebra* of G (with coefficients in $\mathbb{K}$). Note that the group algebra is commutative if and only if G is an abelian group. ◆

Example 4.22 (Algebra of Polynomials). The ring $\mathbb{K}[x_1, \ldots, x_n]$ of polynomials in n indeterminates $x_1, \ldots, x_n$ is also a $\mathbb{K}$-vector space. The ring and the vector space structures are compatible in the sense that $\mathbb{K}[x_1, \ldots, x_n]$ is actually a commutative algebra (the *algebra of polynomials*). ◆

Example 4.23 (Function Algebra). Let X be a set, the $\mathbb{K}$-vector space $\mathbb{K}^X$ of functions $X \to \mathbb{K}$ is also a ring with the point-wise sum and product (see Example 1.9). With all its operations, $\mathbb{K}^X$ is a commutative algebra. ◆

We now show that there is a functor

$$\mathbf{Alg}_{\mathbb{K}} \to \mathbf{ssVect}_{\mathbb{K}}$$

from algebras to semi-simplicial $\mathbb{K}$-vector spaces. Let A be a $\mathbb{K}$-algebra. Consider the family of vector spaces $A^{\otimes} = (A^{\otimes n+1})_{n \in \mathbb{N}_0}$, where we put

$$A^{\otimes k} := \underbrace{A \otimes_{\mathbb{K}} \cdots \otimes_{\mathbb{K}} A}_{k\text{-times}}.$$

The family $A^{\otimes}$ can be given the structure of a semi-simplicial $\mathbb{K}$-vector space

$$\cdots \Rrightarrow A^{\otimes 3} \Rrightarrow A^{\otimes 2} \rightrightarrows A$$

with faces $d = (d_i : A^{\otimes n+1} \to A^{\otimes n})_{0 \le i \le n \in \mathbb{N}}$ uniquely defined by

$$d_i(\alpha_0 \otimes \cdots \otimes \alpha_n) := \begin{cases} \alpha_0 \otimes \cdots \otimes \alpha_i \alpha_{i+1} \otimes \cdots \otimes \alpha_n & \text{if } 0 \le i < n \\ \alpha_n \alpha_0 \otimes \alpha_1 \otimes \cdots \otimes \alpha_{n-1} & \text{if } i = n \end{cases}, \quad (4.27)$$

$\alpha_0, \ldots, \alpha_n \in A$. As the maps $(\alpha_0, \ldots, \alpha_n) \mapsto d_i(\alpha_0 \otimes \cdots \otimes \alpha_n)$ are $\mathbb{K}$-linear in all their arguments $\alpha_0, \ldots, \alpha_n$, it follows from the universal property of the tensor product that the d_i are indeed well-defined linear maps. We leave it to the reader to check the semi-simplicial identities as (part of) Exercise 4.11.

Now, let $f : A \to B$ be an algebra homomorphism. We define $f^{\otimes} : A^{\otimes} \to B^{\otimes}$ to be the family of linear maps $f^{\otimes} = (f^{\otimes n+1} : A^{\otimes n+1} \to B^{\otimes n+1})_{n \in \mathbb{N}_0}$ given by

$$f^{\otimes n+1}(\alpha_0 \otimes \cdots \otimes \alpha_n) := f(\alpha_0) \otimes \cdots \otimes f(\alpha_n), \quad \alpha_0, \ldots, \alpha_n \in A.$$

As the expression $f(\alpha_0) \otimes \cdots \otimes f(\alpha_n)$ is $\mathbb{K}$-linear in all its arguments $\alpha_0, \ldots, \alpha_n$, it follows that the $f^{\otimes n+1}$ are well-defined linear maps. It is easy to see that $f^{\otimes}$ is a semi-simplicial map. Additionally, the assignment $\mathbf{Alg}_{\mathbb{K}} \to \mathbf{ssVect}_{\mathbb{K}}$ that maps an algebra A to the semi-simplicial vector space $A^{\otimes}$ and an algebra homomorphism f to the semi-simplicial module homomorphism $f^{\otimes}$ is a functor (Exercise 4.11).

> **Exercise 4.11.** Prove that the linear maps d_i defined in (4.27) satisfy the semi-simplicial identities. Prove also that, for any algebra homomorphism $f : A \to B$, the family $f^{\otimes} : A^{\otimes} \to B^{\otimes}$ defined above is a semi-simplicial module homomorphism. Finally, prove that the assignment $\mathbf{Alg}_{\mathbb{K}} \to \mathbf{ssVect}_{\mathbb{K}}$ defined in this way is a functor.

Composing with the usual functor $\mathbf{ssVect}_{\mathbb{K}} \to \mathbf{Ch}_{\mathbb{K}}$ from Theorem 4.7 we get a chain complex $(HC_{\bullet}(A), D)$. Explicitly

$$0 \longleftarrow A \xleftarrow{D} A^{\otimes 2} \longleftarrow \cdots \xleftarrow{D} A^{\otimes n} \xleftarrow{D} A^{\otimes n+1} \longleftarrow \cdots$$

where D acts as follows

$$D(\alpha_0 \otimes \cdots \otimes \alpha_n)$$
$$= (d_0 - d_1 + \cdots + (-)^n d_n)(\alpha_0 \otimes \cdots \otimes \alpha_n)$$
$$= \sum_{i=0}^{n-1} (-)^i \alpha_0 \otimes \cdots \otimes \alpha_i \alpha_{i+1} \otimes \cdots \otimes \alpha_n + (-)^n \alpha_n \alpha_0 \otimes \alpha_1 \otimes \cdots \otimes \alpha_{n-1},$$

on decomposable elements $\alpha_0 \otimes \cdots \otimes \alpha_{n+1} \in A^{\otimes n+1}$.

Definition 4.24 (Hochschild Homology). The image $(HC_\bullet(A), D)$ of a $\mathbb{K}$-algebra A under the composition of functors $\mathbf{Alg}_\mathbb{K} \to \mathbf{ssVect}_\mathbb{K} \to \mathbf{Ch}_\mathbb{K}$ is called the *Hochschild chain complex* of A (with coefficients in A) and its homology $HH_\bullet(A) := H_\bullet(HC(A), D)$ is called the *Hochschild homology* of A. The differential D is called the *Hochschild differential*. Cycles in $(HC_\bullet(A), D)$ are denoted $HZ_\bullet(A)$ (*Hochschild cycles*) and boundaries are denoted $HB_\bullet(A)$ (*Hochschild boundaries*).

The n-th Hochschild homology is a functor $HH_n : \mathbf{Alg}_\mathbb{K} \to \mathbf{Vect}_\mathbb{K}$ obtained composing the Hochschild chain complex functor $\mathbf{Alg}_\mathbb{K} \to \mathbf{Ch}_\mathbb{K}$ with the n-th homology functor $H_n : \mathbf{Ch}_\mathbb{K} \to \mathbf{Vect}_\mathbb{K}$. As an immediate consequence we get that isomorphic algebras have isomorphic Hochschild homologies.

As a simple example, consider the 0-th Hochschild homology $HH_0(A)$ of an associative algebra A. The (Hochschild) differential on 1 chains

$$D : A^{\otimes 2} \to A$$

acts on decomposable elements $\alpha_0 \otimes \alpha_1 \in A \otimes_\mathbb{K} A$, $\alpha_0, \alpha_1 \in A$ as follows:

$$D(\alpha_0 \otimes \alpha_1) = \alpha_0 \alpha_1 - \alpha_1 \alpha_0.$$

We conclude that

$$\mathrm{im}\left(D : A \otimes_\mathbb{K} A \to A\right) = [A, A] := \mathrm{Span}\left(\alpha_0 \alpha_1 - \alpha_1 \alpha_0 : \alpha_0, \alpha_1 \in A\right)$$

and

$$HH_0(A) = \frac{A}{\mathrm{im}\left(D : A \otimes_\mathbb{K} A \to A\right)} = A/[A, A].$$

Example 4.25. Let $A = \mathbb{K}$. In this case, it follows from Proposition 1.58(2), that $HC_n(A) = A^{\otimes n+1} \cong \mathbb{K}$ for all n. The latter isomorphism is simply given by

$$a_0 \otimes \cdots \otimes a_n = a_0 \cdots a_n (1 \otimes \cdots \otimes 1) \mapsto a_0 \cdots a_n,$$

on decomposable elements (do you see it?) and, in what follows, we will use it to identify $HC_n(A)$ with $\mathbb{K}$ for all n. If we do so all the face maps $d_i : A^{\otimes n+1} \to A^{\otimes n}$ boil down to the identity: $d_i = \mathrm{id}_\mathbb{K} : \mathbb{K} \to \mathbb{K}$, for all i. It follows that

$$D = \sum_{i=0}^{n} (-)^i d_i = \begin{cases} \mathrm{id}_\mathbb{K} & \text{if } n \text{ is even} \\ 0 & \text{if } n \text{ is odd} \end{cases}.$$

In other words, the Hochschild chain complex of the 1-dimensional $\mathbb{K}$-algebra $\mathbb{K}$ is

$$0 \longleftarrow \mathbb{K} \xleftarrow{0} \mathbb{K} \xleftarrow{\text{id}} \mathbb{K} \xleftarrow{0} \mathbb{K} \xleftarrow{\text{id}} \mathbb{K} \longleftarrow \cdots$$

whose homology clearly is

$$HH_n(\mathbb{K}) = \begin{cases} \mathbb{K} & \text{if } n = 0 \\ 0 & \text{otherwise} \end{cases}$$

(do you see it?). ◆

From an algebra, one can also construct a cochain complex. For a $\mathbb{K}$-algebra A, consider the family of vector spaces $\text{Mult}(A) := (\text{Mult}_{\mathbb{K}}^n(A, A))_{n \in \mathbb{N}_0}$, where we put

$$\text{Mult}_{\mathbb{K}}^n(A, A) := \text{Mult}_{\mathbb{K}}^n(\underbrace{A, \ldots, A}_{n\text{-times}}; A).$$

The family $\text{Mult}(A)$ can be given the structure of a semi-cosimplicial $\mathbb{K}$-vector space

$$\cdots \Longrightarrow \text{Mult}_{\mathbb{K}}^2(A, A) \Longrightarrow \text{Hom}_{\mathbb{K}}(A, A) \Longleftarrow A$$

with faces $d = (d_i : \text{Mult}_{\mathbb{K}}^{n-1}(A) \to \text{Mult}_{\mathbb{K}}^n(A))_{0 \leq i \leq n \in \mathbb{N}}$ defined by

$$d_i \mu(\alpha_1, \ldots, \alpha_n) := \begin{cases} \alpha_1 \mu(\alpha_2, \ldots, \alpha_n) & \text{if } i = 0 \\ \mu(\alpha_1, \ldots, \alpha_i \alpha_{i+1}, \ldots, \alpha_n) & \text{if } 0 < i < n, \\ \mu(\alpha_1, \ldots, \alpha_{n-1}) \alpha_n & \text{if } i = n \end{cases} \tag{4.28}$$

$\mu \in \text{Mult}_{\mathbb{K}}^n(A)$, $\alpha_1, \ldots, \alpha_n \in A$. It is easy to see that the d_i are indeed linear maps (do you see it?). We leave it to the reader to check the semi-cosimplicial identities as (part of) Exercise 4.12.

We remark that, unfortunately, the construction $A \mapsto \text{Mult}_{\mathbb{K}}(A)$ is *not* a functor between the categories $\mathbf{Alg}_{\mathbb{K}}$ and $\mathbf{sCosVect}_{\mathbb{K}}$ (but it is a functor from the category whose objects are associative $\mathbb{K}$-algebras and whose morphisms are $\mathbb{K}$-algebra *isomorphisms*). Nonetheless, isomorphic algebras give rise to isomorphic semi-cosimplicial vector spaces (Exercise 4.12).

Exercise 4.12. Prove that the linear maps d_i defined in (4.42) satisfy the semi-cosimplicial identities. Prove also that any algebra isomorphism $\phi : A \to B$ induces in a natural way a semi-cosimplicial vector space isomorphism $\mathrm{Mult}(\phi) : \mathrm{Mult}(B) \to \mathrm{Mult}(A)$, in such a way that the construction $\phi \mapsto \mathrm{Mult}(\phi)$ satisfies the usual functorial properties (why doesn't this construction work on plain algebra homomorphisms?).

Acting with the usual functor $\mathbf{sCosVect}_{\mathbb{K}} \to \mathbf{CoCh}_{\mathbb{K}}$ from Theorem 4.7 on the semi-cosimplicial vector space $\mathrm{Mult}(A)$, we get a cochain complex denoted $(HC^{\bullet}(A), D)$. Explicitly,

$$0 \longrightarrow A \xrightarrow{D} \mathrm{Hom}_{\mathbb{K}}(A, A) \xrightarrow{D} \cdots \longrightarrow \mathrm{Mult}^n_{\mathbb{K}}(A, A) \xrightarrow{D} \mathrm{Mult}^{n+1}_{\mathbb{K}}(A, A) \xrightarrow{D} \cdots,$$

where D acts as follows

$$\begin{aligned}
D\mu&(\alpha_1, \ldots, \alpha_{n+1}) \\
&= ((d_0 - d_1 + \cdots + (-)^{n+1} d_{n+1})\mu)(\alpha_1, \ldots, \alpha_{n+1}) \\
&= \alpha_1 \mu(\alpha_2, \ldots, \alpha_{n+1}) + \sum_{i=1}^{n} (-)^i \mu(\alpha_1, \ldots, \alpha_i \alpha_{i+1}, \ldots, \alpha_{n+1}) \\
&\quad + (-)^{n+1} \mu(\alpha_1, \ldots, \alpha_n) \alpha_{n+1},
\end{aligned}$$

for all $\mu \in \mathrm{Mult}^n_{\mathbb{K}}(A, A)$ and all $\alpha_1, \ldots, \alpha_{n+1} \in A$.

Definition 4.26 (Hochschild Cohomology). The cochain complex $(HC^{\bullet}(A), D)$ is called the *Hochschild cochain complex* of A (with coefficients in A) and its cohomology $HH^{\bullet}(A) := H^{\bullet}(HC(A), D)$ is called the *Hochschild cohomology* of A. The differential D is also called the *Hochschild differential*. Cocycles in $(HC^{\bullet}(A), D)$ are denoted $HZ^{\bullet}(A)$ (*Hochschild cocycles*) and coboundaries are denoted $HB^{\bullet}(A)$ (*Hochschild coboundaries*).

Despite the n-th Hochschild cohomology not being a functor from $\mathbf{Alg}_{\mathbb{K}}$ to $\mathbf{Vect}_{\mathbb{K}}$, in view of Exercise 4.12, isomorphic algebras have isomorphic Hochschild cohomologies anyway (do you see it?).

Exercise 4.13. Prove that the Hochschild cohomology of the 1-dimensional $\mathbb{K}$-algebra $A = \mathbb{K}$ is $\mathbb{K}$ in degree 0 and it is 0 in other degrees.

In the rest of this section, we concentrate on providing interpretations for the low degree Hochschild cohomologies. The low degree part of the

Hochschild cochain complex is

$$0 \longrightarrow A \xrightarrow{D} \mathrm{Hom}_{\mathbb{K}}(A, A) \xrightarrow{D} \mathrm{Bil}_{\mathbb{K}}(A, A; A) \xrightarrow{D} \cdots.$$

The differential $D : A \to \mathrm{Hom}_{\mathbb{K}}(A, A)$ acts as follows. For all $\alpha \in A$, the differential $D\alpha : A \to A$ is given by

$$D\alpha(\beta) = \beta\alpha - \alpha\beta =: [\beta, \alpha],$$

for all $\beta \in A$. Hence,

$$HH^0(A) = \ker\left(D : A \to \mathrm{Hom}_{\mathbb{K}}(A, A)\right)$$

$$= Z(A) := \{\alpha \in A : [\alpha, \beta] = 0, \text{ for all } \beta \in A\}.$$

The subspace $Z(A) \subseteq A$ is called the *center* of A. Note that if A is a commutative algebra, then $HH^0(A) = Z(A) = A$.

Next, we describe Hochschild 1-cohomologies. The differential

$$D : \mathrm{Hom}_{\mathbb{K}}(A, A) \to \mathrm{Bil}_{\mathbb{K}}(A, A; A)$$

is given by

$$D\varphi(\alpha_1, \alpha_2) = \alpha_1 \varphi(\alpha_2) - \varphi(\alpha_1\alpha_2) + \varphi(\alpha_1)\alpha_2 \tag{4.29}$$

for all $\varphi : A \to A \in \mathrm{Hom}_{\mathbb{K}}(A, A)$ and all $\alpha_1, \alpha_2 \in A$. We conclude that the kernel $HZ^1(A) = \ker\left(D : \mathrm{Hom}_{\mathbb{K}}(A, A) \to \mathrm{Bil}_{\mathbb{K}}(A, A; A)\right)$ consists of those $\mathbb{K}$-linear maps $\varphi : A \to A$ such that

$$\varphi(\alpha_1\alpha_2) = \alpha_1 \varphi(\alpha_2) + \varphi(\alpha_1)\alpha_2, \tag{4.30}$$

for all $\alpha_1, \alpha_2 \in A$. Any such linear map is called a *derivation* of A, and Identity (4.30) is called the *Leibniz rule*.

Remark 4.27. The terminology *derivation* stems from the fact that the usual partial derivatives $\frac{\partial}{\partial x_i}$ are derivations of the commutative $\mathbb{R}$-algebra $C^\infty(\mathbb{R}^n)$ of smooth real valued functions $f = f(x_1, \dots, x_n) : \mathbb{R}^n \to \mathbb{R}$ (i.e., functions that can be differentiated infinitely many times). Actually, it can be proved that an $\mathbb{R}$-linear operator $X : C^\infty(\mathbb{R}^n) \to C^\infty(\mathbb{R}^n)$ is a derivation of $C^\infty(\mathbb{R}^n)$ if and only if it is of the form

$$X = \sum_{i=1}^{n} X_i(x_1, \dots, x_n) \frac{\partial}{\partial x_i}$$

for some smooth functions $X_i = X_i(x_1, \dots, x_n) \in C^\infty(\mathbb{R}^n)$, see, e.g., Vitagliano (2024). $\diamond$

The space of derivations of A is denoted Der A.

For any $\alpha \in A$ the image $D\alpha \in \text{Hom}_{\mathbb{K}}(A, A)$ is a 1-coboundary, hence a 1-cocycle, hence a derivation. Specifically, according to the computation above, it is the derivation given by

$$D\alpha(\beta) = [\beta, \alpha] = -[\alpha, \beta] = -(\alpha\beta - \beta\alpha), \quad \beta \in A.$$

The expression $[\alpha, \beta]$ is called the *commutator* of α, β, and every derivation of the form $D\alpha$ is called an *inner derivation*. The space of inner derivations of A is denoted InnDer A. We conclude that

$$HH^1(A) = \frac{\ker\left(D : \text{Hom}_{\mathbb{K}}(A, A) \to \text{Bil}_{\mathbb{K}}(A, A; A)\right)}{\text{im}\left(D : A \to \text{Hom}_{\mathbb{K}}(A, A)\right)} = \text{Der } A/\text{InnDer } A,$$

which is sometimes called the space of *outer derivations* of A (or the space of *non-trivial infinitesimal symmetries of A*).

We conclude this section discussing Hochschild 2-cohomologies. We begin with an extremely informal motivating discussion about (infinitesimal) deformations of an algebra A. So, let A be an associative $\mathbb{K}$-algebra. The associative product $A \times A \to A$ is a $\mathbb{K}$-bilinear map. In the following, we denote by $\mu : A \times A \to A$ this bilinear map. Note that it can be seen itself as a 2-cochain in the Hochschild complex $(HC^\bullet(A), D)$ of A. A *deformation* of A is then a family $\mu_t : A \times A \to A$ of $\mathbb{K}$-bilinear maps depending on a parameter $t \in \mathbb{K}$ such that

(1) μ_t is an associative product (giving to the $\mathbb{K}$-vector space A a *new* structure of algebra) for all t;
(2) $\mu_0 = \mu$.

A deformation μ_t of A can be thought of as a *curve departing from μ* in the *space of associative algebra structures* on the $\mathbb{K}$-vector space A (note that the vector space structure is fixed, which is not a big loss of generality as two vector spaces are always isomorphic provided only they have the same dimension). The associativity condition on μ_t clearly reads

$$\mu_t\big(\mu_t(\alpha_1, \alpha_2), \alpha_3\big) - \mu_t\big(\alpha_1, \mu_t(\alpha_2, \alpha_3)\big) = 0, \quad \alpha_1, \alpha_2, \alpha_3 \in A. \tag{4.31}$$

Two deformations μ_t, μ_t' of the same associative algebra A are *equivalent* if there is a family $\Phi_t : A \to A$ of $\mathbb{K}$-linear isomorphisms such that

(1) $\Phi_t : (A, \mu_t) \to (A, \mu_t')$ is an algebra isomorphism for all t;
(2) $\Phi_0 = \text{id}_A$.

The idea behind the latter definition is that two isomorphic algebras should be counted as the same algebra. Condition (1) explicitly reads

$$\mu_t'\big(\Phi_t(\alpha_1), \Phi_t(\alpha_2)\big) - \Phi_t\big(\mu_t(\alpha_1, \alpha_2)\big) = 0, \quad \alpha_1, \alpha_2 \in A. \tag{4.32}$$

Now, suppose that it makes sense to take the "Taylor expansion" in t of μ_t (this does actually make sense in certain cases) and that we want to retain the linear term only. In other words, we write

$$\mu_t = \mu_0 + t\dot{\mu} + O(t^2) = \mu + t\dot{\mu} + O(t^2), \tag{4.33}$$

for some $\mathbb{K}$-bilinear map $\dot{\mu} : A \times A \to A$. In terms of the expansion (4.33), the left hand side of the associativity condition (4.31) reads

$$\begin{aligned}
\mu_t\big(\mu_t(\alpha_1, \alpha_2), \alpha_3\big) &- \mu_t\big(\alpha_1, \mu_t(\alpha_2, \alpha_3)\big) \\
&= t\big(\dot{\mu}(\mu(\alpha_1, \alpha_2), \alpha_3) + \mu(\dot{\mu}(\alpha_1, \alpha_2), \alpha_3) - \dot{\mu}(\alpha_1, \mu(\alpha_2, \alpha_3)) \\
&\quad - \mu(\alpha_1, \dot{\mu}(\alpha_2, \alpha_3))\big) + O(t^2)
\end{aligned} \tag{4.34}$$

for all $\alpha_1, \alpha_2, \alpha_3 \in A$, where the degree 0 term vanishes because (4.31) is satisfied when $t = 0$.

Next, assume that μ_t, μ_t' are equivalent deformations and that the equivalence is realized by the family $\Phi_t : A \to A$ of $\mathbb{K}$-linear isomorphisms. If it also makes sense to take the Taylor expansion of Φ_t then

$$\Phi_t = \Phi_0 + t\dot{\Phi} + O(t^2) = \mathrm{id}_A + t\dot{\Phi} + O(t^2),$$

for some linear map $\dot{\Phi} : A \to A$, and the left hand side of (4.32) reads

$$\begin{aligned}
\mu_t'\big(\Phi_t(\alpha_1), \Phi_t(\alpha_2)\big) &- \Phi_t\big(\mu_t(\alpha_1, \alpha_2)\big) \\
&= t\big(\dot{\mu}'(\alpha_1, \alpha_2) + \mu(\dot{\Phi}(\alpha_1), \alpha_2) + \mu(\alpha_1, \dot{\Phi}(\alpha_2)) \\
&\quad - \dot{\Phi}(\mu(\alpha_1, \alpha_2)) - \dot{\mu}(\alpha_1, \alpha_2)\big) + O(t^2)
\end{aligned} \tag{4.35}$$

for all $\alpha_1, \alpha_2 \in A$, where the degree 0 term vanishes because (4.32) is satisfied when $t = 0$.

Exercise 4.14. Prove Formulas (4.34) and (4.35).

Going back to the usual notation $\alpha\beta$ for the product $\mu(\alpha,\beta)$, $\alpha,\beta \in A$, the coefficient of the linear term in (4.34) can be rewritten

$$\dot{\mu}(\alpha_1,\alpha_2)\alpha_3 + \dot{\mu}(\alpha_1\alpha_2,\alpha_3) - \alpha_1\dot{\mu}(\alpha_2,\alpha_3) - \dot{\mu}(\alpha_1,\alpha_2\alpha_3),$$

while the coefficient of the linear term in (4.35) can be rewritten

$$\dot{\mu}'(\alpha_1,\alpha_2) + \dot{\Phi}(\alpha_1)\alpha_2 + \alpha_1\dot{\Phi}(\alpha_2) - \dot{\Phi}(\alpha_1\alpha_2) - \dot{\mu}(\alpha_1,\alpha_2).$$

All this discussion suggests the following

Definition 4.28 (Infinitesimal Deformation of an Associative Algebra).
An *infinitesimal deformation* of an associative $\mathbb{K}$-algebra A is a $\mathbb{K}$-bilinear map $\nu : A \times A \to A$ such that

$$\nu(\alpha_1,\alpha_2)\alpha_3 + \nu(\alpha_1\alpha_2,\alpha_3) - \alpha_1\nu(\alpha_2,\alpha_3) - \nu(\alpha_1,\alpha_2\alpha_3) = 0$$

for all $\alpha_1,\alpha_2,\alpha_3 \in A$. Two infinitesimal deformations ν,ν' are *equivalent* if there exists a $\mathbb{K}$-linear map $\psi : A \to A$ such that

$$\nu(\alpha_1,\alpha_2) - \nu'(\alpha_1,\alpha_2) = \psi(\alpha_1)\alpha_2 + \alpha_1\psi(\alpha_2) - \psi(\alpha_1\alpha_2)$$

for all $\alpha_1,\alpha_2 \in A$.

We are now ready to describe Hochschild 2-cohomologies. The Hochschild differential

$$D : \mathrm{Bil}_{\mathbb{K}}(A,A;A) \to HC^3(A) = \mathrm{Mult}^3_{\mathbb{K}}(A,A)$$

is given by

$$Dv(\alpha_1,\alpha_2,\alpha_3)$$
$$= d_0\nu(\alpha_1,\alpha_2,\alpha_3) - d_1\nu(\alpha_1,\alpha_2,\alpha_3) + d_2\nu(\alpha_1,\alpha_2,\alpha_3) - d_3\nu(\alpha_1,\alpha_2,\alpha_3)$$
$$= \alpha_1\nu(\alpha_2,\alpha_3) - \nu(\alpha_1\alpha_2,\alpha_3) + \nu(\alpha_1,\alpha_2\alpha_3) - \nu(\alpha_1,\alpha_2)\alpha_3$$

for all $\nu : A \times A \to A \in \mathrm{Bil}_{\mathbb{K}}(A,A;A)$ and all $\alpha_1,\alpha_2,\alpha_3 \in A$. We immediately see that the kernel $HZ^2(A) = \ker\left(D : \mathrm{Bil}_{\mathbb{K}}(A,A;A) \to \mathrm{Mult}^3_{\mathbb{K}}(A,A)\right)$ consists exactly of infinitesimal deformations of A. Two cocycles $\nu,\nu' \in HZ^2(A)$ are cohomologous if they differ by a coboundary $D\psi \in HB^2(A) = \mathrm{im}\left(D : \mathrm{Hom}_{\mathbb{K}}(A,A) \to \mathrm{Bil}_{\mathbb{K}}(A,A;A)\right)$, $\psi \in \mathrm{Hom}_{\mathbb{K}}(A,A)$ and we see from (4.29) that this exactly means that ν,ν' are equivalent infinitesimal deformations. We conclude that "being equivalent" is indeed an equivalence relation on the space of infinitesimal deformations of A and that

$$HH^2(A) = \{\text{equivalence classes of infinitesimal deformations of } A\}.$$

The latter remark is the starting point of an important chapter of current algebra and geometry called *Deformation Theory* (Manetti, 2022).

4.4 Chevalley–Eilenberg (Co)Homology

In this section, we introduce *Lie algebras* and show that any Lie algebra determines both a chain and a cochain complex (in a functorial way). Similarly as for groups and associative algebras, the (co)chain complex of a Lie algebra contains important information on it. Let $\mathbb{K}$ be a field.

Definition 4.29 (Lie Algebra). A *Lie algebra* over $\mathbb{K}$ is a $\mathbb{K}$-vector space $\mathfrak{g}$ equipped with an additional composition law $[-,-] : \mathfrak{g} \times \mathfrak{g} \to \mathfrak{g}$ (*Lie bracket*) such that

- $[-,-]$ is $\mathbb{K}$-bilinear;
- $[-,-]$ is *alternating*;
- $[-,-]$ satisfies the following *Jacobi identity*:

$$[u,[v,w]] + [v,[w,u]] + [w,[u,v]] = 0, \quad u,v,w \in \mathfrak{g}.$$

A *Lie subalgebra* in a Lie algebra $\mathfrak{g}$ is a vector subspace $\mathfrak{k} \subseteq \mathfrak{g}$ which is preserved by the Lie bracket, i.e., $[v,w] \in \mathfrak{k}$ for all $v,w \in \mathfrak{k}$. If $\mathfrak{g},\mathfrak{h}$ are Lie algebras, a *Lie algebra homomorphism* between them is a $\mathbb{K}$-linear map $f : \mathfrak{g} \to \mathfrak{h}$ preserving the Lie brackets, i.e., for every $v,w \in \mathfrak{g}$

$$f([v,w]) = [f(v),f(w)].$$

A Lie algebra *isomorphism* is an invertible Lie algebra homomorphism. Lie algebras over $\mathbb{K}$ and their homomorphisms form a category denoted $\mathbf{Lie}_{\mathbb{K}}$ (Exercise 4.15).

We remark that every Lie subalgebra is a Lie algebra itself with the restricted operations.

Exercise 4.15. Show that Lie algebras over $\mathbb{K}$ and Lie algebra homomorphisms form a category whose isomorphisms are Lie algebra isomorphisms.

Example 4.30 (Abelian Lie Algebra). Let V be a $\mathbb{K}$-vector space. The zero bracket $0 : V \times V \to V$, $(v,w) \mapsto 0$ is a Lie bracket on V. Hence, with this trivial bracket, V is a Lie algebra: the *abelian Lie algebra*.   ◆

Example 4.31 (Commutator). Let A be an associative $\mathbb{K}$-algebra. We define on A the following bracket (already encountered in the previous section) called the *commutator*:

$$[-,-] : A \times A \to A, \quad (\alpha, \beta) \mapsto [\alpha, \beta] := \alpha\beta - \beta\alpha.$$

A direct computation shows that the commutator is a Lie bracket, hence $(A, [-,-])$ is a Lie algebra (Exercise 4.16). In other words, every associative algebra, equipped with the commutator, is a Lie algebra. ◆

Exercise 4.16. Show that the commutator in an associative algebra is a Lie bracket. Show also that the assignment $\mathbf{Alg}_{\mathbb{K}} \to \mathbf{Lie}_{\mathbb{K}}$ mapping an associative algebra A to the Lie algebra $(A, [-,-])$ and an algebra homomorphism $f : A \to B$ to itself is a well-defined functor.

Example 4.32 (General Linear Lie Algebra). Consider the associative algebra $M_n(\mathbb{K})$ of $n \times n$ matrices. According to Example 4.31, $M_n(\mathbb{K})$ is also a Lie algebra when equipped with the commutator $[-,-]$. The Lie algebra $(M_n(\mathbb{K}), [-,-])$ is called the *general linear Lie algebra* of order n over the field $\mathbb{K}$ and it is denoted by $\mathfrak{gl}_n(\mathbb{K})$. More generally, given a $\mathbb{K}$-vector space V, the Lie algebra $(\mathrm{End}_{\mathbb{K}} V, [-,-])$ of $\mathbb{K}$-linear endomorphisms $f : V \to V$ (with the commutator) is called the *general linear Lie algebra* of V and it is denoted $\mathfrak{gl}(V)$. ◆

Example 4.33 (Special Linear Lie Algebra). Let $\mathfrak{sl}_n(\mathbb{K}) \subseteq M_n(\mathbb{K})$ be the vector subspace of trace-free matrices:

$$\mathfrak{sl}_n(\mathbb{K}) := \big\{ A = (a_{ij}) \in M_n(\mathbb{K}) : \mathrm{tr}\, A := \textstyle\sum_{i=1}^n a_{ii} = 0 \big\}.$$

It is easy to see that $\mathfrak{sl}_n(\mathbb{K})$ is a Lie subalgebra (beware, not an *associative* subalgebra) in $\mathfrak{gl}_n(\mathbb{K})$ (Exercise 4.17). Accordingly, it is a Lie algebra itself called the *special linear Lie algebra*. ◆

Example 4.34 (Special Orthogonal Lie Algebra). Let $\mathfrak{so}_n(\mathbb{K}) \subseteq M_n(\mathbb{K})$ be the vector subspace of skew-symmetric matrices:

$$\mathfrak{so}_n(\mathbb{K}) := \{ A \in M_n(\mathbb{K}) : A^T = -A \}.$$

It is easy to see that $\mathfrak{so}_n(\mathbb{K})$ is a Lie subalgebra (beware, not an *associative* subalgebra) in $\mathfrak{gl}_n(\mathbb{K})$ (Exercise 4.17). Accordingly, it is a Lie algebra itself called the *special orthogonal Lie algebra*. ◆

Exercise 4.17. Prove that the vector subspaces $\mathfrak{sl}_n(\mathbb{K}), \mathfrak{so}_n(\mathbb{K}) \subseteq M_n(\mathbb{K})$ of trace-free and skew-symmetric matrices respectively are Lie subalgebras of the general linear Lie algebra $\mathfrak{gl}_n(\mathbb{K})$.

Example 4.35 (Lie Algebra of Derivations). Let A be an associative $\mathbb{K}$-algebra. We know that derivations of A form a vector subspace Der A in $\mathrm{End}_{\mathbb{K}}(A)$ (indeed Der $A = HZ^1(A)$ the vector space of Hochschild 1-cocycles). A direct computation shows that Der A is also a Lie subalgebra of the general linear Lie algebra $\mathfrak{gl}(A)$ of A (Exercise 4.18). Hence, it is a Lie algebra itself.   ◆

Exercise 4.18. Prove that the space Der A of derivations of an associative $\mathbb{K}$-algebra A is a Lie subalgebra of the general linear Lie algebra $\mathfrak{gl}(A)$.

Example 4.36 (2-Dimensional Non-abelian Lie Algebra). Consider the associative (commutative) $\mathbb{R}$-algebra $C^\infty(\mathbb{R})$ of smooth functions $f = f(t) : \mathbb{R} \to \mathbb{R}$. The operators

$$\frac{d}{dt}, \quad \text{and} \quad t\frac{d}{dt}$$

are linearly independent derivations of $C^\infty(\mathbb{R})$. Hence, they span a 2-dimensional vector subspace V in Der $C^\infty(\mathbb{R})$. It is easy to see that V is actually a Lie subalgebra. For instance,

$$\left[\frac{d}{dt}, t\frac{d}{dt}\right] = \frac{d}{dt} \in V.$$

    ◆

Exercise 4.19. Prove all the unproven claims in Example 4.36.

We now define a functor

$$CE : \mathbf{Lie}_{\mathbb{K}} \to \mathbf{Ch}_{\mathbb{K}}$$

from Lie algebras over $\mathbb{K}$ to chain complexes of $\mathbb{K}$-vector spaces. Let $\mathfrak{g}$ be a Lie algebra over $\mathbb{K}$. For all $n \in \mathbb{Z}$, put

$$C_n(\mathfrak{g}) := \begin{cases} \wedge^n \mathfrak{g} & \text{if } n \geq 0 \\ 0 & \text{otherwise} \end{cases}.$$

We also define arrows

$$\delta : C_n(\mathfrak{g}) \to C_{n-1}(\mathfrak{g}),$$

declaring how do they act on decomposable vectors. Namely, we put

$$\delta(v_1 \wedge \cdots \wedge v_n) := \sum_{i<j} (-)^{i+j} [v_i, v_j] \wedge v_1 \wedge \cdots \wedge \widehat{v_i} \wedge \cdots \wedge \widehat{v_j} \wedge \cdots \wedge v_n,$$

$$(4.36)$$

for all $v_1, \ldots, v_n \in \mathfrak{g}$, where, as usual, a hat "$\hat{}$" denotes omission. A direct check reveals that the right hand side of (4.36) is multilinear alternating in its arguments $v_1, \ldots, v_n$. Hence, from the universal property of the exterior product, Definition (4.36) can be uniquely extended to a linear map $\delta : \wedge^n \mathfrak{g} \to \wedge^{n-1} \mathfrak{g}$.

Proposition 4.37. *The sequence*

$$0 \longleftarrow C_0(\mathfrak{g}) \overset{\delta}{\longleftarrow} C_1(\mathfrak{g}) \overset{\delta}{\longleftarrow} C_2(\mathfrak{g}) \overset{\delta}{\longleftarrow} \cdots \qquad (4.37)$$

is a chain complex of vector spaces.

Proof. The proof is a long but straightforward computation using that

(1) the exterior product is alternating;
(2) the Lie bracket $[-, -]$ satisfies the Jacobi identity.

We omit the details but invite the brave reader to check everything themselves. □

Now, let $f : \mathfrak{g} \to \mathfrak{h}$ be a Lie algebra homomorphism. We define $\wedge^\bullet f : C_\bullet(\mathfrak{g}) \to C_\bullet(\mathfrak{h})$ to be the family of maps $\wedge^\bullet f := (\wedge^n f : \wedge^n \mathfrak{g} \to \wedge^n \mathfrak{h})_{n \in \mathbb{Z}}$ given by

$$\wedge^n f(v_1 \wedge \cdots \wedge v_n) := f(v_1) \wedge \cdots \wedge f(v_n), \quad v_1, \ldots, v_n \in \mathfrak{g}. \qquad (4.38)$$

As the right hand side of (4.38) is multilinear alternating in its arguments $v_1, \ldots, v_n$, this definition extends uniquely to a linear map $\wedge^n f : \wedge^n \mathfrak{g} \to \wedge^n \mathfrak{h}$. It is easy to see that $\wedge^\bullet f : (C_\bullet(\mathfrak{g}), \delta) \to (C_\bullet(\mathfrak{h}), \delta)$ is a chain map. Additionally the assignment $CE : \mathbf{Lie}_\mathbb{K} \to \mathbf{Ch}_\mathbb{K}$ that maps a Lie algebra $\mathfrak{g}$ to the chain complex $(C_\bullet(\mathfrak{g}), \delta)$ and a Lie algebra homomorphism f to the chain map $\wedge^\bullet f$ is a functor (Exercise 4.20).

Exercise 4.20. Prove that, for a Lie algebra homomorphism $f : \mathfrak{g} \to \mathfrak{h}$, the family $\wedge^\bullet f : (C_\bullet(\mathfrak{g}), \delta) \to (C_\bullet(\mathfrak{h}), \delta)$ defined above is a chain map. Prove also that the assignment $CE : \mathbf{Lie}_\mathbb{K} \to \mathbf{Ch}_\mathbb{K}$ defined by putting $CE(\mathfrak{g}) = (C_\bullet(\mathfrak{g}), \delta)$ and $CE(f) = \wedge^\bullet f$, for every Lie algebra $\mathfrak{g}$ and any Lie algebra homomorphism f, is a functor.

Definition 4.38 (Chevalley–Eilenberg Homology). The image $(C_\bullet(\mathfrak{g}), \delta)$ of a Lie algebra $\mathfrak{g}$ over $\mathbb{K}$ under the functor $CE : \mathbf{Lie}_\mathbb{K} \to \mathbf{Ch}_\mathbb{K}$ is called the *Chevalley–Eilenberg chain complex* of $\mathfrak{g}$ (with trivial coefficients) and its homology $H_\bullet(\mathfrak{g}) := H_\bullet(C(\mathfrak{g}), \delta)$ is called the *Chevalley–Eilenberg homology* of $\mathfrak{g}$. The differential δ is called the *Chevalley–Eilenberg differential*. Cycles in $(C_\bullet(\mathfrak{g}), \delta)$ are denoted $Z_\bullet(\mathfrak{g})$ (*Chevalley–Eilenberg cycles*) and boundaries are denoted $B_\bullet(\mathfrak{g})$ (*Chevalley–Eilenberg boundaries*).

Note that the Chevalley–Eilenberg differential δ vanishes when $\mathfrak{g}$ is an abelian Lie algebra (so what is the Chevalley–Eilenberg homology of an abelian Lie algebra?).

The n-th Chevalley–Eilenberg homology is a functor $\mathbf{Lie}_{\mathbb{K}} \to \mathbf{Vect}_{\mathbb{K}}$ obtained composing the Chevalley–Eilenberg chain complex functor $\mathbf{Lie}_{\mathbb{K}} \to \mathbf{Ch}_{\mathbb{K}}$ with the n-th homology functor $H_n : \mathbf{Ch}_{\mathbb{K}} \to \mathbf{Vect}_{\mathbb{K}}$. Hence, isomorphic Lie algebras have isomorphic Chevalley–Eilenberg homologies.

Let's look at 0-th and 1-st Chevalley–Eilenberg homologies. In low degree, the Chevalley–Eilenberg chain complex reads

$$0 \xleftarrow{\quad} \mathbb{K} \xleftarrow{\;0\;} \mathfrak{g} \xleftarrow{\;\delta\;} \wedge^2 \mathfrak{g} \xleftarrow{\;\delta\;} \cdots$$

(do you see it?). It immediately follows that $H_0(\mathfrak{g}) = \mathbb{K}$. As for the first homology, we have

$$\mathrm{im}\left(\delta : \wedge^2 \mathfrak{g} \to \mathfrak{g}\right) = [\mathfrak{g}, \mathfrak{g}] := \mathrm{Span}\left([v_1, v_2] : v_1, v_2 \in \mathfrak{g}\right)$$

(do you see it?). Hence,

$$H_1(\mathfrak{g}) = \mathfrak{g}/[\mathfrak{g}, \mathfrak{g}] .$$

Example 4.39 (Chevalley–Eilenberg Homology of $\mathfrak{so}_3(\mathbb{K})$). In this example we compute by hands the Chevalley–Eilenberg homology of the order 3 special orthogonal Lie algebra $\mathfrak{so}_3(\mathbb{K})$. Specifically, we prove that

$$H_n\left(\mathfrak{so}_3(\mathbb{K})\right) \cong \begin{cases} \mathbb{K} & \text{if } n = 0, 3 \\ 0 & \text{otherwise} \end{cases} .$$

We discuss more examples in Exercise 4.21 and in the End-of-Chapter Problems. In the following, we denote $\mathfrak{g} = \mathfrak{so}_3(\mathbb{K})$. We begin choosing an appropriate basis of $\mathfrak{g}$. Recall that $\mathfrak{g}$ consists of skew-symmetric 3×3 matrices, i.e., matrices of the form

$$\begin{pmatrix} 0 & a_1 & a_3 \\ -a_1 & 0 & a_2 \\ -a_2 & -a_2 & 0 \end{pmatrix}, \quad a_1, a_2, a_3 \in \mathbb{K}.$$

So, it is a 3-dimensional vector space spanned by the matrices

$$E_1 := \begin{pmatrix} 0 & 1 & 0 \\ -1 & 0 & 0 \\ 0 & 0 & 0 \end{pmatrix}, \quad E_2 := \begin{pmatrix} 0 & 0 & 0 \\ 0 & 0 & 1 \\ 0 & -1 & 0 \end{pmatrix}, \quad E_3 := \begin{pmatrix} 0 & 0 & 1 \\ 0 & 0 & 0 \\ -1 & 0 & 0 \end{pmatrix}.$$

A direct computation (that we invite the reader to perform in details) shows that

$$[E_1, E_2] = E_3, \quad [E_2, E_3] = E_1, \quad [E_3, E_1] = E_2.$$

From Proposition 1.72, $\wedge^2 \mathfrak{g}$ is a 3-dimensional vector space spanned by

$$E_{12} := E_1 \wedge E_2, \quad E_{13} := E_1 \wedge E_3, \quad E_{23} := E_2 \wedge E_3.$$

Similarly, $\wedge^3 \mathfrak{g}$ is a 1-dimensional vector space spanned by

$$E_{123} := E_1 \wedge E_2 \wedge E_3,$$

while $\wedge^n \mathfrak{g} = 0$ for $n > 3$. It follows that the Chevalley–Eilenberg chain complex of $\mathfrak{g}$ is concentrated in degree $0, 1, 2, 3$:

$$0 \longleftarrow \mathbb{K} \overset{0}{\longleftarrow} \mathfrak{g} \overset{\delta}{\longleftarrow} \wedge^2 \mathfrak{g} \overset{\delta}{\longleftarrow} \wedge^3 \mathfrak{g} \longleftarrow 0.$$

Let's compute $\delta : \wedge^2 \mathfrak{g} \to \mathfrak{g}$. On the basis elements, we clearly have

$$\delta E_{12} = [E_1, E_2] = E_3, \quad \delta E_{13} = [E_1, E_3] = -E_2, \quad \delta E_{23} = E_1.$$

This shows that $\delta : \wedge^2 \mathfrak{g} \to \mathfrak{g}$ maps a basis to a basis. Hence, it is a vector space isomorphism. It follows that $\mathrm{im}\left(\delta : \wedge^2 \mathfrak{g} \to \mathfrak{g}\right) = \mathfrak{g}$ and

$$H_1(\mathfrak{g}) = \frac{\mathfrak{g}}{\mathrm{im}\left(\delta : \wedge^2 \mathfrak{g} \to \mathfrak{g}\right)} = \mathfrak{g}/\mathfrak{g} = 0.$$

It also follows that $\ker\left(\delta : \wedge^2 \mathfrak{g} \to \mathfrak{g}\right) = 0$. Hence, $\mathrm{im}\left(\delta : \wedge^3 \mathfrak{g} \to \wedge^2 \mathfrak{g}\right) = 0$ as well, i.e., $\delta : \wedge^3 \mathfrak{g} \to \wedge^2 \mathfrak{g}$ is the zero map. We conclude that

$$H_2(\mathfrak{g}) = \frac{\ker\left(\delta : \wedge^2 \mathfrak{g} \to \mathfrak{g}\right)}{\mathrm{im}\left(\delta : \wedge^3 \mathfrak{g} \to \wedge^2 \mathfrak{g}\right)} = 0/0 = 0$$

and

$$H_3(\mathfrak{g}) = \frac{\ker\left(\delta : \wedge^3 \mathfrak{g} \to \wedge^2 \mathfrak{g}\right)}{0} = \wedge^3 \mathfrak{g}/0 \cong \mathbb{K},$$

as claimed. ◆

Exercise 4.21. Compute the Chevalley–Eilenberg homologies of the order 2 special linear Lie algebra $\mathfrak{sl}_2(\mathbb{K})$ and of the 2-dimensional non-abelian real Lie algebra of Example 4.36.

There is also a functor

$$\mathbf{Lie}_{\mathbb{K}} \rightarrow \mathbf{CoCh}_{\mathbb{K}}$$

from Lie algebras over $\mathbb{K}$ to cochain complexes of $\mathbb{K}$-vector spaces. For a Lie algebra $\mathfrak{g}$, put

$$C^n(\mathfrak{g}) := \begin{cases} \mathrm{Alt}_{\mathbb{K}}^n(\mathfrak{g}; \mathbb{K}) & \text{if } n \geq 0 \\ 0 & \text{otherwise} \end{cases}.$$

We also define arrows

$$d : C^n(\mathfrak{g}) \rightarrow C^{n+1}(\mathfrak{g})$$

by putting

$$d\omega(v_1, \ldots, v_{n+1}) := \sum_{i<j} (-)^{i+j} \omega\left([v_i, v_j], v_1, \ldots, \widehat{v_i}, \ldots, \widehat{v_j}, \ldots, v_{n+1}\right),$$

for all $\omega \in \mathrm{Alt}_{\mathbb{K}}^n(\mathfrak{g}; \mathbb{K})$ and all $v_1, \ldots, v_{n+1} \in \mathfrak{g}$. A direct check reveals that $d\omega$ is a well-defined multilinear alternating map. It is then obvious that $d : \mathrm{Alt}_{\mathbb{K}}^n(\mathfrak{g}; \mathbb{K}) \rightarrow \mathrm{Alt}_{\mathbb{K}}^{n+1}(\mathfrak{g}; \mathbb{K})$ defined in this way is a linear map (do you see it?).

Proposition 4.40. *The sequence*

$$0 \longrightarrow C^0(\mathfrak{g}) \xrightarrow{d} C^1(\mathfrak{g}) \xrightarrow{d} C^2(\mathfrak{g}) \xrightarrow{d} \cdots \qquad (4.39)$$

is a cochain complex of vector spaces.

Proof. It is easy to see that the sequence (4.39) can be obtained from the sequence (4.37) applying the duality functor $* : \mathbf{Vect}_{\mathbb{K}} \rightarrow \mathbf{Vect}_{\mathbb{K}}$ (first, and then the natural isomorphisms $\mathrm{Hom}_{\mathbb{K}}(\wedge^n \mathfrak{g}, \mathbb{K}) \cong \mathrm{Alt}_{\mathbb{K}}^n(\mathfrak{g}; \mathbb{K})$). The statement now follows from Proposition 4.37. $\qquad \square$

Now, let $f : \mathfrak{g} \rightarrow \mathfrak{h}$ be a Lie algebra homomorphism. We define $\mathrm{Alt}(f) : C^\bullet(\mathfrak{h}) \rightarrow C^\bullet(\mathfrak{g})$ to be the family of maps $\mathrm{Alt}(f) := (\mathrm{Alt}^n(f) : \mathrm{Alt}_{\mathbb{K}}^n(\mathfrak{h}; \mathbb{K}) \rightarrow \mathrm{Alt}_{\mathbb{K}}^n(\mathfrak{g}; \mathbb{K}))_{n \in \mathbb{Z}}$ defined by

$$\mathrm{Alt}^n(f)(\omega)(v_1, \ldots, v_n) := \omega(f(v_1), \ldots, f(v_n))$$

for all $\omega \in \mathrm{Alt}_{\mathbb{K}}^n(\mathfrak{h}; \mathbb{K})$, $v_1, \ldots, v_n \in \mathfrak{g}$. It is easy to see that $\mathrm{Alt}(f)$ is a cochain map. Additionally, the assignment $\mathbf{Lie}_{\mathbb{K}} \rightarrow \mathbf{CoCh}_{\mathbb{K}}$ that maps a Lie algebra $\mathfrak{g}$ to the cochain complex $(C^\bullet(\mathfrak{g}), d)$ and a Lie algebra homomorphism f to the cochain map $\mathrm{Alt}(f)$ is a contravariant functor (Exercise 4.22).

Exercise 4.22. Prove that, for a Lie algebra homomorphism $f : \mathfrak{g} \to \mathfrak{h}$, the family $\mathrm{Alt}(f) : (C^\bullet(\mathfrak{h}), d) \to (C^\bullet(\mathfrak{g}), d)$ defined above is a cochain map. Prove also that the assignment $\mathbf{Lie}_\mathbb{K} \to \mathbf{CoCh}_\mathbb{K}$ defined in this way is a contravariant functor.

Definition 4.41 (Chevalley–Eilenberg Cohomology). The image $(C^\bullet(\mathfrak{g}), d)$ of a Lie algebra $\mathfrak{g}$ over $\mathbb{K}$ under the functor $\mathbf{Lie}_\mathbb{K} \to \mathbf{CoCh}_\mathbb{K}$ is called the *Chevalley–Eilenberg cochain complex* of $\mathfrak{g}$ (with trivial coefficients) and its cohomology $H^\bullet(\mathfrak{g}) := H^\bullet(C^\bullet(\mathfrak{g}), d)$ is called the *Chevalley–Eilenberg cohomology* of $\mathfrak{g}$. The differential d is also called the *Chevalley–Eilenberg differential*. Cocycles in $(C^\bullet(\mathfrak{g}), d)$ are denoted $Z^\bullet(\mathfrak{g})$ (*Chevalley–Eilenberg cocycles*) and coboundaries are denoted $B^\bullet(\mathfrak{g})$ (*Chevalley–Eilenberg coboundaries*).

The n-th Chevalley–Eilenberg cohomology is a contravariant functor $\mathbf{Lie}_\mathbb{K} \to \mathbf{Vect}_\mathbb{K}$ obtained composing the Chevalley–Eilenberg cochain complex functor $\mathbf{Lie}_\mathbb{K} \to \mathbf{CoCh}_\mathbb{K}$ and the n-th cohomology functor $H^n : \mathbf{CoCh}_\mathbb{K} \to \mathbf{Vect}_\mathbb{K}$. Hence isomorphic Lie algebras have isomorphic Chevalley–Eilenberg cohomologies.

We conclude this chapter discussing low degree Chevalley–Eilenberg cohomologies. In low degree, the Chevalley–Eilenberg cochain complex reads

$$0 \longrightarrow \mathbb{K} \xrightarrow{\ 0\ } \mathrm{Hom}_\mathbb{K}(\mathfrak{g}, \mathbb{K}) \xrightarrow{\ d\ } \mathrm{Alt}^2_\mathbb{K}(\mathfrak{g}; \mathbb{K}) \xrightarrow{\ d\ } \cdots.$$

So, $H^0(\mathfrak{g}) = \mathbb{K}$, and the first cohomology is

$$H^1(\mathfrak{g}) = \ker\left(d : \mathrm{Hom}_\mathbb{K}(\mathfrak{g}, \mathbb{K}) \to \mathrm{Alt}^2_\mathbb{K}(\mathfrak{g}; \mathbb{K})\right).$$

Now, a linear map $f : \mathfrak{g} \to \mathbb{K}$ is in the kernel of d if and only if, for all $v_1, v_2 \in \mathfrak{g}$

$$0 = df(v_1, v_2) = f([v_1, v_2]),$$

i.e., $f \in \mathrm{Ann}\left([\mathfrak{g}, \mathfrak{g}]\right)$ (the annihilator subspace of the subspace $[\mathfrak{g}, \mathfrak{g}]$). We conclude that

$$H^1(\mathfrak{g}) = \mathrm{Ann}\left([\mathfrak{g}, \mathfrak{g}]\right).$$

Remark 4.42. There is an equivalent description of $H^1(\mathfrak{g})$ more similar to the first group cohomology. Namely, we can rephrase the property of a linear map $f : \mathfrak{g} \to \mathbb{K}$ of being in the annihilator of $[\mathfrak{g}, \mathfrak{g}]$ by saying that f is a Lie algebra homomorphism from $\mathfrak{g}$ to the abelian Lie algebra $\mathbb{K}$ (do you see it?). Hence, we also have

$$H^1(\mathfrak{g}) = \mathrm{Hom}_{\mathbf{Lie}_{\mathbb{K}}}(\mathfrak{g}, \mathbb{K}).$$

$\diamond$

Finally, we briefly describe Chevalley–Eilenberg 2-cohomologies. The reader is invited to notice the similarity between this situation and the group 2-cohomologies. Let $\mathfrak{g}, \mathfrak{k}$ be Lie algebras over $\mathbb{K}$. A *Lie algebra extension* of the Lie algebra $\mathfrak{g}$ by the Lie algebra $\mathfrak{k}$ is (another Lie algebra $\mathfrak{h}$ together with) a *short exact sequence of Lie algebras*

$$0 \longrightarrow \mathfrak{k} \xrightarrow{\alpha} \mathfrak{h} \xrightarrow{\beta} \mathfrak{g} \rightarrow 0. \tag{4.40}$$

This means that (4.40) is a short exact sequence of vector spaces and, additionally, α and β are Lie algebra homomorphisms. Two Lie algebra extensions $0 \rightarrow \mathfrak{k} \rightarrow \mathfrak{h} \rightarrow \mathfrak{g} \rightarrow 0$ and $0 \rightarrow \mathfrak{k} \rightarrow \mathfrak{h}' \rightarrow \mathfrak{g} \rightarrow 0$ of $\mathfrak{g}$ by $\mathfrak{k}$ are *equivalent* if there exists a Lie algebra isomorphism $\Phi : \mathfrak{h} \rightarrow \mathfrak{h}'$ such that the diagram

$$
\begin{array}{ccccccccc}
0 & \longrightarrow & \mathfrak{k} & \longrightarrow & \mathfrak{h} & \longrightarrow & \mathfrak{g} & \longrightarrow & 0 \\
 & & \| & & \downarrow{\scriptstyle\Phi} & & \| & & \\
0 & \longrightarrow & \mathfrak{k} & \longrightarrow & \mathfrak{h}' & \longrightarrow & \mathfrak{g} & \longrightarrow & 0
\end{array}
$$

commutes (the vertical "$=$" denote the identity maps). "Equivalence" is indeed an equivalence relation on the collection of Lie algebra extensions of $\mathfrak{g}$ by $\mathfrak{k}$. A Lie algebra extension (4.40) is called *central* if im α is in the *center* of H, i.e., for all $k \in \mathfrak{k}$ and all $h \in \mathfrak{h}$ we have $[k, h] = 0$. Every Lie algebra extension equivalent to a central extension is also central. For more on Lie algebra extensions see Hilgert and Neeb (2012).

Theorem 4.43. *Let $\mathfrak{g}$ be a Lie algebra over a field $\mathbb{K}$. Then central extensions of $\mathfrak{g}$ by the abelian Lie algebra $\mathbb{K}$ are* classified by the second Chevalley–Eilenberg cohomology $H^2(\mathfrak{g})$, *i.e., there exists a natural bijection between $H^2(\mathfrak{g})$ and equivalence classes of central extensions of $\mathfrak{g}$ by $\mathbb{K}$.*

Proof. The proof is similar in spirit to that of Theorem 4.16 and we only sketch it leaving the details as Exercise 4.23.

A 2-cocycle $c \in Z^2(\mathfrak{g})$ determines a central extension of $\mathfrak{g}$ by $\mathbb{K}$ as follows. First of all recall that $c : \mathfrak{g} \times \mathfrak{g} \rightarrow \mathbb{K}$ is a bilinear alternating map. Its differential is the 3-multilinear alternating map $dc : \mathfrak{g} \times \mathfrak{g} \times \mathfrak{g} \rightarrow \mathbb{K}$ given by

$$dc(v_1, v_2, v_3) = -c([v_1, v_2], v_3) + c([v_1, v_3], v_2) - c([v_2, v_3], v_1),$$

which vanishes iff

$$-c([v_1, v_2], v_3) + c([v_1, v_3], v_2) - c([v_2, v_3], v_1) = 0$$

for all $v_1, v_2, v_3 \in \mathfrak{g}$. Now, consider the vector space $\mathbb{K} \oplus \mathfrak{g}$ and define a bracket

$$[-, -]_c : \mathbb{K} \oplus \mathfrak{g} \times \mathbb{K} \oplus \mathfrak{g} \to \mathbb{K} \oplus \mathfrak{g},$$

by putting

$$[(a_1, v_1), (a_2, v_2)]_c := (c(v_1, v_2), [v_1, v_2]).$$

It is easy to see that $[-, -]_c$ is a Lie bracket precisely because c is a cocycle. So, $(\mathbb{K} \oplus \mathfrak{g}, [-, -]_c)$ is a Lie algebra that we denote $\mathfrak{h}_c$. The sequence

$$
\begin{array}{ccccccccc}
0 & \longrightarrow & \mathbb{K} & \stackrel{\alpha}{\longrightarrow} & \mathfrak{h}_c & \stackrel{\beta}{\longrightarrow} & \mathfrak{g} & \longrightarrow & 0 \\
& & a & \longmapsto & (a, 0) & & & & \\
& & & & (a, v) & \longmapsto & v & &
\end{array}
$$
(4.41)

is a central extension of $\mathfrak{g}$ by $\mathbb{K}$ as desired.

If $c, c' \in Z^2(\mathfrak{g})$ are cohomologous, the associated central extensions $0 \to \mathfrak{k} \to \mathfrak{h}_c \to \mathfrak{g} \to 0$ and $0 \to \mathfrak{k} \to \mathfrak{h}_{c'} \to \mathfrak{g} \to 0$ are equivalent. An explicit equivalence is provided by the isomorphism

$$\Phi_\varphi : \mathfrak{h}_c \to \mathfrak{h}_{c'}, \quad (a, v) \mapsto \Phi_\varphi(a, v) := (a + \varphi(v), v),$$

where $\varphi \in C^1(\mathfrak{g}) = \mathrm{Hom}_\mathbb{K}(\mathfrak{g}, \mathbb{K})$ is a 1-cochain such that $c - c' = d\varphi$ (which in turn means that $c(v_1, v_2) = c'(v_1, v_2) - \varphi([v_1, v_2])$ for all $v_1, v_2 \in \mathfrak{g}$, do you see it?). So, we have a well-defined map

$$H^2(\mathfrak{g}) \to \{\text{equivalence classes of central extensions of } \mathfrak{g} \text{ by } \mathbb{K}\}$$
$$[c] \mapsto \text{equivalence class of } 0 \to \mathbb{K} \to \mathfrak{h}_c \to \mathfrak{g} \to 0.$$

It remains to prove that this map is bijective. For the injectivity, take two cocycles $c, c' \in Z^2(\mathfrak{g})$ and assume that the associated central extensions $0 \to \mathfrak{k} \to \mathfrak{h}_c \to \mathfrak{g} \to 0$ and $0 \to \mathfrak{k} \to \mathfrak{h}_{c'} \to \mathfrak{g} \to 0$ are equivalent. Let $\Phi : \mathfrak{h}_c \to \mathfrak{h}_{c'}$ be an isomorphism realizing the equivalence. Then it is easy to see that Φ is necessarily of the form Φ_φ for some 1-cochain $\varphi \in C^1(\mathfrak{g})$ such that $c - c' = d\varphi$ and we conclude that $[c] = [c']$ as desired.

For the surjectivity, let $0 \to \mathbb{K} \to \mathfrak{h} \to \mathfrak{g} \to 0$ be any central extension of $\mathfrak{g}$ by $\mathbb{K}$. In particular, it is a short exact sequence of vector spaces. Hence, it splits. Choose any splitting $s : \mathfrak{g} \to \mathfrak{h}$ (beware that s is a linear map but it *needs not* be a Lie algebra homomorphism). We know from Example 1.43 that the splitting s induces a vector space isomorphism $\Phi : \mathfrak{h} \to \mathbb{K} \oplus \mathfrak{g}$ that we can use to transport the Lie algebra structure from the domain to the codomain. We can also transport the maps $\mathbb{K} \to \mathfrak{h}$, and $\mathfrak{h} \to \mathfrak{g}$ getting an equivalent central extension $0 \to \mathbb{K} \to \mathbb{K} \oplus \mathfrak{g} \to \mathfrak{g} \to 0$. A closer inspection (implementing all the properties of a central extension) reveals that the latter is necessarily of the form $0 \to \mathbb{K} \to \mathfrak{h}_c \to \mathfrak{g} \to 0$ for some 2-cocycle $c \in Z^2(\mathfrak{g})$. This concludes the proof. $\square$

> **Exercise 4.23.** Fill all the gaps in the proof of Theorem 4.43.

Example 4.44. The central extensions corresponding to the zero cohomology class are called *trivial*. Suppose that $H^2(\mathfrak{g}) = 0$. One can then use Theorem 4.43 to conclude that there are no non-trivial central extensions of $\mathfrak{g}$ by $\mathbb{K}$. This is the case, e.g., for $\mathfrak{g} = \mathfrak{so}_3(\mathbb{K}), \mathfrak{sl}_2(\mathbb{K})$ if the characteristic of the field $\mathbb{K}$ is different from 2 (Exercise 4.24).   ◆

> **Exercise 4.24.** Prove that there are no non-trivial central extensions of the Lie algebras $\mathfrak{sl}_2(\mathbb{K})$ (when the characteristic of the field $\mathbb{K}$ is not 2), $\mathfrak{so}_3(\mathbb{K})$ by $\mathbb{K}$ (**Hint:** *Prove that the 2-nd Chevalley–Eilenberg cohomology vanishes in these two cases and then apply Theorem* 4.43).

4.5   End-of-Chapter Problems

> **Problem 4.1.** Let R be a commutative ring with unit, and let M be an R-module. Show that the constant family of R-modules $(M_n)_{n \in \mathbb{N}_0}$, with $M_n = M$ for all n, can be given a structure of semi-simplicial module
>
> $$\cdots \rightrightarrows M \Rrightarrow M \rightrightarrows M$$
>
> by letting all the faces be equal to the identity $\mathrm{id}_M : M \to M$. Then compute the associated chain complex $(C_\bullet(M), D)$ and its homologies.

> **Problem 4.2.** Compute the group homologies and cohomologies of the trivial group $G = \{1\}$.

Problem 4.3. Show that any group homomorphism $f : G \to H$ preserves the derived subgroups: $f(G') \subseteq H'$, and conclude that f defines a linear map $f^{ab} : G^{ab} \to H^{ab}$ between the abelianization via $f^{ab}(gG') := f(g)H'$, for all $g \in G$. Show also that the assignment $ab : \mathbf{Gr} \to \mathbf{Ab}$ mapping

(1) a group G to its abelianization G^{ab} and
(2) a group homomorphism $f : G \to H$ to the corresponding linear map $f^{ab} : G^{ab} \to H^{ab}$

is a well-defined functor. Finally, prove that the assignment $G \mapsto \Phi_G$, where $\Phi_G : H_1(G) \to G^{ab}$ is the isomorphism of Theorem 4.11, is a natural isomorphism between the functors $H_1 : \mathbf{Gr} \to \mathbf{Ab}$ and $ab : \mathbf{Gr} \to \mathbf{Ab}$.

Problem 4.4. Let G be a group. A *G-module* is an abelian group $\mathfrak{A}$ together with a group homomorphism $\alpha : G \to \operatorname{Aut}\mathfrak{A}$, where $\operatorname{Aut}\mathfrak{A}$ is the group of invertible linear maps $\mathfrak{A} \to \mathfrak{A}$. The homomorphism α is also called a *linear action* of G on $\mathfrak{A}$ and we denote $\alpha(g)(a) =: g.a$ for all $g \in G$ and $a \in \mathfrak{A}$. Prove that the family $(\mathfrak{A}^{G^{\times n}})_{n \in \mathbb{N}_0}$ can be given the structure of a semi-cosimplicial abelian group

$$\cdots \Longrightarrow \mathfrak{A}^{G \times G} \Lleftarrow \mathfrak{A}^{G} \Longleftarrow \mathfrak{A}$$

with cofaces $(d_i : \mathfrak{A}^{G^{\times(n-1)}} \to \mathfrak{A}^{G^{\times n}})_{0 \le i \le n \in \mathbb{N}}$ given by

$$d_i f(g_1, \ldots, g_n) := \begin{cases} g_1 . f(g_2, \ldots, g_n) & \text{if } i = 0 \\ f(g_1, \ldots, g_{i-1}, g_i g_{i+1}, g_{i+2}, \ldots, g_n) & \text{if } i = 1, \ldots, n-1, \\ f(g_1, \ldots, g_{n-1}) & \text{if } i = n \end{cases}$$

$f \in \mathfrak{A}^{G^{\times(n-1)}}$, $g_1, \ldots, g_n \in G$ (here $\mathfrak{A}^{G^{\times n}} = \mathcal{F}(G^{\times n}, \mathfrak{A})$ is the abelian group of maps $G^{\times n} \to \mathfrak{A}$). The associated cochain complex $C^{\bullet}(G; \mathfrak{A})$ is called the *group cochain complex with coefficients in* $\mathfrak{A}$ and its cohomology $H^{\bullet}(G; \mathfrak{A})$ is the *group cohomology with coefficients in* $\mathfrak{A}$. Provide an interpretation for $H^0(G; \mathfrak{A})$ and $H^1(G; \mathfrak{A})$.

Problem 4.5. Show that the assignment $\mathbf{Gr} \to \mathbf{Alg}_{\mathbb{K}}$ mapping a group G to its group algebra $\mathbb{K}G$ (Example 4.21) and a group homomorphism $f : G \to H$ to the unique $\mathbb{K}$-linear map $\mathbb{K}f : \mathbb{K}G \to \mathbb{K}H$ extending f is a well-defined functor from the category of groups (and group homomorphisms) to the category of associative $\mathbb{K}$-algebra (and their algebra homomorphisms).

Problem 4.6. Let G be a group, let $\mathbb{K}$ be a field and let $\mathbb{K}G$ be the group algebra of G. Moreover, let $\mathfrak{A}$ be a $\mathbb{K}$-vector space and a G-module (Problem 4.4), in such a way that the linear G-action $\alpha : G \to \operatorname{Aut}\mathfrak{A}$ takes values in $\mathbb{K}$-linear automorphisms $\operatorname{Aut}_{\mathbb{K}}\mathfrak{A} \subseteq \operatorname{Aut}\mathfrak{A}$. Prove that, under these hypotheses, $\alpha : G \to \operatorname{Aut}_{\mathbb{K}}\mathfrak{A}$ can be uniquely extended to a homomorphism of associative $\mathbb{K}$-algebras $\alpha' : \mathbb{K}G \to \operatorname{End}_{\mathbb{K}}\mathfrak{A}$.

Problem 4.7. Let $\mathbb{K}$ be a field and let V be a $\mathbb{K}$-vector space. Consider the vector space

$$\operatorname{Mult}(V;\mathbb{K}) := \bigoplus_{i>0} \operatorname{Mult}^i(V;\mathbb{K}), \quad \operatorname{Mult}^i(V;\mathbb{K}) := \operatorname{Mult}^i_{\mathbb{K}}(\underbrace{V,\ldots,V}_{i\text{-times}};\mathbb{K}),$$

and the bilinear composition law

$$\star : \operatorname{Mult}(V;\mathbb{K}) \times \operatorname{Mult}(V;\mathbb{K}) \to \operatorname{Mult}(V;\mathbb{K})$$

uniquely defined by $\operatorname{Mult}^i(V;\mathbb{K}) \star \operatorname{Mult}^j(V;\mathbb{K}) \subseteq \operatorname{Mult}^{i+j}(V;\mathbb{K})$ and

$$(\mu \star \nu)(v_1,\ldots,v_{i+j}) := \mu(v_1,\ldots,v_i)\nu(v_{i+1},\ldots,v_{i+j}),$$

$\mu \in \operatorname{Mult}^i(V;\mathbb{K}), \nu \in \operatorname{Mult}^j(V;\mathbb{K}), i,j > 0$. Prove that $(\operatorname{Mult}(V;\mathbb{K}),\star)$ is a well-defined associative $\mathbb{K}$-algebra.

Problem 4.8. Let $\mathbb{K}$ be a field and let V be a $\mathbb{K}$-vector space. Consider the vector space

$$\mathcal{T}(V) := \bigoplus_{i>0} V^{\otimes i}, \quad V^{\otimes i} := \underbrace{V \otimes \cdots \otimes V}_{i\text{-times}},$$

and the bilinear composition law

$$\star : \mathcal{T}(V) \times \mathcal{T}(V) \to \mathcal{T}(V)$$

uniquely defined by $V^{\otimes i} \star V^{\otimes j} \subseteq V^{\otimes(i+j)}$ and

$$(v_1 \otimes \cdots \otimes v_i) \otimes (v_{i+1} \otimes \cdots \otimes v_{i+j}) := v_1 \otimes \cdots \otimes v_{i+j},$$

$v_1, \ldots, v_{i+j} \in V$, $i, j > 0$. Prove that $(\mathcal{T}(V), \star)$ is a well-defined associative $\mathbb{K}$-algebra. The algebra $(\mathcal{T}(V), \star)$ is called the *tensor algebra over* V.

Problem 4.9. Let $\mathbb{K}$ be a field and let V be a $\mathbb{K}$-vector space. The tensor algebra over $\mathbb{K}$ of Problem 4.8, together with the natural injection

$$\iota : V \to \mathcal{T}(V),$$

is also called the *free algebra over* V because it satisfies the following universal property: for every other $\mathbb{K}$-algebra A and every $\mathbb{K}$-linear map $\alpha : V \to A$, there exists a unique algebra homomorphism $\mathcal{T}(V) \to A$ such that the diagram

$$
\begin{array}{ccc}
V & \xrightarrow{\alpha} & A \\
{\scriptstyle \iota} \downarrow & \nearrow_{\exists!} & \\
\mathcal{T}(V) & &
\end{array}
$$

commutes. Prove the universal property of the tensor algebra. Prove also that the tensor algebra is the unique algebra with such universal property up to unique isomorphisms. Finally, prove that the universal property of the tensor algebra says that $\mathcal{T}(V)$ is an initial object in an appropriate category (Problem 3.24).

Problem 4.10. Find "symmetric" and "alternating versions" of the algebras in Problems 4.7 and 4.8.

Problem 4.11. Let $\mathbb{K}$ be a field and let A be an associative $\mathbb{K}$-algebra. An *A-bimodule* is a $\mathbb{K}$-vector space $\mathfrak{M}$ together with $\mathbb{K}$-linear maps $\rho, \sigma : A \to \mathrm{End}_{\mathbb{K}} \mathfrak{M}$ such that

(1) ρ is a *left A-action*, i.e., an algebra homomorphism;
(2) σ is a *right A-action*, i.e., an algebra *anti-homomorphism*, i.e., $\sigma(\alpha\beta) = \sigma(\beta) \circ \sigma(\alpha)$ for all $\alpha, \beta \in A$;
(3) ρ and σ *commute*, i.e., $\rho(\alpha) \circ \sigma(\beta) = \sigma(\beta) \circ \rho(\alpha)$ for all $\alpha, \beta \in A$.

In this situation, we also denote $\rho(\alpha)(p) = \alpha.p$, and $\sigma(\beta)(p) =: p.\beta$ for all $\alpha, \beta \in A$, and $p \in \mathfrak{M}$. Finally, put

$$\mathrm{Mult}_{\mathbb{K}}^{n}(A; \mathfrak{M}) := \mathrm{Mult}_{\mathbb{K}}(\underbrace{A, \ldots, A}_{n\text{-times}}; \mathfrak{M}), \quad n \geq 0.$$

Show that the family $\mathrm{Mult}(A; \mathfrak{M}) := (\mathrm{Mult}_{\mathbb{K}}^{n}(A; \mathfrak{M}))_{n \in \mathbb{N}_0}$ can be given the structure of a semi-cosimplicial $\mathbb{K}$-vector space

$$\cdots \Longrightarrow \mathrm{Bil}_{\mathbb{K}}^{2}(A; \mathfrak{M}) \Longrightarrow \mathrm{Hom}_{\mathbb{K}}(A, \mathfrak{M}) \Longleftarrow \mathfrak{M}.$$

with cofaces $d = (d_i : \mathrm{Mult}_{\mathbb{K}}^{n-1}(A; \mathfrak{M}) \to \mathrm{Mult}_{\mathbb{K}}^{n}(A; \mathfrak{M}))_{0 \leq i \leq n \in \mathbb{N}}$ defined by

$$d_i\mu(\alpha_1, \ldots, \alpha_n) := \begin{cases} \alpha_1.\mu(\alpha_2, \ldots, \alpha_n) & \text{if } i = 0 \\ \mu(\alpha_1, \ldots, \alpha_i\alpha_{i+1}, \ldots, \alpha_n) & \text{if } 0 < i < n, \quad (4.42) \\ \mu(\alpha_1, \ldots, \alpha_{n-1}).\alpha_n & \text{if } i = n \end{cases}$$

$\mu \in \mathrm{Mult}_{\mathbb{K}}^{n-1}(A; \mathfrak{M})$, $\alpha_1, \ldots, \alpha_n \in A$. The associated cochain complex $HC^{\bullet}(A; \mathfrak{M})$ is called the *Hochschild cochain complex with coefficients in* $\mathfrak{M}$ and its cohomology $HH^{\bullet}(A; \mathfrak{M})$ is the *Hochschild cohomology with coefficients in* $\mathfrak{M}$. Provide an interpretation for $H^0(A; \mathfrak{M})$ and $H^1(A; \mathfrak{M})$.

Problem 4.12. Let $\mathbb{K}$ be a field and let $\mathfrak{sp}_n(\mathbb{K}) \subseteq M_{2n}(\mathbb{K})$ be the vector subspace consisting of matrices A such that

$$JA + A^T J = 0,$$

with

$$J = \begin{pmatrix} 0_n & I_n \\ -I_n & 0_n \end{pmatrix} \in M_{2n}(\mathbb{K}),$$

where $0_n, I_n \in M_n(\mathbb{K})$ are the zero and the identical matrix, respectively, $n \geq 1$. Prove that $\mathfrak{sp}_n(\mathbb{K})$ is a Lie subalgebra in $\mathfrak{gl}_{2n}(\mathbb{K})$. The Lie algebra $\mathfrak{sp}_n(\mathbb{K})$ is called the *symplectic Lie algebra*.

Problem 4.13. Compute the Chevalley–Eilenberg homology and the Chevalley–Eilenberg cohomology of the general linear Lie algebra $\mathfrak{gl}_2(\mathbb{K})$.

Problem 4.14. Let $\mathbb{K}$ be a field and let $\mathfrak{g}$ be a Lie algebra over $\mathbb{K}$. A *right $\mathfrak{g}$-module* is a $\mathbb{K}$-vector space W together with a *Lie algebra antihomomorphism* $\sigma : \mathfrak{g} \to \mathrm{End}_{\mathbb{K}} V$, i.e., a linear map such that $\sigma([v_1, v_2]) = -[\sigma(v_1), \sigma(v_2)]$ for all $v_1, v_2 \in \mathfrak{g}$. The map σ is called a *right action* of $\mathfrak{g}$. Let W be a right $\mathfrak{g}$-module, and put

$$C_n(\mathfrak{g}; W) := \begin{cases} W \otimes \wedge^n \mathfrak{g} & \text{if } n \geq 0 \\ 0 & \text{otherwise} \end{cases}.$$

We also define arrows

$$\delta : C_n(\mathfrak{g}) \to C_{n-1}(\mathfrak{g}),$$

declaring how do they act on decomposable vectors. Namely, we put

$$\begin{aligned} \delta(w \otimes v_1 \wedge \cdots \wedge v_n) \\ := \sum_i (-)^{i+1} \sigma(v_i)(w) \otimes v_1 \wedge \cdots \wedge \widehat{v_i} \wedge \cdots \wedge v_n \\ + \sum_{i<j} (-)^{i+j} w \otimes [v_i, v_j] \wedge v_1 \wedge \cdots \wedge \widehat{v_i} \wedge \cdots \wedge \widehat{v_j} \wedge \cdots \wedge v_n, \end{aligned} \tag{4.43}$$

for all $w \in W$ and $v_1, \ldots, v_n \in \mathfrak{g}$. Prove that $(C_{\bullet}(\mathfrak{g}; W), \delta)$ is a well-defined chain complex. The chain complex $(C_{\bullet}(\mathfrak{g}; W), \delta)$ is called the *Chevalley–Eilenberg chain complex of $\mathfrak{g}$ with coefficients in W* and its homology $H_{\bullet}(\mathfrak{g}; W)$ is the *Chevalley–Eilenberg homology with coefficients in W*. Provide an interpretation for $H_0(\mathfrak{g}; W)$ and $H_1(\mathfrak{g}; W)$.

Problem 4.15. Let $\mathbb{K}$ be a field and let $\mathfrak{g}$ be a Lie algebra over $\mathbb{K}$. A *left $\mathfrak{g}$-module* is a $\mathbb{K}$-vector space V together with a Lie algebra homomorphism $\rho : \mathfrak{g} \to \mathrm{End}_{\mathbb{K}} V$. A left $\mathfrak{g}$-module is also called a *representation* of $\mathfrak{g}$ and the structure map ρ is called a *left action* of $\mathfrak{g}$. Let V be a left $\mathfrak{g}$-module, and put

$$C^n(\mathfrak{g}; V) := \begin{cases} \mathrm{Alt}_{\mathbb{K}}^n(\mathfrak{g}; V) & \text{if } n \geq 0 \\ 0 & \text{otherwise} \end{cases}.$$

We also define arrows

$$d : C^n(\mathfrak{g}; V) \to C^{n+1}(\mathfrak{g}; V)$$

by putting

$$\begin{aligned} d\omega(v_1, &\ldots, v_{n+1}) \\ &:= \sum_i (-)^{i+1} \rho(v_i) \omega(v_1, \ldots, \widehat{v}_i, \ldots, v_{n+1}) \\ &\quad + \sum_{i<j} (-)^{i+j} \omega\left([v_i, v_j], v_1, \ldots, \widehat{v}_i, \ldots, \widehat{v}_j, \ldots, v_{n+1}\right), \end{aligned}$$

for all $\omega \in \mathrm{Alt}_{\mathbb{K}}^n(\mathfrak{g}; V)$ and all $v_1, \ldots, v_{n+1} \in \mathfrak{g}$. Prove that $(C^\bullet(\mathfrak{g}; V), d)$ is a well-defined cochain complex. The cochain complex $(C^\bullet(\mathfrak{g}; V), d)$ is called the *Chevalley–Eilenberg cochain complex of $\mathfrak{g}$ with coefficients in V* and its cohomology is the *Chevalley–Eilenberg cohomology with coefficients in V*. Provide an interpretation for $H^0(\mathfrak{g}; V)$ and $H^1(\mathfrak{g}; V)$.

Chapter 5

Singular Homology

In this chapter, we show that (co)chain complexes pop up naturally in Topology as well. In particular, every topological space X gives rise to a chain complex (actually one for each ring) in a functorial way. The associated homology contains relevant information on X. We will analyze this case more thoroughly than those in Chapter 4 and the reader will see homotopies and short exact sequences of chain complexes in action. For much deeper presentations of Singular Homology and other homology theories attached to appropriate classes of topological spaces, the interested reader may consult, e.g., Hatcher (2002) and Rotman (1998).

5.1 Singular (Co)Chains and Singular (Co)Homology

In this section, we show that every topological space X functorially defines both a chain complex and a cochain complex (actually one for each ring) called the complex of *singular (co)chains* in X. The (co)homology of such complex is the *singular (co)homology* of X and contains information on the topology of X. More precisely, due to its functorial properties, singular homology is a *topological invariant*, i.e., it doesn't change when replacing X by a homeomorphic space and one can use it to separate homeomorphism classes of topological spaces. We will actually show that singular homology is a *homotopical invariant*, i.e., it doesn't change when replacing X by a space which is only *homotopy equivalent* to it (see Definition 5.19 below). So, it can be used to separate homotopy equivalence classes of topological spaces. Singular homology (together with the *fundamental group* and the other *homotopy groups*, Hatcher, 2002; Kosniowski, 1980; Rotman, 1998) is one of the starting points of a whole branch of Geometry that applies

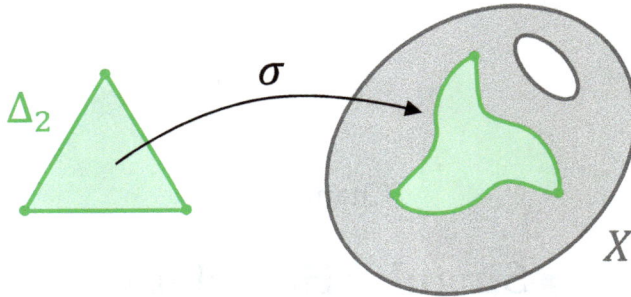

Figure 5.1.   A singular 2-simplex σ in the topological space X.

algebraic methods to study topological spaces and, for this reason, is called *Algebraic Topology*.

Let X be a topological space. The complex of singular (co)chains of X arises as the (co)chain complex associated to an appropriate semi-simplicial set (Section 4.1).

Definition 5.1 (Singular Simplex). A *singular n-simplex* in X is a continuous map

$$\sigma : \Delta_n \to X$$

where Δ_n is the standard n-simplex (with the subspace topology induced from the standard topology on $\mathbb{R}^{n+1}$, see Figure 5.1). The set of singular n-simplexes in X is denoted $S_n(X)$.

Let n be a non-negative integer and let $\sigma : \Delta_n \to X$ be a singular n-simplex in X. For any $i = 0, \ldots, n$, the composition $\sigma \circ d_i : \Delta_{n-1} \to X$ of the i-th coface of the standard simplex $(\Delta_\bullet, d)$ (see Example 4.4) followed by σ is a singular $(n-1)$-simplex in X (see Figure 5.2). This follows from the fact that both d_i and σ are continuous maps and that the composition of continuous maps is continuous. In this way, we get maps

$$d_i^\sharp : S_n(X) \to S_{n-1}(X), \quad \sigma \mapsto d_i^\sharp \sigma := \sigma \circ d_i.$$

Lemma 5.2. *The pair $(S_\bullet(X), d^\sharp)$ with $S_\bullet(X) := (S_n(X))_{n \in \mathbb{N}_0}$ and*

$$d^\sharp = (d_i^\sharp : S_n(X) \to S_{n-1}(X))_{0 \leq i \leq n \in \mathbb{N}}$$

is a semi-simplicial set.

Proof.   The semi-simplicial identities for the $d_i^\sharp$ easily follow from the semi-cosimplicial identities for the d_i. We leave the details to the reader as Exercise 5.1. $\square$

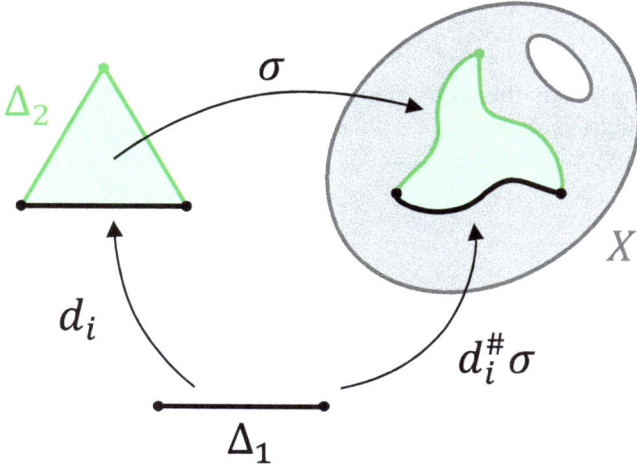

Figure 5.2.   A face map in $(S_\bullet(X), d^\sharp)$.

Exercise 5.1. Prove Lemma 5.2.

Next, we fix a ring R and apply the constructions described in Section 4.1 to produce, out of the semi-simplicial set $(S_\bullet(X), d^\sharp)$, a semi-simplicial R-module

$$(RS_\bullet(X), Rd^\sharp)$$

and a semi-cosimplicial R-module

$$(R_\bullet^{S(X)}, R^{d^\sharp}).$$

In other words, as usual, we are applying to $(S_\bullet(X), d^\sharp)$ the functors **ssFree** : **ssSet** $\to$ **ssMod**$_R$ and **ssFun** : **ssSet** $\to$ **sCosMod**$_R$. Remember that $RS_n(X)$ is the free module spanned by $S_n(X)$, so its elements are formal finite linear combinations of singular n-simplexes with coefficients in R:

$$\sum_{i \in I} a_i \sigma_i, \quad a_i \in R, \quad \sigma_i \in S_n(X),$$

where I is a set of indexes in bijection with $S_n(X)$ so that we can understand $S_n(X)$ as a family $(\sigma_i)_{i \in I}$, and the a_i are all zero but finitely many. While $R_n^{S(X)} = R^{S_n(X)}$ is the function module consisting of maps a : $S_n(X) \to R$. In their turn, $(RS_\bullet(X), Rd^\sharp)$ and $(R_\bullet^{S(X)}, R^{d^\sharp})$ determine a chain and a cochain complex that we denote

$$(C_\bullet(X, R), \partial) \quad \text{and} \quad (C^\bullet(X, R), \delta),$$

respectively. From the definition, $(C_\bullet(X, R), \partial)$ is concentrated in non-negative degrees. For all $n \geq 0$, we have $C_n(X, R) = RS_n(X)$ and the differential $\partial : C_n(X, R) \to C_{n-1}(X, R)$ is given by

$$\partial = \sum_{i=0}^{n} (-)^i R d_i^{\sharp}.$$

Hence, it acts on a basis element $\sigma \in S_n(X) \subseteq RS_n(X)$ as follows:

$$\partial\sigma = \sum_{i=0}^{n} (-)^i R d_i^{\sharp}\sigma = \sum_{i=0}^{n} (-)^i d_i^{\sharp}\sigma = d_0^{\sharp}\sigma - d_1^{\sharp}\sigma + d_2^{\sharp}\sigma + \cdots + (-)^n d_n^{\sharp}\sigma.$$

In the following, we often abuse the notation and denote simply by $d_i^{\sharp}$ the maps $R d_i^{\sharp} : C_n(X, R) \to C_{n-1}(X, R)$. Then the differential ∂ simply reads $\partial = \sum_{i=0}^{n}(-)^i d_i^{\sharp}$.

Definition 5.3 (Singular Homology). The elements of $C_n(X, R)$ are called *singular n-chains* in X with coefficients in R, and the differential ∂ is called the *boundary operator* (because it is essentially the alternating sum of face maps). The *n-cycles* in the chain complex $(C_\bullet(X, R), \partial)$ are denoted $Z_n(X, R)$ and the *n-boundaries* $B_n(X, R)$. The homology of $(C_\bullet(X, R), \partial)$ is called the *singular homology* of X with coefficients in R, and it is denoted

$$H_\bullet(X, R) := H_\bullet(C(X, R), \partial).$$

When $R = \mathbb{Z}$, we simply write $C_\bullet(X)$, $Z_\bullet(X)$, $B_\bullet(X)$ and $H_\bullet(X)$ (instead of $C_\bullet(X, \mathbb{Z})$, $Z_\bullet(X, \mathbb{Z})$, $B_\bullet(X, \mathbb{Z})$ and $H_\bullet(X, \mathbb{Z})$) and call $H_\bullet(X)$ simply the *singular homology* of X.

As for $(C_\bullet(X, R), \delta)$, from the definition, it is concentrated in non-negative degrees as well. For all $n \geq 0$, we have $C^n(X, R) = R^{S_n(X)}$ and the differential $\delta : C_n(X, R) \to C_{n-1}(X, R)$ is the alternating sum

$$\delta = \sum_{i=1}^{n} (-)^i R d_i^{\sharp}$$

of the pull-backs along the face maps $d_i^{\sharp}$, hence it acts on a function $a : S_n(X) \to R$ as follows:

$$\delta a = \sum_{i=1}^{n} (-)^i R d_i^{\sharp} a = \sum_{i=1}^{n} (-)^i a \circ d_i^{\sharp}$$

$$= a \circ d_0^{\sharp} - a \circ d_1^{\sharp} + a \circ d_2^{\sharp} + \cdots + (-)^n a \circ d_n^{\sharp}.$$

In other words, $\delta a \in R^{S_{n+1}(X)}$ is the function $\delta a : S_{n+1}(X) \to R$ given by

$$\delta a(\sigma) = \sum_{i=1}^{n} (-)^i a(d_i^\sharp \sigma).$$

Definition 5.4 (Singular Cohomology). Elements in $C^n(X, R)$ are called *singular n-cochains* in X with coefficients in R, and the differential δ is called the *coboundary operator*. The n-cocycles in the cochain complex $(C^\bullet(X, R), \delta)$ are denoted $Z^n(X, R)$ and the n-coboundaries $B^n(X, R)$. The cohomology of $(C^\bullet(X, R), \delta)$ is called the *singular cohomology* of X with coefficients in R, and it is denoted

$$H^\bullet(X, R) := H^\bullet(C(X, R), \delta).$$

When $R = \mathbb{Z}$, we simply write $C^\bullet(X)$, $Z^\bullet(X)$, $B^\bullet(X)$ and $H^\bullet(X)$ (instead of $C^\bullet(X, \mathbb{Z})$, $Z^\bullet(X, \mathbb{Z})$, $B^\bullet(X, \mathbb{Z})$ and $H^\bullet(X, \mathbb{Z})$), and call $H^\bullet(X)$ simply the *singular cohomology* of X.

Remark 5.5. Using that $R^{S_n(X)}$ is naturally isomorphic to the dual module of $RS_n(X)$, we can identify $C^n(X, R)$ with the dual module of $C_n(X, R)$ for all n. It turns out that, if we do this, then the coboundary operator δ identifies with the transpose of the boundary operator ∂. In other words, the cochain complex $(C^\bullet(X, R), \delta)$ is obtained form the chain complex $(C_\bullet(X, R), \partial)$ applying the duality functor $*$ (to the chains and to the differential). See also Problem 3.38. $\diamond$

In this chapter, we mainly (if not *only*) concentrate on singular *homology with coefficients in* $\mathbb{Z}$. In this section, we discuss the singular n-homology $H^n(X)$ of a topological space X in two easy cases: $X = \{*\}$, the one point space, but n arbitrary, and X any topological space but $n = 0$. Before proving anything, we discuss singular chains in degree 0 and 1, with a special emphasis on cycles and boundaries. This will help us gaining some intuition on what singular homologies really are. The discussion also motivates the terms "cycle" and "boundary" that we have been using since our presentation of chain complexes in Chapter 2.

Let us begin with singular 0-simplexes. Having a singular 0-simplex in a topological space X is actually equivalent to having a point in X. Indeed, a singular 0-simplex is a continuous map $\sigma : \Delta_0 \to X$. But $\Delta_0 = \{1\}$ is a one point space. So σ is completely determined by its value $\sigma(1)$. Conversely, given a point $x \in X$, we can consider the map $\sigma_x : \Delta_0 \to X$ defined by $\sigma_x(1) = x$. As Δ_0 is a one point topological space, σ_x is automatically continuous (see Figure 5.3).

In the following, we often use the notation σ_x for the singular 0-simplex mapping 1 to $x \in X$. If we interpret singular 0-simplexes as points,

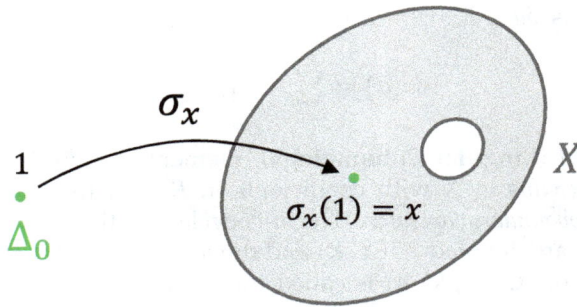

Figure 5.3.   A singular 0-simplex is just a point.

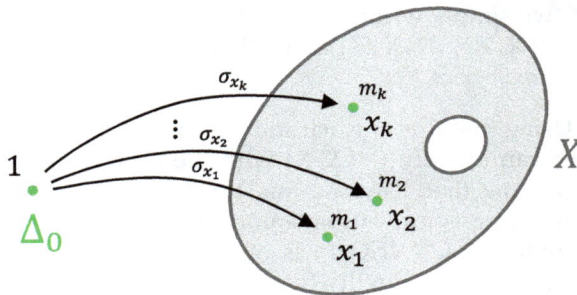

Figure 5.4.   A singular 0-chain.

a singular 0-chain $c = \sum_{i=1}^{k} m_i \sigma_{x_i}$ becomes a finite set of points $x_1, \ldots, x_k$ weighted by integer numbers $m_1, \ldots, m_k$, see Figure 5.4 (remember that the linear combination is purely formal).

Clearly, every singular 0-chain is also a 0-cycle. Before discussing 0-boundaries, we discuss 1-chains. To do this, we need a brief digression on *paths* that we make in the next remark (where we also collect some facts about *path connectedness* that will be useful in the sequel).

Remark 5.6. Let X be a topological space. A *path* in X is a continuous map $\gamma : [0, 1] \to X$ (where the closed interval $[0, 1] \subseteq \mathbb{R}$ is equipped with the subspace topology induced from the standard topology of $\mathbb{R}$). Two points $x_0, x_1 \in X$ are *connected by a path* if there exists a path $\gamma : [0, 1] \to X$ such that $\gamma(0) = x_0$ and $\gamma(1) = x_1$ (Figure 5.5).

Being connected by a path is an equivalence relation on X. Indeed, any point x_0 is connected to itself by the constant path $\gamma : [0, 1] \to X, t \mapsto x_0$, showing that being connected by a path is a reflexive relation. If x_0, x_1 are two points connected by a path $\gamma : [0, 1] \to X$, then x_1, x_0 are connected by

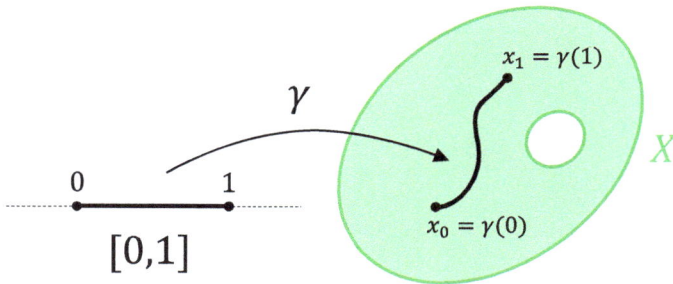

Figure 5.5. Two points connected by a path in the topological space X.

the path $\overline{\gamma} : [0,1] \to X$ defined by

$$\overline{\gamma}(t) := \gamma(1-t),$$

showing that being connected by a path is a symmetric relation. Note that $\overline{\gamma}$ is a well-defined path. Indeed, it is the composition of the continuous map $[0,1] \to [0,1], t \mapsto 1-t$, followed by γ which is continuous, hence $\overline{\gamma}$ is a continuous map as well. Finally, suppose that x_0, x_1, x_2 are three points in X such that x_0, x_1 are connected by the path γ and x_1, x_2 are connected by the path γ'. Then x_0, x_2 are connected by the path (Figure 5.6)

$$\gamma * \gamma' : [0,1] \to X$$

defined by

$$\gamma * \gamma'(t) := \begin{cases} \gamma(2t) & \text{if } t \leq 1/2 \\ \gamma'(2t-1) & \text{if } t > 1/2 \end{cases}.$$

This shows that being connected by a path is also a transitive relation. In order to see that $\gamma * \gamma'$ is indeed a well-defined path, first consider the maps

$$\Gamma : [0, 1/2] \to X, \quad t \mapsto \gamma(2t)$$

and

$$\Gamma' : [1/2, 1] \to X, \quad t \mapsto \gamma'(2t-1).$$

They are both continuous. For instance, the first one is the continuous composition of the continuous map $[0,1/2] \to [0,1], t \mapsto 2t$ followed by γ. Additionally, Γ and Γ' agree on the intersection of their domains: $\Gamma(1/2) = \gamma(1) = x_1 = \gamma'(0) = \Gamma'(1/2)$. Note that the domains $[0,1/2], [1/2,1]$ of

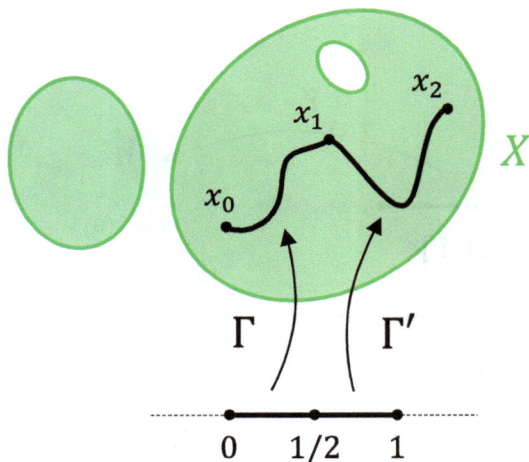

Figure 5.6.   Concatenation of paths.

Γ, Γ' are closed subsets in $[0,1]$ such that $[0,1/2] \cup [1/2,1] = [0,1]$. By the *Gluing Lemma*, Γ, Γ' glue to a well-defined continuous map $\Gamma^* : [0,1] \to X$ (uniquely defined by the conditions $\Gamma^*|_{[0,1/2]} = \Gamma$ and $\Gamma^*|_{[1/2,1]} = \Gamma'$). But it is clear that $\Gamma^* = \gamma * \gamma'$. We conclude that $\gamma * \gamma'$ is continuous as well. The path $\gamma * \gamma'$ is sometimes called the *concatenation* of γ and γ'.

Each equivalence class with respect to the equivalence relation "being connected by a path" is called a *path connected component of X*. The *path connected component* of a point $x \in X$ is the equivalence class of x. In other words, it consists of all points in X that can be connected to x by a path and will be denote by X_x. The set of path connected components of X is sometimes denoted $\pi_0(X)$ (or simply π_0 if it is clear what topological space we are talking about). The topological space X is called *path connected* if there is just one element in $\pi_0(X)$. In other words, every two points in X are connected by a path. $\Diamond$

We are ready to provide a new interpretation for singular 1-simplexes. Namely, singular 1-simplexes in a topological space X are essentially the same as paths in X. To see this, first note that there is a canonical homeomorphism

$$h : [0,1] \to \Delta_1, \quad t \mapsto h(t) := (t, 1-t).$$

Indeed, h is continuous because it is the restriction to the subspace $[0,1]$ in the domain and to the subspace Δ_1 in the codomain of a continuous map $\mathbb{R} \to \mathbb{R}^2, t \mapsto (t, 1-t)$ (Figure 5.7).

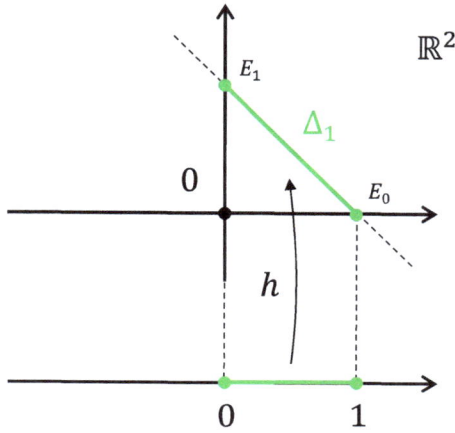

Figure 5.7. The homeomorphism $h : [0,1] \to \Delta_1$.

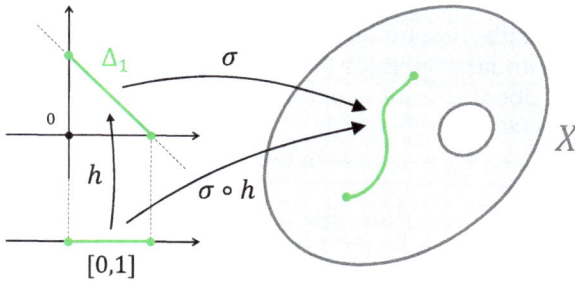

Figure 5.8. A singular 1-simplex is just a path.

Additionally, h is invertible with inverse

$$h^{-1} : \Delta_1 \to [0,1], \quad (x_0, x_1) \mapsto x_0.$$

The inverse h^{-1} is also continuous: it is the restriction to subspaces in both the domain and the codomain of the continuous map $\mathbb{R}^2 \to \mathbb{R}$, $(x,y) \mapsto x$. Now, given a singular 1-simplex $\sigma : \Delta_1 \to X$ in the topological space X, we can build a path γ by right composition with h, $\gamma = \sigma \circ h$, and vice-versa, given a path $\gamma : [0,1] \to X$ in X we can build a singular 1-simplex σ by right composition with h^{-1}: $\sigma = \gamma \circ h^{-1}$ (Figure 5.8).

It is clear that these two constructions invert each other, giving a canonical bijection between $S_1(X)$ and the set of paths in X. If we use this bijection to interpret singular 1-simplexes as paths, then a singular 1-chain becomes a finite set of paths weighted by integer numbers, see Figure 5.9.

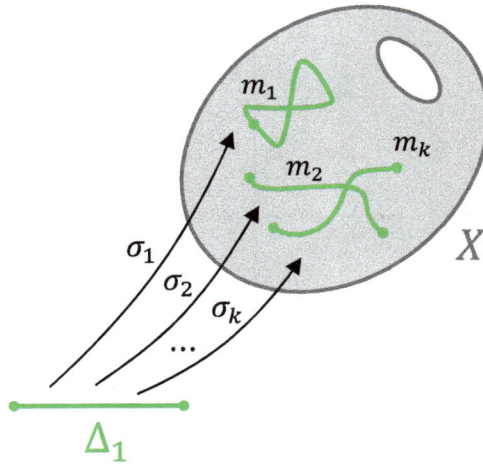

Figure 5.9.   A singular 1-chain.

Next, we describe 0-boundaries. To do that we provide an explicit formula for the boundary operator $\partial : C_1(X) \to C_0(X)$. By $\mathbb{Z}$-linearity it is enough to describe it on singular 1-simplexes. Take a singular 1-simplex $\sigma : \Delta_1 \to X$, and denote $x_0 = \sigma(0,1)$ and $x_1 = \sigma(1,0)$ its *extremal points*. We claim that $\partial\sigma = \sigma_{x_0} - \sigma_{x_1}$. To see this, compute

$$\partial\sigma = d_0^{\#}\sigma - d_1^{\#}\sigma = \sigma \circ d_0 - \sigma \circ d_1.$$

But the maps $\sigma \circ d_0, \sigma \circ d_1 : \Delta_0 \to X$ are given by

$$(\sigma \circ d_0)(1) = \sigma(0,1) = x_0 = \sigma_{x_0}(1) \quad \text{and}$$
$$(\sigma \circ d_1)(1) = \sigma(1,0) = x_1 = \sigma_{x_1}(1).$$

This shows that $\sigma \circ d_0 = \sigma_{x_0}$ and $\sigma \circ d_1 = \sigma_{x_1}$, hence

$$\partial\sigma = \sigma_{x_0} - \sigma_{x_1} \tag{5.1}$$

as claimed (see Figure 5.10). This is already a good motivation for the term "boundary" attributed to ∂ and its image.

With a description of $\partial : C_1(X) \to C_0(X)$ at hand, we can also discuss 1-cycles. For simplicity, we check first what does it mean for a singular 1-simplex to be in the kernel of ∂. So, let $\sigma : \Delta_1 \to X$ be a singular 1-simplex in the topological space X, and let $x_0 = \sigma(0,1)$ and $x_1 = \sigma(1,0)$ be its extremal points. We know that $\partial\sigma = \sigma_{x_0} - \sigma_{x_1}$ and this linear combination

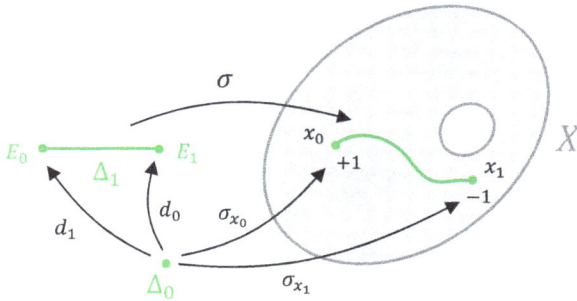

Figure 5.10.   The boundary $\partial\sigma$ of a singular 1-simplex.

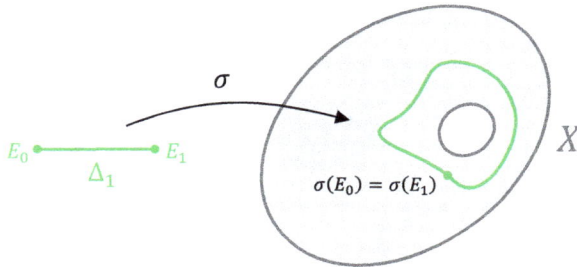

Figure 5.11.   A 1-cycle.

vanishes if and only if $\sigma_{x_1} = \sigma_{x_0}$, i.e., $x_1 = x_0$. In other words, σ identifies with a *closed path* in X, and this should motivate the term "cycle" attributed to a chain in the kernel of ∂ (Figure 5.11).

Similar remarks hold for a generic singular 1-chain. In Figure 5.12, we illustrate an example of a singular 1-cycle c: we have a specific linear combination

$$c = m_1\rho_1 + m\left(\rho_2 + \rho_3 + \rho_4\right)$$

of four singular 1-simplexes $\rho_1, \rho_2, \rho_3, \rho_4 : \Delta_1 \to X$ whose extremal points $(x_{0,i}, x_{1,i})$, $i = 1, \ldots, 4$ satisfy $x_{0,1} = x_{1,1}$, $x_{1,2} = x_{0,3}$, $x_{1,3} = x_{0,4}$ and $x_{1,4} = x_{0,2}$, $m_1, m \in \mathbb{Z}$ so that

$$\partial c = m_1\partial\rho_1 + m\left(\partial\rho_2 + \partial\rho_3 + \partial\rho_4\right)$$
$$= m_1\left(\sigma_{x_{0,1}} - \sigma_{x_{1,1}}\right) + m\left(\sigma_{x_{0,2}} - \sigma_{x_{1,2}} + \sigma_{x_{0,3}} - \sigma_{x_{1,3}} + \sigma_{x_{0,4}} - \sigma_{x_{1,4}}\right) = 0.$$

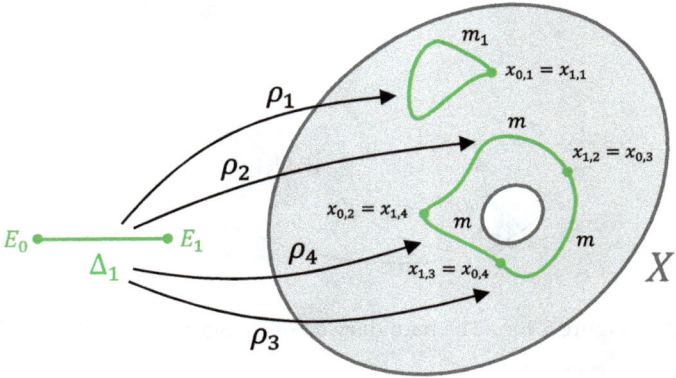

Figure 5.12.   A more complicated 1-cycle c (actually c is the sum of two 1-cycles, do you see it?).

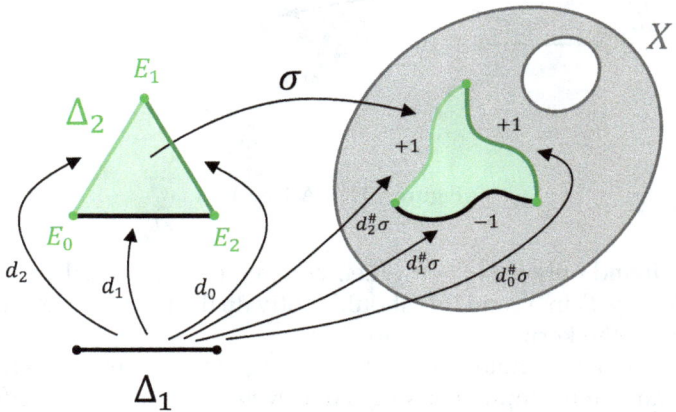

Figure 5.13.   The boundary of a singular 2-simplex.

Finally, in Figure 5.13, we represent the boundary operator acting on a singular 2-simplex. This should reinforce our motivation for the term "boundary".

Proposition 5.7 (Singular Homology of a Point). *The singular homology (with coefficient in $\mathbb{Z}$) of the 1-point space $\{*\}$ is given by*

$$H_n(\{*\}) = \begin{cases} \mathbb{Z} & \text{if } n = 0 \\ 0 & \text{otherwise} \end{cases}.$$

Proof. Note that, for every $n \geq 0$, there is only one singular n-simplex in $\{*\}$, namely the constant map

$$\sigma_n : \Delta_n \to \{*\}, \quad x \mapsto *.$$

So, for each n, the abelian group $C_n(\{*\})$ of singular n-chains possesses a one element basis, hence it is canonically isomorphic to $\mathbb{Z}$, where the isomorphism $C_n(\{*\}) \cong \mathbb{Z}$ is the only linear map $S_n(\{*\}) \to \mathbb{Z}$ mapping σ_n to 1. If we understand this isomorphisms, the chain complex $(C_\bullet(\{*\}), \partial)$ looks like

$$0 \longleftarrow \underset{0}{\mathbb{Z}} \overset{\partial}{\longleftarrow} \underset{1}{\mathbb{Z}} \overset{\partial}{\longleftarrow} \underset{2}{\mathbb{Z}} \overset{\partial}{\longleftarrow} \underset{3}{\mathbb{Z}} \longleftarrow \cdots .$$

We now describe the boundary operator on n-chains. We begin remarking that, for all $n > 0$,

$$d_i^\sharp \sigma_n = \sigma_n \circ d_i = \sigma_{n-1}.$$

As the generic n-chain is $m\sigma_n$ with $m \in \mathbb{Z}$, we get

$$\partial(m\sigma_n) = m\partial\sigma_n = m \sum_{i=0}^{n} (-)^i d_i^\sharp \sigma_n = m \sum_{i=0}^{n} (-)^i \sigma_{n-1}$$

$$= \begin{cases} 0 & \text{if } n \text{ is odd} \\ m\sigma_{n-1} & \text{if } n \text{ is even} \end{cases} .$$

If we understand the isomorphisms $S_n(\{*\}) \cong \mathbb{Z}$ again, then $(C_\bullet(\{*\}), \partial)$ now reads

$$0 \longleftarrow \underset{0}{\mathbb{Z}} \overset{0}{\longleftarrow} \underset{1}{\mathbb{Z}} \overset{\text{id}}{\longleftarrow} \underset{2}{\mathbb{Z}} \overset{0}{\longleftarrow} \underset{3}{\mathbb{Z}} \longleftarrow \cdots$$

and the claim immediately follows. Indeed,

$$H_0(\{*\}) = \frac{\ker(0 : \mathbb{Z} \to 0)}{\text{im}(0 : 0 \to \mathbb{Z})} = \frac{\mathbb{Z}}{0} = \mathbb{Z}.$$

When n is positive odd,

$$H_n(\{*\}) = \frac{\ker(0 : \mathbb{Z} \to \mathbb{Z})}{\text{im}(\text{id} : \mathbb{Z} \to \mathbb{Z})} = \frac{\mathbb{Z}}{\mathbb{Z}} = 0.$$

Finally, when n is positive even,

$$H_n(\{*\}) = \frac{\ker(\text{id} : \mathbb{Z} \to \mathbb{Z})}{\text{im}(0 : \mathbb{Z} \to \mathbb{Z})} = \frac{0}{0} = 0.$$

$\square$

We conclude this section by showing that the 0-th singular homology $H_0(X)$ of a topological space *counts the number of path connected components of X*.

Proposition 5.8 (0-th Singular Homology). *Let X be a topological space. Denote by π_0 the set of path connected components of X. The 0-th singular homology $H_0(X)$ of X is canonically isomorphic to the free module $\mathbb{Z}\pi_0$ spanned by π_0. In particular, X is path connected if and only if $H_0(X) \cong \mathbb{Z}$.*

Proof.   We can define a map

$$\varphi_0 : \pi_0 \to H_0(X)$$

as follows. We map the path connected component X_x of a point $x \in X$ to the homology class of the singular 0-chain σ_x. We have to show that this map is well defined, namely that if $x' \in X$ is another point in the same path connected component X_x then $\sigma_{x'}$ is homologous to σ_x so that they have the same homology class. But $x' \in X_x$ if and only if x, x' are connected by a path $\gamma : [0,1] \to X$. Consider the singular 1-simplex $\sigma = \gamma \circ h^{-1} : \Delta_1 \to X$ (in particular, it is a singular 1-chain). Then $\sigma(1,0) = \gamma(h^{-1}(1,0)) = \gamma(1) = x'$. Similarly, $\sigma(0,1) = x$ so that

$$\partial \sigma = \sigma_x - \sigma_{x'},$$

i.e., $\sigma_x, \sigma_{x'}$ are homologous as claimed. We conclude that the map $\varphi_0 : \pi_0 \to H_0(X)$ is well defined.

From the universal property of free modules, there is a unique $\mathbb{Z}$-linear map

$$\varphi : \mathbb{Z}\pi_0 \to H_0(X)$$

such that $\varphi|_{\pi_0} = \varphi_0$. The map φ is the isomorphism we are looking for. In order to show that it is bijective, we construct its inverse

$$\psi : H_0(X) \to \mathbb{Z}\pi_0$$

explicitly. We begin defining a linear map

$$\Psi : Z_0(X) = C_0(X) \to \mathbb{Z}\pi_0.$$

To do this, it is enough to define a map

$$\Psi_0 : S_0(X) \to \mathbb{Z}\pi_0$$

and then use the universal property of free modules. So, let σ_x be the singular 0-simplex corresponding to the point $x \in X$. It is natural to put $\Psi_0(\sigma_x) = X_x \in \pi_0 \subseteq \mathbb{Z}\pi_0$ the path connected component of x. Next, we prove that Ψ annihilates the 0-boundaries, i.e., $\Psi(\partial b) = 0$ for all $b \in C_1(X)$. Actually, by linearity, it is enough to show that $\Psi(\partial\sigma) = 0$ for all singular 1-simplexes σ (do you agree?). So, let $\sigma \in S_1(X)$, and denote $y_0 = \sigma(0,1)$, $y_1 = \sigma(1,0)$. Notice that y_0, y_1 are connected by the path $\sigma \circ h$ (do you see it? If not check it explicitly) and we find

$$\Psi(\partial\sigma) = \Psi(\sigma_{y_0} - \sigma_{y_1}) = \Psi(\sigma_{y_0}) - \Psi(\sigma_{y_1})$$
$$= \Psi_0(\sigma_{y_0}) - \Psi_0(\sigma_{y_1}) = X_{y_0} - X_{y_1} = 0$$

where, in the last step, we used that y_0, y_1 belong to the same path connected component. Summarizing 0-boundaries belong to the kernel of Ψ, hence Ψ descends to a well-defined linear map

$$\psi : \frac{Z_0(X)}{B_0(X)} = H_0(X) \to \mathbb{Z}\pi_0, \quad [c] \mapsto \psi(c).$$

We leave it to the reader to check that ψ inverts φ as Exercise 5.2. This concludes the proof. $\qquad \square$

Exercise 5.2. Complete the proof of Proposition 5.8 showing that the linear maps $\varphi : \mathbb{Z}\pi_0 \to H_0(X)$ and $\psi : H_0(X) \to \mathbb{Z}\pi_0$ described in the proof invert each other (**Hint:** *It is enough to work on generators*).

5.2 Geometric Homotopies

Homotopies were first defined in Topology and only later they were defined for (co)chain complexes. In this section, we show how a homotopy between continuous maps gives rise to an algebraic homotopy between singular chains with integer coefficients (the cases of singular cochains and of arbitrary coefficients are similar and we leave them to the reader). We begin showing that a continuous map $F : X \to Y$ between topological spaces determines a chain map $F_\sharp : C_\bullet(X) \to C_\bullet(Y)$ between the associated singular chains in a functorial way. First of all, from F and a singular n-simplex $\sigma : \Delta_n \to X$ in X we can construct a singular n-simplex in Y by left composition with F:

$$F_\sharp\sigma := F \circ \sigma : \Delta_n \to Y.$$

As both σ and F are continuous, $F \circ \sigma$ is continuous as well, hence it is a singular chain. In this way, we get a map

$$F_\sharp : S_n(X) \to S_n(Y), \quad \sigma \mapsto F_\sharp\sigma.$$

We claim that the family $(F_\sharp : S_n(X) \to S_n(Y))_{n \in \mathbb{N}_0}$ is a semi-simplicial map. To see this, we have to prove that $F_\sharp \circ d_i^\sharp = d_i^\sharp \circ F_\sharp$ for all i. So, take a singular n-simplex $\sigma \in S_n(X)$ and compute

$$F_\sharp \circ d_i^\sharp(\sigma) = F_\sharp(d_i^\sharp \sigma) = F \circ (\sigma \circ d_i) = (F \circ \sigma) \circ d_i$$

$$= F_\sharp(\sigma) \circ d_i = d_i^\sharp F_\sharp(\sigma) = d_i^\sharp \circ F_\sharp(\sigma).$$

The assignment $S_\bullet : \mathbf{Top} \to \mathbf{ssSet}$ mapping a topological space X to the semi-simplicial set $(S_\bullet(X), d^\sharp)$ and a continuous map of topological spaces $F : X \to Y$ to the semi-simplicial map $F_\sharp : (S_\bullet(X), d^\sharp) \to (S_\bullet(Y), d^\sharp)$ is actually a functor. This easily follows from the semi-cosimplicial identities for the d_i. We leave the details to the reader as the following

---

Exercise 5.3. Prove that the assignment $S_\bullet : \mathbf{Top} \to \mathbf{ssSet}$ defined above is a functor.

Now consider the sequence of functors

$$\mathbf{Top} \xrightarrow{\;S_\bullet\;} \mathbf{ssSet} \xrightarrow{\;\mathrm{ssFree}\;} \mathbf{ssAb} \xrightarrow{\;\mathrm{Thm.\ 4.7}\;} \mathbf{Ch}_\mathbb{Z} \; .$$

Their composition

$$C_\bullet : \mathbf{Top} \to \mathbf{Ch}_\mathbb{Z}$$

is again a functor mapping a topological space X to its complex $(C_\bullet(X), \partial)$ of singular chains and a continuous map between topological spaces $F : X \to Y$ to the chain map

$$F_\sharp : (C_\bullet(X), \partial) \to (C_\bullet(Y), \partial),$$

defined as follows. Take $c \in C_n(X) = \mathbb{Z}S_n(X)$. Then c is a formal linear combination of singular n-simplexes with integer coefficients:

$$c = \sum_{j=1}^k m_j \sigma_j,$$

$m_j \in \mathbb{Z}, \sigma_j \in S_n(X)$, for all $j = 1, \ldots, k$, and

$$F_\sharp(c) = \sum_{j=1}^k m_j F_\sharp(\sigma_j) = \sum_{j=1}^k m_j (F \circ \sigma_j).$$

Composing further the functor $C_\bullet : \mathbf{Top} \to \mathbf{Ch}_\mathbb{Z}$ with the n-th homology functor

$$H_n : \mathbf{Ch}_\mathbb{Z} \to \mathbf{Ab}$$

we get a new functor also denoted

$$H_n : \mathbf{Top} \to \mathbf{Ab}$$

and called the *n-th singular homology functor*. Given a continuous map $F : X \to Y$, the linear map $H_n(F) : H_n(X) \to H_n(Y)$, $[c] \mapsto [F_\# c]$ associated with it via the functor H_n is also called the *map induced by F in the n-th singular homology*. It immediately follows from the functorial properties of the singular *n*-th homology that *homeomorphic topological spaces have isomorphic singular homologies*.

Example 5.9 (Map Induced in Singular Homology by a Constant Map).
Let X, Y be topological spaces (with $Y \neq \varnothing$). Take a point $y_0 \in Y$ and consider the constant map $c_{y_0} : X \to Y$ mapping every point $x \in X$ to y_0. We want to compute the induced map in singular homology $H_n(c_{y_0}) :$ $H_n(X) \to H_n(Y)$ for all $n \in \mathbb{Z}$. We use a trick. We consider the one point topological space $\{y_0\}$ and interpret c_{y_0} as the composition

$$X \xrightarrow{c} \{y_0\} \xrightarrow{in} Y$$

of the only (necessarily constant) map $c : X \to \{y_0\}$ and the inclusion in : $\{y_0\} \to Y$. Both c and in are continuous hence, from the functorial properties of the *n*-th singular homology,

$$H_n(c_{y_0}) = H_n(c) \circ H_n(in).$$

But, from Proposition 5.7, $H_n(\{y_0\}) = 0$ for all $n \neq 0$. It follows that $H_n(c) : H_n(X) \to H_n(\{y_0\})$ and $H_n(in) : H_n(\{y_0\}) \to H_n(Y)$ are both the zero maps for all $n \neq 0$ so that $H_n(c_{y_0}) = 0$ for all $n \neq 0$. It remains to compute

$$H_0(c_{y_0}) : H_0(X) \to H_0(Y).$$

Denote $\pi_0(X), \pi_0(Y)$ the sets of path connected components of X, Y respectively, denote also by Y_0 the path connected component of y_0 in Y. From Proposition 5.8, we have $H_0(X) \cong \mathbb{Z}\pi_0(X)$ and $H_0(Y) \cong \mathbb{Z}\pi_0(Y)$. We claim that the map $H_0(c_{y_0})$ is the unique abelian group homomorphism mapping any path connected component of X to Y_0. In other words,

$$H_0(c_{y_0}) \left(\sum_{j=1}^{k} m_j X_{x_j} \right) = \sum_{j=1}^{k} m_j H_0(c_{y_0})(X_{x_j}) = \left(\sum_{j=1}^{k} m_j \right) Y_0,$$

for any k points $x_1, \ldots, x_k \in X$, and any k integers $m_1, \ldots, m_k \in \mathbb{Z}$. To see this, remember that the isomorphism $H_0(X) \cong \mathbb{Z}\pi_0(X)$ identifies the homology class of the constant 0-cycle σ_x with the path connected component X_x of x, for all $x \in X$. Now,

$$H_0(c_{y_0})([\sigma_x]) = [c_{y_0} \circ \sigma_x] = [\sigma_{y_0}],$$

which identifies with Y_0 (under the isomorphism $H_0(Y) \cong \mathbb{Z}\pi_0(Y)$). This concludes the proof. $\qquad\qquad\qquad\qquad\qquad\qquad\qquad\qquad\qquad\quad$ ◆

We now come to (geometric) homotopies. Let X, Y be topological spaces and let $F, G : X \to Y$ be continuous maps.

Definition 5.10 (Geometric Homotopy). A *homotopy* (more precisely a *geometric homotopy*) between the continuous maps $F, G : X \to Y$ is a continuous map $\mathcal{H} : [0, 1] \times X \to Y$ such that

$$\mathcal{H}(0, x) = F(x) \quad \text{and} \quad \mathcal{H}(1, x) = G(x)$$

for all $x \in X$. Two continuous maps F, G are said to be *homotopic* if there exists a homotopy $\mathcal{H}$ between them. In this case, we write $F \sim_{\mathcal{H}} G$. A continuous map F is *null-homotopic* if it is homotopic to a constant map.

Given a homotopy $\mathcal{H} : [0, 1] \times X \to Y$ and a point $t \in [0, 1]$, we usually denote by $\mathcal{H}_t : X \to Y$ the map given by $\mathcal{H}_t(x) := \mathcal{H}(t, x)$ for all $x \in X$. It is a continuous map, indeed it is the composition of the map $\text{in}_t : X \to [0, 1] \times X$, $x \mapsto \text{in}_t(x) := (t, x)$ (which is continuous because it has continuous components, do you see it?) followed by $\mathcal{H}$. Note that the homotopy $\mathcal{H}$ can be reconstructed from the family $(\mathcal{H}_t)_{t \in [0,1]}$ reading the definition $\mathcal{H}_t(x) = \mathcal{H}(t, x)$ from the right to the left. In terms of $(\mathcal{H}_t)_{t \in [0,1]}$, the condition $F \sim_{\mathcal{H}} G$ reads $F = \mathcal{H}_0$ and $G = \mathcal{H}_1$. In other words, a homotopy between continuous maps $F, G : X \to Y$ can be seen as a continuous deformation of F into G (along the family $(\mathcal{H}_t)_{t \in [0,1]}$, see Figure 5.14).

Remark 5.11. A homotopy between two continuous maps $F, G : X \to Y$ of topological spaces can also be seen as a path connecting F and G in the *space of continuous maps* $X \to Y$. $\qquad\qquad\qquad\qquad\qquad\qquad\qquad$ ◇

Proposition 5.12. *"Being homotopic" is an equivalence relation on the set of continuous maps (between given topological spaces). More precisely, if $F, G, L : X \to Y$ are continuous maps such that $F \sim_{\mathcal{H}} G$ and $G \sim_{\mathcal{K}} L$ for some homotopies $\mathcal{H}, \mathcal{K}$, then there are homotopies $\mathcal{O}, \overline{\mathcal{H}}, \mathcal{H} * \mathcal{K}$ such that*

- $F \sim_{\mathcal{O}} F$ *(reflexivity)*,
- $G \sim_{\overline{\mathcal{H}}} F$ *(symmetry)*,
- $F \sim_{\mathcal{H} * \mathcal{K}} L$ *(transitivity)*.

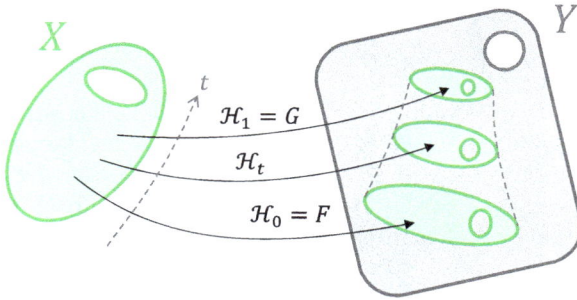

Figure 5.14. A homotopy $\mathcal{H}$ between the continuous maps $F, G : X \to Y$.

Proof. For the reflexivity, denote by $\mathcal{O} : [0,1] \times X \to X$ the *"constant homotopy"* defined by $\mathcal{O}(t,x) = F(x)$ for all $(t,x) \in [0,1] \times X$. It is clear that $\mathcal{O}$ is a continuous map, hence a homotopy, and that $F \sim_\mathcal{O} F$ (do you see it?).

For the symmetry, given a homotopy $\mathcal{H} : [0,1] \times X \to Y$ between F and G, we define a new homotopy $\overline{\mathcal{H}} : [0,1] \times X \to Y$ between G and F by putting $\overline{\mathcal{H}}(t,x) := \mathcal{H}(1-t,x)$. We leave it to the reader to check that $\overline{\mathcal{H}}$ is a continuous map, hence a homotopy (see Exercise 5.4). The rest is obvious.

For the transitivity, define a homotopy $\mathcal{H} * \mathcal{K} : [0,1] \times X \to Y$ between F and L by putting

$$\mathcal{H} * \mathcal{K}(t,x) := \begin{cases} \mathcal{H}(2t,x) & \text{if } t \leq 1/2 \\ \mathcal{K}(2t-1,x) & \text{if } t > 1/2 \end{cases}.$$

Show that $\mathcal{H} * \mathcal{K}$ is a continuous map, hence a homotopy, as part of Exercise 5.4. The rest is clear. This concludes the proof. $\qquad\square$

Exercise 5.4. Fill the gaps in the proof of Proposition 5.12 proving that $\overline{\mathcal{H}}$ and $\mathcal{H} * \mathcal{K}$ are continuous maps, hence homotopies (**Hint:** *Get inspired by Remark* 5.6).

We now present some examples, including a trivial but somewhat *universal* example. More examples are given toward the end of this section, after showing how does geometric homotopies help computing singular (co)homologies of topological spaces.

Example 5.13 (The Tautological Homotopy). Let X be a topological space. Consider the maps $\mathrm{in}_0 : X \to [0,1] \times X, x \mapsto (0,x)$ and $\mathrm{in}_1 : X \to [0,1] \times X, x \mapsto (1,x)$. As already remarked they are continuous injections. There is an obvious homotopy between in_0, in_1, namely

$$\mathcal{H}_{\text{can}} := \text{id}_{[0,1] \times X} : [0,1] \times X \to [0,1] \times X, \quad (t, x) \mapsto (t, x).$$

It is also clear that $(\mathcal{H}_{\text{can}})_t = \text{in}_t$ for all $t \in [0,1]$. The homotopy $\mathcal{H}_{\text{can}}$ might well be called the *tautological homotopy* and it is *universal* in the sense that every homotopy $F : [0,1] \times X \to Y$ can be seen as the composition of $\mathcal{H}_{\text{can}}$ followed by a continuous map, namely F itself. This might seem tricky and trivial but it has interesting consequences (see, e.g., the proof of Theorem 5.16). ◆

Example 5.14. Let $F, G : X \to \mathbb{R}^d$ be continuous maps from an arbitrary topological space X to the standard d-dimensional Euclidean space. Then F, G are automatically homotopic. Indeed, we can easily define a homotopy $\mathcal{H} : [0,1] \times X \to \mathbb{R}^d$ between them by putting

$$\mathcal{H}(t, x) = tG(x) + (1 - t)F(x)$$

(do you see that $\mathcal{H}$ is a continuous map? If not, prove it in details). In particular, any $\mathbb{R}^d$-valued continuous map is null-homotopic. More generally, recall that a subset $Y \subseteq \mathbb{R}^d$ is said to be *convex* if, for every $x_0, x_1 \in Y$, the segment

$$\overline{x_0 x_1} := \{ t x_1 + (1 - t) x_0 : t \in [0,1] \} \subseteq \mathbb{R}^d$$

is entirely contained into Y: $\overline{x_0 x_1} \subseteq Y$. It is clear that any two continuous maps $F, G : X \to Y \subseteq \mathbb{R}^d$ with values in a convex subspace Y of $\mathbb{R}^d$ are homotopic (just define a homotopy as above). So, every continuous map with values in a convex subspace of $\mathbb{R}^d$ is null-homotopic. ◆

Proposition 5.15. *Homotopies respect the composition of continuous maps. More precisely if*

$$X \underset{G}{\overset{F}{\rightrightarrows}} Y \underset{G'}{\overset{F'}{\rightrightarrows}} Z$$

are continuous maps such that $F \sim_{\mathcal{H}} G$ and $F' \sim_{\mathcal{H}'} G'$ for some homotopies $\mathcal{H}, \mathcal{H}'$, then there exists a homotopy $\mathbb{H}$ (to be specified in the proof) such that $F' \circ F \sim_{\mathbb{H}} G' \circ G$.

Proof. We define a homotopy $\mathbb{H} : [0,1] \times X \to Z$ between $F' \circ F$ and $G' \circ G$ by putting

$$\mathbb{H}(t, x) := \mathcal{H}'(t, \mathcal{H}(t, x)).$$

Clearly,

$$\mathbb{H}(0, x) := \mathcal{H}'(0, \mathcal{H}(0, x)) = \mathcal{H}'(0, F(x)) = F'(F(x)),$$

and similarly,

$$\mathbb{H}(1, x) := \mathcal{H}'(1, \mathcal{H}(1, x)) = \mathcal{H}'(1, G(x)) = G'(G(x)),$$

for all $(t, x) \in [0, 1] \times X$. It remains to check that $\mathbb{H}$ is a continuous map. But $\mathbb{H}$ is the composition of the map $[0, 1] \times X \to [0, 1] \times Y$, $(t, x) \mapsto (t, \mathcal{H}(t, x))$, which is continuous because it has continuous components, followed by $\mathcal{H}'$, which is continuous by hypothesis. Hence, $\mathbb{H}$ is continuous as well. □

Theorem 5.16. *Let $F, G : X \to Y$ be homotopic continuous maps between topological spaces. Then F, G induce the same map in singular homology:*

$$H_n(F) = H_n(G), \quad \text{for all } n \in \mathbb{Z}.$$

Proof. The tautological homotopy $\mathcal{H}_{\text{can}} : [0, 1] \times X \to [0, 1] \times X$ allows us to consider the case $Y = [0, 1] \times X$, $F = \text{in}_0$ and $G = \text{in}_1$ only. Indeed, suppose preliminarily that in_0 and in_1 induce the same map in singular homology:

$$H_n(\text{in}_0) = H_n(\text{in}_1), \quad \text{for all } n \in \mathbb{Z}.$$

Then, if $\mathcal{H}$ is a homotopy between F and G, we have $\mathcal{H}_t = \mathcal{H} \circ \text{in}_t$ for all $t \in [0, 1]$. Hence,

$$H_n(F) = H_n(\mathcal{H}_0) = H_n(\mathcal{H} \circ \text{in}_0) = H_n(\mathcal{H}) \circ H_n(\text{in}_0)$$
$$= H_n(\mathcal{H}) \circ H_n(\text{in}_1) = H_n(\mathcal{H} \circ \text{in}_1) = H_n(\mathcal{H}_1) = H_n(G).$$

It remains to check that $H_n(\text{in}_0) = H_n(\text{in}_1)$ for all n. To do this, we explicitly construct an algebraic homotopy

$$h = (h_n : C_n(X) \to C_{n+1}([0, 1] \times X))_{n \in \mathbb{Z}}$$

between the chain maps $(\text{in}_0)_\sharp, (\text{in}_1)_\sharp : (C_\bullet(X), \partial) \to (C_\bullet([0, 1] \times X), \partial)$.
For every $n \in \mathbb{N}_0$ and every $i = 0, \ldots, n$, consider the map

$$P_i^n : \Delta_{n+1} \to [0, 1] \times \Delta_n,$$

defined by

$$P_i^n(x_0, \ldots, x_{n+1}) := \left(1 - \sum_{j=0}^{i} x_j, (x_0, \ldots, x_{i-1}, x_i + x_{i+1}, x_{i+2}, \ldots, x_{n+1})\right).$$

The family of maps $P = (P_i^n)_{0 \le i \le n \in \mathbb{N}_0}$ is sometimes called the *prism map*. The reason is illustrated in Figure 5.15. We often denote P_i^n simply by P_i if it is clear which simplex it acts on.

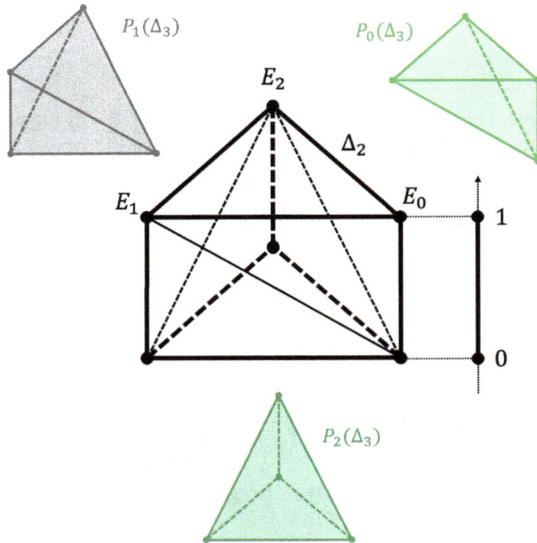

Figure 5.15.   The prism maps $P_i^n : \Delta_{n+1} \to [0,1] \times \Delta_n$ embed the $(n+1)$-simplex into the $(n+1)$-dimensional prism $[0,1] \times \Delta_n$ in $n+1$ different ways. Here we depict the case $n = 2$. In this case, P_0 is the only affine map such that $P_0(E_0) = (0, E_0)$, $P_0(E_1) = (1, E_0)$, $P_0(E_2) = (1, E_1)$, and $P_0(E_3) = (1, E_2)$. Similarly for higher i.

Inspired by the prism map, we define maps

$$P_i^\sharp : S_n(X) \to S_{n+1}([0,1] \times X)$$

by putting

$$P_i^\sharp(\sigma)(x_0, \ldots, x_{n+1})$$

$$:= \left(1 - \sum_{j=0}^i x_j, \sigma\left(x_0, \ldots, x_{i-1}, x_i + x_{i+1}, x_{i+2}, \ldots, x_{n+1}\right)\right)$$

for all singular n-simplexes $\sigma : \Delta_n \to X$. The map $P_i^\sharp(\sigma) : \Delta_{n+1} \to [0,1] \times X$ defined in this way is continuous (do you see it?). Hence, it is a singular $(n+1)$-simplex in $[0,1] \times X$ as desired. The maps $P_i^\sharp$ interact with the face maps $d_j^\sharp$ on $S_\bullet(X)$ and $S_\bullet([0,1] \times X)$ as follows:

$$d_j^\sharp \circ P_i^\sharp = \begin{cases} P_i^\sharp \circ d_{j-1}^\sharp & \text{if } 0 \le i < j-1 \le n \\ d_j^\sharp \circ P_j^\sharp & \text{if } 0 \le i = j-1 < n. \\ P_{i-1}^\sharp \circ d_j^\sharp & \text{if } 0 \le j < i \le n \end{cases} \qquad (5.2)$$

We leave it to the reader to check the *Prism Identities* (5.2) as Exercise 5.5. Now, for each i, the map $P_i^\sharp$ can be uniquely extended to a linear map, also denoted

$$P_i^\sharp : C_n(X) = \mathbb{Z}S_n(X) \to C_{n+1}([0,1] \times X) = \mathbb{Z}S_{n+1}([0,1] \times X).$$

Define

$$h := \sum_{i=0}^{n}(-)^i P_i^\sharp : C_n(X) \to C_{n+1}([0,1] \times X), \quad n \in \mathbb{Z}.$$

Finally, we use the Prism Identities (5.2) to check that h is the desired algebraic homotopy:

$$\partial \circ h = \sum_{j=0}^{n+1}(-)^j d_j^\sharp \circ \sum_{i=0}^{n}(-)^i P_i^\sharp$$

$$= \sum_{j=0}^{n+1}\sum_{i=0}^{n}(-)^{i+j}d_j^\sharp \circ P_i^\sharp$$

$$= \sum_{j=2}^{n+1}\sum_{i=0}^{j-2}(-)^{i+j}d_j^\sharp \circ P_i^\sharp - \sum_{j=1}^{n+1}d_j^\sharp \circ P_{j-1}^\sharp$$

$$\quad + \sum_{j=0}^{n}d_j^\sharp \circ P_j^\sharp + \sum_{j=0}^{n-1}\sum_{i=j+1}^{n}(-)^{i+j}d_j^\sharp \circ P_i^\sharp$$

$$= \sum_{j=2}^{n+1}\sum_{i=0}^{j-2}(-)^{i+j}P_i^\sharp \circ d_{j-1}^\sharp - d_{n+1}^\sharp \circ P_n^\sharp - \sum_{j=1}^{n}d_j^\sharp \circ P_{j-1}^\sharp + \sum_{j=0}^{n}d_j^\sharp \circ P_j^\sharp$$

$$\quad + \sum_{j=0}^{n-1}\sum_{i=j+1}^{n}(-)^{i+j}P_{i-1}^\sharp \circ d_j^\sharp$$

$$= -\sum_{j=1}^{n}\sum_{i=0}^{j-1}(-)^{i+j}P_i^\sharp \circ d_j^\sharp - d_{n+1}^\sharp \circ P_n^\sharp - \sum_{j=1}^{n}d_j^\sharp \circ P_j^\sharp + \sum_{j=0}^{n}d_j^\sharp \circ P_j^\sharp$$

$$\quad - \sum_{j=0}^{n-1}\sum_{i=j}^{n-1}(-)^{i+j}P_i^\sharp \circ d_j^\sharp$$

$$= d_0^\sharp \circ P_0^\sharp - d_{n+1}^\sharp \circ P_n^\sharp - \sum_{j=0}^{n}\sum_{i=0}^{n-1}(-)^{i+j}P_i^\sharp \circ d_j^\sharp$$

$$= d_0^\sharp \circ P_0^\sharp - d_{n+1}^\sharp \circ P_n^\sharp - \sum_{i=0}^{n-1}(-)^i P_i^\sharp \circ \sum_{j=0}^{n}(-)^j \circ d_j^\sharp$$

$$= d_0^\sharp \circ P_0^\sharp - d_{n+1}^\sharp \circ P_n^\sharp - h \circ \partial.$$

But, for any $\sigma \in S_n(X)$ and any $(x_0, \ldots, x_n) \in \Delta_n$

$$\left((d_0^\sharp \circ P_0^\sharp)\sigma \right)(x_0, \ldots, x_n)$$

$$= P_0^\sharp \sigma\left(d_0(x_0, \ldots, x_n)\right) = P_0^\sharp \sigma(0, x_0, \ldots, x_n)$$

$$= (1, \sigma(x_0, \ldots, x_n)) = \text{in}_1 \circ \sigma(x_0, \ldots, x_n) = (\text{in}_1)_\sharp \sigma(x_0, \ldots, x_n),$$

and

$$\left((d_{n+1}^\sharp \circ P_n^\sharp)\sigma \right)(x_0, \ldots, x_n)$$

$$= P_n^\sharp \sigma\left(d_{n+1}(x_0, \ldots, x_n)\right) = P_n^\sharp \sigma(x_0, \ldots, x_n, 0)$$

$$= (1 - x_0 - \cdots - x_n, \sigma(x_0, \ldots, x_n)) = (0, \sigma(x_0, \ldots, x_n))$$

$$= \text{in}_0 \circ \sigma(x_0, \ldots, x_n) = (\text{in}_0)_\sharp \sigma(x_0, \ldots, x_n),$$

so that

$$d_0^\sharp \circ P_0^\sharp - d_{n+1}^\sharp \circ P_n^\sharp = (\text{in}_1)_\sharp - (\text{in}_0)_\sharp.$$

We conclude that

$$\partial \circ h = (\text{in}_1)_\sharp - (\text{in}_0)_\sharp - h \circ \partial,$$

as desired. $\qquad\qquad\qquad\qquad\qquad\qquad\qquad\qquad\qquad\qquad\qquad \square$

Exercise 5.5. Prove the *Prism Identities* (5.2).

Exercise 5.6. Let $F, G : X \to Y$ be continuous maps between topological spaces, let $\mathcal{H} : [0,1] \times X \to Y$ be a homotopy such that $F \sim_\mathcal{H} G$, and let

$$h = (h : C_n(X) \to C_{n+1}([0,1] \times X))_{n \in \mathbb{Z}}$$

be the algebraic homotopy constructed in the proof of Theorem 5.16. Show that

$$h_\mathcal{H} := (\mathcal{H}_\sharp \circ h : C_n(X) \to C_{n+1}(Y))_{n \in \mathbb{Z}}$$

is an algebraic homotopy between the chain maps $F_\sharp, G_\sharp : (C_\bullet(X), \partial) \to (C_\bullet(Y), \partial)$.

Corollary 5.17. *If* $F : X \rightarrow Y$ *is a null-homotopic continuous map, then* $H_n(F) = 0$ *for all* $n \neq 0$, *and*

$$H_0(F) : H_0(X) \cong \mathbb{Z}\pi_0(X) \rightarrow H_0(Y) \cong \mathbb{Z}\pi_0(Y)$$

maps every path connected component of X to a single path connected component of Y.

Example 5.18. It immediately follows from Corollary 5.17 and Example 5.14 that, for any continuous map $F : X \rightarrow Y \subseteq \mathbb{R}^d$ with values in a convex subspace Y of $\mathbb{R}^d$ we have $H_n(F) = 0$ for all $n \neq 0$. ◆

We now discuss a new notion that formalizes the idea of *continuously deforming a topological space X into another topological space X'.*

Definition 5.19 (Homotopy Equivalence of Topological Spaces). A continuous map $F : X \rightarrow X'$ between topological spaces is a *homotopy equivalence* if there exists a continuous map in the other direction $G : X' \rightarrow X$ such that $G \circ F$ is homotopic to the identity of X and $F \circ G$ is homotopic to the identity of X':

$$G \circ F \sim_{\mathcal{H}} \mathrm{id}_X \quad \text{and} \quad F \circ G \sim_{\mathcal{H}'} \mathrm{id}_{X'},$$

for some homotopies $\mathcal{H}, \mathcal{H}'$. In this situation, G is clearly a homotopy equivalence as well. We also say that G is a *homotopy inverse* of F (and viceversa) or that G *inverts F up to homotopy*. If X, X' are topological spaces connected by a homotopy equivalence, we say that they are *homotopy equivalent*.

Proposition 5.20. *Let* $F : X \rightarrow X'$ *be a homotopy equivalence between the topological spaces* X, X', *and let* $G : X' \rightarrow X$ *be a homotopy inverse of F. Then F, G induce mutually inverse abelian group isomorphisms in singular homology, i.e.,* $H_n(F) : H_n(X) \rightarrow H_n(X')$ *and* $H_n(G) : H_n(X') \rightarrow H_n(X)$ *are abelian group isomorphisms and*

$$H_n(F)^{-1} = H_n(G) \quad \text{for all } n \in \mathbb{Z}.$$

In particular, homotopy equivalent topological spaces have isomorphic singular homologies.

Proof. We have

$$H_n(F) \circ H_n(G) = H_n(F \circ G) = H_n(\mathrm{id}_{X'}) = \mathrm{id}_{H_n(X')}$$

for all n, where, in the first and the last step, we used the functorial properties of singular homology and, in the second step, we used Theorem 5.16. Swapping the roles of F and G we get $H_n(G) \circ H_n(F) = \mathrm{id}_{H_n(X)}$. This concludes the proof. □

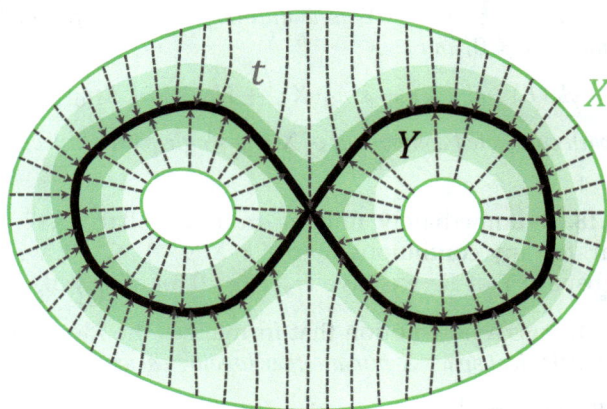

Figure 5.16. A deformation retract with a deformation retraction.

Sometimes a topological space X is homotopy equivalent to a subspace $Y \subseteq X$.

Definition 5.21 (Deformation Retract). A subspace $Y \subseteq X$ in a topological space X is a *deformation retract* of X if there exists a continuous map $r : X \to Y$, called a *deformation retraction*, inverting the inclusion $i_Y : Y \to X$ on the left and such that $i_Y \circ r : X \to X$ is homotopic to the identity of X (Figure 5.16). A topological space X is *contractible* if there exists a point $x_0 \in X$ such that the one point subspace $\{x_0\} \subseteq X$ is a deformation retract of X.

Proposition 5.22. *Let X be a topological space and let $Y \subseteq X$ be a deformation retract of X. Then the inclusion $i_Y : Y \to X$ induces an isomorphism in singular homology: $H_n(i_Y) : H_n(Y) \to H_n(X)$ is an isomorphism for all $n \in \mathbb{Z}$. Hence, if X is a contractible space, then*

$$H_n(X) \cong \begin{cases} \mathbb{Z} & \text{if } n = 0 \\ 0 & \text{otherwise} \end{cases}.$$

Proof. Let r be a deformation retraction for $Y \subseteq X$. Then, $r \circ i_Y = \mathrm{id}_Y$ is clearly homotopic to the identity of Y itself. At the same time, $i_Y \circ r$ is homotopic to the identity of X by definition of deformation retraction. This shows that i_Y and r are mutually homotopy inverse homotopy equivalences between X and Y. So they both induce an isomorphism in singular homology by Proposition 5.20. The last part of the statement now follows from Proposition 5.7. $\square$

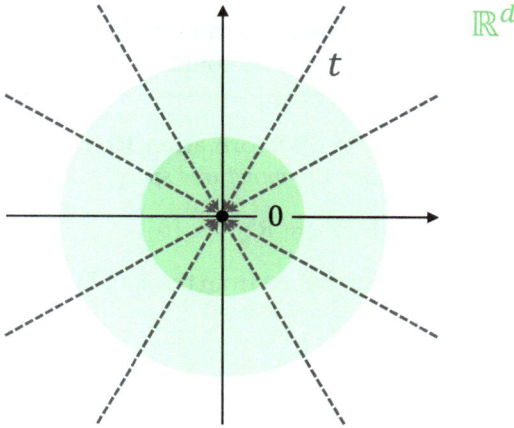

Figure 5.17. The standard Euclidean space $\mathbb{R}^d$ is contractible.

We finally come to concrete examples.

Example 5.23 ($\mathbb{R}^d$ is Contractible). The standard Euclidean space $\mathbb{R}^d$ is contractible for every d. More precisely the constant map

$$r : \mathbb{R}^d \to \{0\}, \quad x \mapsto 0$$

is a deformation retraction. Indeed, the continuous map

$$\mathcal{H} : [0,1] \times \mathbb{R}^d \to \mathbb{R}^d, \quad (t, x) \mapsto \mathcal{H}(t, x) := tx$$

is clearly a homotopy between $i_{\{0\}} \circ r$ and $\mathrm{id}_{\mathbb{R}^d}$ (do you see it? See also Figure 5.17). Restricting $\mathcal{H}$ to the *standard n-dimensional disk*

$$D^d := \left\{ x \in \mathbb{R}^d : \|x\| \le 1 \right\} \subseteq \mathbb{R}^d,$$

we see that D^d is contractible as well. We conclude that

$$H_n(\mathbb{R}^d) = H_n(D^d) = \begin{cases} \mathbb{Z} & \text{if } n = 0 \\ 0 & \text{otherwise} \end{cases}. \tag{5.3}$$

$\blacklozenge$

Exercise 5.7. Recall that a subset $X \subseteq \mathbb{R}^d$ is said to be *star-shaped* if there exists a point $x_0 \in X$ such that, for every other point $x \in X$, the segment

$$\overline{x_0 x} := \left\{ tx_0 + (1-t)x \in \mathbb{R}^d : t \in [0,1] \right\} \subseteq \mathbb{R}^d$$

> is entirely contained into X. Show that any star-shaped subspace of $\mathbb{R}^d$, in particular any convex subspace, is contractible.

Example 5.24 (The Sphere is a Deformation Retract of the Punctured Space). For all $d \geq 0$, consider the *punctured space*

$$\mathbb{R}^{d+1}_\times := \mathbb{R}^{d+1} \smallsetminus \{0\},$$

(with the subspace topology induced from the standard topology of $\mathbb{R}^{d+1}$).
  The d-dimensional sphere

$$S^d := \{x \in \mathbb{R}^{d+1} : \|x\| = 1\} \subseteq \mathbb{R}^{d+1}_\times$$

is a deformation retract of $\mathbb{R}^{d+1}_\times$, with deformation retraction $r : \mathbb{R}^{d+1}_\times \to S^d$ given by

$$r(x) := \frac{x}{\|x\|}.$$

Indeed, the continuous map

$$\mathcal{H} : [0,1] \times \mathbb{R}^{d+1}_\times \to \mathbb{R}^{d+1}_\times, \quad (t,x) \mapsto \mathcal{H}(t,x) := tx + (1-t)\frac{x}{\|x\|}$$

is clearly a homotopy between $i_{S^d} \circ r$ and the identity of $\mathbb{R}^{d+1}_\times$ (do you see it? see also Figure 5.18). Hence, the punctured space $\mathbb{R}^{d+1}_\times$ and the d-dimensional sphere have isomorphic singular homology. We compute the singular homology of spheres in the following section. Note also that $\mathbb{R}^{d+1}_\times$ is homeomorphic to the *cylinder* $\mathbb{R} \times S^d$. An explicit homeomorphism is given by

$$\Phi : \mathbb{R}^{d+1}_\times \to \mathbb{R} \times S^d, \quad x \mapsto (\log\|x\|, x/\|x\|)$$

whose inverse is

$$\Phi^{-1} : \mathbb{R} \times S^d \to \mathbb{R}^{d+1}_\times, \quad (s,y) \mapsto e^s y.$$

We conclude that the cylinder $\mathbb{R} \times S^d$ has the same singular homology as $\mathbb{R}^{d+1}_\times$ and S^d (see also Exercise 5.8).   ◆

Exercise 5.8. Find an explicit homotopy equivalence between the cylinder $\mathbb{R} \times S^d$ and the sphere S^d (thus confirming that they have isomorphic singular homology).

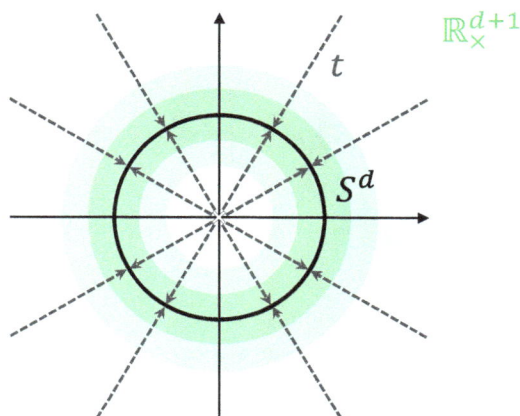

Figure 5.18. The sphere S^d is a deformation retract of the punctured space $\mathbb{R}_\times^{d+1}$.

Example 5.25 (The Eight Figure is a Deformation Retract of the 2-Punctured Plane). In the standard Euclidean plane $\mathbb{R}^2$ consider the two points $x_\pm := (\pm 1, 0)$. The 2-*punctured plane* is the subspace

$$X := \mathbb{R}^2 \setminus \{x_-, x_+\} \subseteq \mathbb{R}^2.$$

Consider the *eight figure*, i.e., the subspace

$$Y = C_- \cup C_+ \subseteq X,$$

where

$$C_\pm = \{x \in \mathbb{R}^2 : \|x - x_\pm\| = 1\}$$

is the unit circle centered in $x_\pm$. So, Y consists of two circles with a common point. The eight figure Y is a deformation retract of the 2-punctured plane X, with deformation retraction $r : X \to Y$ given by (see Figure 5.19)

$$r(x) := \begin{cases} x_+ + \dfrac{x - x_+}{\|x - x_+\|} & \text{if } \|x - x_+\| \leq 1 \\[2ex] x_- + \dfrac{x - x_-}{\|x - x_-\|} & \text{if } \|x - x_-\| \leq 1 \\[2ex] \dfrac{2x_1 x}{\|x\|^2} & \text{if } \|x - x_+\| \geq 1, \\ & \quad x \neq 0,\ x_1 \geq 0 \\[2ex] -\dfrac{2x_1 x}{\|x\|^2} & \text{if } \|x - x_-\| \geq 1, \\ & \quad x \neq 0,\ x_1 \leq 0 \end{cases} \quad \text{for all } x = (x_1, x_2) \in X.$$

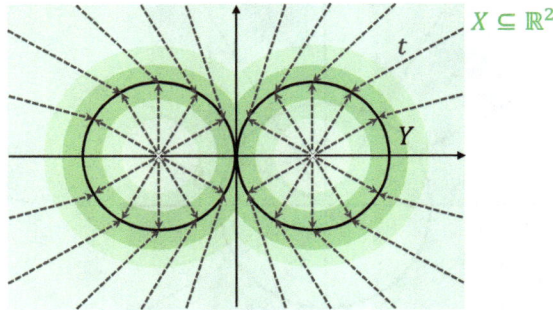

Figure 5.19.   The eight figure Y is a deformation retract of the 2-punctured plane $X \subseteq \mathbb{R}^2$.

We leave it to the reader to check that r is indeed a well-defined continuous map. To see that it is a deformation retraction, it is enough to notice that the continuous map

$$\mathcal{H} : [0,1] \times X \to X, \quad (t,x) \mapsto \mathcal{H}(t,x) := tx + (1-t)r(x)$$

is a well-defined homotopy between $i_Y \circ r$ and id_X. We conclude that the eight figure and the 2-punctured plane have isomorphic singular homologies. In the following section, we compute the singular homologies of the eight figure, hence of the 2-punctured plane.   ◆

Exercise 5.9. Prove all unproven claims in Example 5.25.

Proposition 5.26. *Homotopy equivalence of topological spaces is an equivalence relation.*

Proof.   The identity map $\mathrm{id}_X : X \to X$ of a topological space is a homotopy equivalence, with homotopy inverse itself. The involved homotopies are both given by the map $\mathcal{H} : [0,1] \times X \to X, (t,x) \mapsto x$ (do you see it?). Hence, homotopy equivalence is a reflexive relation. It is also clear that it is a symmetric relation and it remains to prove that it is transitive. So, let

$$X \xrightarrow[\;G\;]{\;F\;} X' \xrightarrow[\;G'\;]{\;F'\;} X''$$

be homotopy equivalences between topological spaces with their homotopy inverses. We want to show that $F' \circ F$ is a homotopy equivalence with homotopy inverse given by $G \circ G'$. So let $\mathcal{H}, \mathcal{H}'$ be homotopies such that $G \circ F \sim_{\mathcal{H}} \mathrm{id}_X$ and $G' \circ F' \sim_{\mathcal{H}'} \mathrm{id}_{X'}$. Then we have

$$G \circ G' \circ F' \sim_{G \circ \mathcal{H}'} G$$

(check it explicitly). Hence,

$$G \circ G' \circ F' \circ F \sim_{\mathbb{H}'} G \circ F \sim_{\mathcal{H}} \mathrm{id}_X,$$

where $\mathbb{H}' : [0,1] \times X \to X$ is the homotopy given by

$$\mathbb{H}'(t,x) := G \circ \mathcal{H}'(t, F(x)), \quad \text{for all } (t,x) \in [0,1] \times X,$$

and, from Proposition 5.12,

$$G \circ G' \circ F' \circ F \sim_{\mathbb{H}' * \mathcal{H}} \mathrm{id}_X.$$

Similarly, there is a homotopy $\mathcal{K}$ such that $F' \circ F \circ G \circ G' \sim_{\mathcal{K}} \mathrm{id}_{X''}$. This concludes the proof. $\qquad\qquad\square$

Remark 5.27. Consider the category **Top** of topological spaces. Define a new category **hTop** as follows. The objects in **hTop** are topological spaces themselves, i.e., $\mathrm{Ob}_{\mathbf{hTop}} = \mathrm{Ob}_{\mathbf{Top}}$. In order to define morphisms, recall that "being homotopic" is an equivalence relation on the set $\mathrm{Hom}_{\mathbf{Top}}(X,Y)$ of continuous maps between the topological spaces X, Y (Proposition 5.12). Denote by $\sim$ this equivalence relation and, for any two topological spaces X, Y, put

$$\mathrm{Hom}_{\mathbf{hTop}}(X,Y) := \mathrm{Hom}_{\mathbf{Top}}(X,Y)/\!\!\sim,$$

the set of *homotopy classes* of continuous maps. Given a continuous map $F : X \to Y$ we will denote by $[F]_\sim \in \mathrm{Hom}_{\mathbf{hTop}}(X,Y)$ its homotopy class. The composition law of morphisms in **hTop** is defined as follows. Let

$$X \xrightarrow{F} Y \xrightarrow{G} Z$$

be continuous maps between topological spaces. We put

$$[G]_\sim \circ [F]_\sim := [G \circ F]_\sim.$$

As homotopies respect the composition of continuous maps (Proposition 5.15), this is well defined (do you see it?). The composition law in **hTop** defined in this way is clearly associative. The units are the homotopy classes of the identity maps. The isomorphisms in **hTop** are the (homotopy classes of) homotopy equivalences of topological spaces (do you see it?). The category **hTop** is called the *homotopy category of topological spaces*.

Exercise 5.6 now shows that the singular chain complex construction can also be seen as a functor

$$C_\bullet : \mathbf{hTop} \to \mathbf{hCh}_{\mathbb{Z}}$$

from the homotopy category of topological spaces to the homotopy category of chain complexes (of abelian groups). Similarly, for all $n \in \mathbb{Z}$, the n-th singular homology functor can be seen as a functor

$$H_n : \mathbf{hTop} \rightarrow \mathbf{Ab}. \qquad \diamondsuit$$

5.3   Mayer–Vietoris Sequence

Let X be a topological space. Sometimes, the singular (co)homology of X can be computed from the singular (co)homology of appropriate pieces of X. Namely, let $\{U, V\}$ be an open cover of X, i.e., $U, V \subseteq X$ are open subspaces such that $X = U \cup V$. Then $U \cap V \subseteq X$ is also an open subspace and we have a commuting diagram of continuous maps:

$$
\begin{array}{ccc}
& X & \\
{}^{i_U}\nearrow & & \nwarrow^{i_V} \\
U & & V \\
{}_{j_U}\nwarrow & & \nearrow_{j_V} \\
& U \cap V &
\end{array}
\qquad (5.4)
$$

where the arrows are the inclusions. Applying the singular chain complex functor to Diagram (5.4) we get a commuting diagram of chain maps:

$$
\begin{array}{ccc}
& (C_\bullet(X), \partial) & \\
{}^{i_{U\sharp}}\nearrow & & \nwarrow^{i_{V\sharp}} \\
(C_\bullet(U), \partial) & & (C_\bullet(V), \partial) \\
{}_{j_{U\sharp}}\nwarrow & & \nearrow_{j_{V\sharp}} \\
& (C_\bullet(U \cap V), \partial) &
\end{array}
\qquad (5.5)
$$

We can combine the top chain maps $i_{U\sharp}, i_{V\sharp}$ in the latter diagram in one single chain map. In order to explain this at a conceptual level, we have to explain how to take *direct sums of (co)chain complexes*. We take this opportunity to present this construction in full generality (although we will only need the case of the direct sum of only *two* chain complexes). So, let $\left(({}^i C_\bullet, {}^i \partial) \right)_{i \in I}$ be a family of chain complexes parameterized by some index set I. Out of such family we construct a new chain complex $(C_\bullet^\oplus, d^\oplus)$ as follows. For any $n \in \mathbb{Z}$, put

$$C_n^\oplus = \bigoplus_{i \in I} {}^iC_n.$$

and define maps $d^\oplus : C_n^\oplus \to C_{n-1}^\oplus$ by putting

$$d^\oplus({}^ic)_{i \in I} := ({}^id^ic)_{i \in I}$$

for all $({}^ic)_{i \in I} \in C_n^\oplus$. It is clear that $d^\oplus$ is a linear map such that $d^\oplus \circ d^\oplus = 0$ (do you see it? If not, check all the details), hence $(C_\bullet^\oplus, d^\oplus)$ is a chain complex also denoted

$$\left(\bigoplus_{i \in I} {}^iC_\bullet, d^\oplus \right)$$

and called the *direct sum* of the family of chain complexes $\left(({}^iC_\bullet, {}^id) \right)_{i \in I}$. Note that, by definition of $d^\oplus$, the usual inclusion

$$\iota_j : ({}^jC, {}^jd) \to \left(\bigoplus_{i \in I} {}^iC_\bullet^\oplus, d^\oplus \right)$$

is a chain map for all $j \in I$.

Lemma 5.28. *The homology of $\left(\bigoplus_{i \in I} {}^iC_\bullet, d^\oplus \right)$ is the direct sum of the homologies of the chain complexes $({}^iC_\bullet, {}^id)$. More precisely, for each $n \in \mathbb{Z}$,*

$$H_n \left(\bigoplus_{i \in I} {}^iC_\bullet, d^\oplus \right)$$

together with the linear maps

$$H_n(\iota_j) : H_n({}^jC, {}^jd) \to H_n \left(\bigoplus_{i \in I} {}^iC, d^\oplus \right), \quad j \in I,$$

is a direct sum of $\left(H_n({}^iC, {}^id) \right)_{i \in I}$:

$$H_n \left(\bigoplus_{i \in I} {}^iC_\bullet, d^\oplus \right) \cong \bigoplus_{i \in I} H_n \left({}^iC_\bullet, {}^id \right).$$

Proof.   We construct the isomorphism

$$\Phi : H_n \left(\bigoplus_{i \in I} {}^i C_\bullet, d^\oplus \right) \to \bigoplus_{i \in I} H_n \left({}^i C_\bullet, {}^i d \right)$$

as follows. A cycle $c \in Z_n(\bigoplus_{i \in I} {}^i C_\bullet, d^\oplus)$ is a family $({}^i c)_{i \in I}$ with ${}^i c \in {}^i C_n$ for all $i \in I$. The cycle condition reads

$$0 = d^\oplus c = ({}^i d {}^i c)_{i \in I}.$$

As the 0 element in $\bigoplus_{i \in I} {}^i C_{n-1}$ is the constant zero family, we get ${}^i d {}^i c = 0$ for all $i \in I$, i.e., ${}^i c \in Z_n({}^i C, {}^i d)$, and we can consider the cohomology class $[{}^i c] \in H_n({}^i C, {}^i d)$. We put

$$\Phi[c] := \left([{}^i c] \right)_{i \in I} \in \bigoplus_{i \in I} H_n({}^i C, {}^i d).$$

We leave it to the reader to check that, defined in this way, Φ is indeed an isomorphism as Exercise 5.10. $\square$

Exercise 5.10. Complete the proof of Lemma 5.28 proving that, for all n, the map Φ is an isomorphism identifying the maps $H_n(\iota_j)$ with the inclusions $H_n({}^j C, {}^j d) \to \bigoplus_{i \in I} H_n({}^i C, {}^i d)$.

Now, Diagram (5.5) gives linear maps $-i_{U\sharp} : C_n(U) \to C_n(X)$ (beware the sign!) and $i_{V\sharp} : C_n(V) \to C_n(X)$, for all n, that (from the universal property of the direct sum) we can combine into one single linear map

$$i_\sharp : C_n(U) \oplus C_n(V) \to C_n(X), \quad (c_U, c_V) \mapsto i_{V\sharp}(c_V) - i_{U\sharp}(c_U).$$

The family

$$\left(i_\sharp : C_n(U) \oplus C_n(V) \to C_n(X) \right)_{n \in \mathbb{Z}}$$

is a chain map between $(C_\bullet(U) \oplus C_\bullet(V), \partial^\oplus)$ and $(C_\bullet(X), \partial)$. Indeed, for all $n \in \mathbb{Z}$ and all $(c_U, c_V) \in C_n(U) \oplus C_n(V)$,

$$i_\sharp(\partial^\oplus(c_U, c_V)) = i_\sharp(\partial c_U, \partial c_V) = i_{V\sharp}(\partial c_V) - i_{U\sharp}(\partial c_U) = \partial i_{V\sharp}(c_V) - \partial i_{U\sharp}(c_U)$$

$$= \partial(i_{V\sharp}(c_V) - i_{U\sharp}(c_U)) = \partial i_\sharp(c_U, c_V),$$

where we used that $i_{U\sharp}, i_{V\sharp}$ are both chain maps.

The maps $j_{U\sharp}, j_{V\sharp}$ in (5.5) can also be combined in one chain map. To explain this conceptually, we need to explain *direct products of (co)chain complexes*. So, let $\left(({}^iC_\bullet, {}^id)\right)_{i\in I}$ be again a family of chain complexes. For any $n \in \mathbb{Z}$, put

$$C_n^\Pi = \Pi_{i\in I}{}^iC_n.$$

and define a map $d^\Pi : C_n^\Pi \to C_{n-1}^\Pi$ by putting

$$d^\Pi({}^ic)_{i\in I} := ({}^id{}^ic)_{i\in I}$$

for all $({}^ic)_{i\in I} \in C_n^\Pi$. It is clear that d^Π is a linear map such that $d^\Pi \circ d^\Pi = 0$ (do you see it? If not, check all the details), hence $(C_\bullet^\Pi, d^\Pi)$ is a chain complex also denoted

$$\left(\Pi_{i\in I}{}^iC_\bullet, d^\Pi\right)$$

and called the *direct product* of the family of chain complexes $\left(({}^iC_\bullet, {}^id)\right)_{i\in I}$. By definition of d, the usual projection

$$\pi_j : \left(\Pi_{i\in I}{}^iC_\bullet, d^\Pi\right) \to ({}^jC, {}^jd)$$

is a chain map for all $j \in I$.

Lemma 5.29. *The homology of $\left(\Pi_{i\in I}{}^iC_\bullet, d^\Pi\right)$ is the direct product of the homologies of the chain complexes $({}^iC_\bullet, {}^id)$. More precisely, for each $n \in \mathbb{Z}$,*

$$H_n\left(\Pi_{i\in I}{}^iC_\bullet, d^\Pi\right)$$

together with the linear maps

$$H_n(\pi_j) : H_n\left(\Pi_{i\in I}{}^iC, d^\Pi\right) \to H_n({}^jC, {}^jd), \quad j \in I,$$

is a direct product of $\left(H_n({}^iC, {}^id)\right)_{i\in I}$:

$$H_n\left(\Pi_{i\in I}{}^iC_\bullet, d^\Pi\right) \cong \Pi_{i\in I}H_n\left({}^iC_\bullet, {}^id\right).$$

Proof. Left as Exercise 5.11. $\qquad\qquad\square$

Exercise 5.11. Prove Lemma 5.29.

Example 5.30. Remember from Example 1.47 that the direct product of finitely many modules agrees, as a module, with their direct sum. It immediately follows that, when $\left(({}^{i}C_{\bullet}, {}^{i}d)\right)_{i \in I}$ is a family of chain complexes parameterized by a *finite* index set, then the direct product and the direct sum of $\left(({}^{i}C_{\bullet}, {}^{i}d)\right)_{i \in I}$ do actually agree:

$$\left(\Pi_{i \in I}{}^{i}C_{\bullet}, d^{\Pi}\right) = \left(\bigoplus_{i \in I}{}^{i}C_{\bullet}, d^{\oplus}\right).$$

In the following, we freely use this simple fact without further comments. ◆

Now, from the universal property of direct products, the maps $j_{U\sharp}, j_{V\sharp}$ give linear maps

$$j_{\sharp} : C_{n}(U \cap V) \to C_{n}(U) \oplus C_{n}(V), \quad c \mapsto j_{\sharp}(c) = (j_{U\sharp}(c), j_{V\sharp}(c)),$$

$n \in \mathbb{Z}$. The family

$$\left(j_{\sharp} : C_{n}(U \cap V) \to C_{n}(U) \oplus C_{n}(V)\right)_{n \in \mathbb{Z}}$$

is a chain map between $(C_{\bullet}(U \cap V), \partial)$ and $(C_{\bullet}(U) \oplus C_{\bullet}(V), \partial^{\oplus})$. Indeed, for all $n \in \mathbb{Z}$ and all $c \in C_{n}(U \cap V)$,

$$j_{\sharp}(\partial c) = (j_{U\sharp}(\partial c), j_{V\sharp}(\partial c)) = (\partial j_{U\sharp}(c), \partial j_{V\sharp}(c))$$
$$= \partial^{\oplus}(j_{U\sharp}(c), j_{V\sharp}(c)) = \partial^{\oplus}j_{\sharp}(c),$$

where we used that $j_{U\sharp}, j_{V\sharp}$ are both chain maps.

Summarizing, from Diagram (5.5), we get a sequence of chain maps

$$0 \longrightarrow (C_{\bullet}(U \cap V), \partial) \xrightarrow{j_{\sharp}} (C_{\bullet}(U) \oplus C_{\bullet}(V), \partial^{\oplus}) \xrightarrow{i_{\sharp}} (C_{\bullet}(X), \partial). \quad (5.6)$$

Lemma 5.31. *The sequence (5.6) is exact, in the sense that*

(1) $j_{\sharp}$ *is injective and*
(2) $\ker i_{\sharp} = \operatorname{im} j_{\sharp}$.

Proof. We begin explaining what do the linear maps $j_{U\sharp}, j_{V\sharp}, i_{U\sharp}, i_{V\sharp}$ really do. Let $\sigma \in S_n(U)$ be a singular n-simplex in U. Then $j_{U\sharp}(\sigma) = j_U \circ \sigma$. Recall that $j_U : U \cap V \to U$ is just the inclusion. So $j_{U\sharp}(\sigma)$ agrees with σ but seen as an n-simplex in U rather that in $U \cap V$. Similarly, for an n-chain $c \in C_n(U \cap V)$, the chain $j_{U\sharp}(c)$ is just the same singular chain but seen as a singular chain in U. So, the image of $j_{U\sharp}$ consists of singular chains in U that do actually take values in $U \cap V$ (more precisely, they are linear combinations of singular simplexes taking values in $U \cap V$). Likewise, for $j_{V\sharp}, i_{U\sharp}, i_{V\sharp}$.

Now, for item (1), let $c \in C_n(U \cap V)$ be in the kernel of $j_\sharp$. This means that

$$j_\sharp(c) = (j_{U\sharp}(c), j_{V\sharp}(c)) = (0,0).$$

As both $j_{U\sharp}(c), j_{V\sharp}(c)$ agree with c (but seen as a singular n-chain in U, V respectively) we conclude that $c = 0$. From the kernel criterion, $j_\sharp$ is injective.

For item (2), let $(c_U, c_V) \in C_n(U) \oplus C_n(V)$ be in the kernel of $i_\sharp$. Then

$$i_\sharp(c_U, c_V) = i_{V\sharp}(c_V) - i_{U\sharp}(c_U) = 0,$$

i.e., $i_{V\sharp}(c_V) = i_{U\sharp}(c_U)$. Denote $c = i_{V\sharp}(c_V) = i_{U\sharp}(c_U) \in C_n(X)$. In other words, c takes values both in U and in V (more precisely, it is a linear combination of singular n-simplexes taking values both in U and in V). The only possibility is that c takes values in $U \cap V$, i.e., there exists $c_{U\cap V} \in C_n(U \cap V)$ such that $c = j_{U\sharp}(c_{U\cap V}) = j_{V\sharp}(c_{U\cap V})$. We conclude that

$$(c_U, c_V) = (c,c) = (j_{U\sharp}(c_{U\cap V}), j_{V\sharp}(c_{U\cap V})) = j_\sharp(c_{U\cap V}).$$

This shows that $\ker i_\sharp \subseteq \operatorname{im} j_\sharp$. For the converse inclusion, let $c_{U\cap V} \in C_n(U \cap V)$, and compute

$$i_\sharp \circ j_\sharp(c_{U\cap V}) = i_\sharp \big(j_{U\sharp}(c_{U\cap V}), j_{V\sharp}(c_{U\cap V})\big)$$

$$= i_{V\sharp} \circ j_{V\sharp}(c_{U\cap V}) - i_{U\sharp} \circ j_{U\sharp}(c_{U\cap V}) = 0,$$

where, in the last step, we used that the compositions $i_V \circ j_V$ and $i_U \circ j_U$ agree (they are both just the inclusion $U \cap V \to X$) hence $i_{V\sharp} \circ j_{V\sharp}$ and $i_{U\sharp} \circ j_{U\sharp}$ agree as well. $\qquad\square$

Note however that the chain map $i_\sharp : C_\bullet(U) \oplus C_\bullet(V) \to C_\bullet(X)$ is not surjective in general. The image of $i_\sharp$ consists of linear combinations of singular simplexes taking values either in U or in V (do you see this?), and it is a subcomplex in $(C_\bullet(X), \partial)$ (see Example 2.32) that we denote $(C_\bullet(X; U, V), \partial)$. The homology of $(C_\bullet(X; U, V), \partial)$ will be denoted $H_\bullet(X; U, V)$. Clearly, we have a short exact sequence of chain complexes

$$0 \longrightarrow (C_\bullet(U \cap V), \partial) \xrightarrow{j_\sharp} (C_\bullet(U) \oplus C_\bullet(V), \partial^\oplus) \xrightarrow{i_\sharp} (C_\bullet(X; U, V), \partial) \longrightarrow 0$$

and an associated long exact sequence in homology

$$\cdots$$

$$\longrightarrow H_{n+1}(U \cap V) \xrightarrow{H(j_\sharp)} H_{n+1}(U) \oplus H_{n+1}(V) \xrightarrow{H(i_\sharp)} H_{n+1}(X; U, V) \xrightarrow{\Delta}$$

$$\longrightarrow H_n(U \cap V) \xrightarrow{H(j_\sharp)} H_n(U) \oplus H_n(V) \xrightarrow{H(i_\sharp)} H_n(X; U, V) \xrightarrow{\Delta}$$

$$\longrightarrow H_{n-1}(U \cap V) \xrightarrow{H(j_\sharp)} H_{n-1}(U) \oplus H_{n-1}(V) \xrightarrow{H(i_\sharp)} H_{n-1}(X; U, V) \xrightarrow{\Delta}$$

$$\longrightarrow \quad \cdots$$

$$(5.7)$$

Proposition 5.32. *The inclusion* $\mathfrak{J} : (C_\bullet(X; U, V), \partial) \to (C_\bullet(X), \partial)$ *is a quasi-isomorphism.*

Proof (a sketch).   The proof is technical and it is based on a construction called *barycentric subdivision* which is also useful for different purposes but we will not explain (see, e.g., Hatcher, 2002; Lee, 2011). We only discuss the basic ideas. There exists a chain map

$$S : (C_\bullet(X), \partial) \to (C_\bullet(X), \partial)$$

with the following properties:

(1) S preserves the subcomplex $C_\bullet(X; U, V) \subseteq C_\bullet(X)$, i.e., $S(C_\bullet(X; U, V)) \subseteq C_\bullet(X; U, V)$;

(2) for every n and every singular n-chain $c \in C_n(X)$, there exists a $k \in \mathbb{N}_0$ such that

$$S^k(c) := \underbrace{S \circ \cdots \circ S}_{k\text{-times}}(c) \in C_n(X; U, V);$$

(3) for every n and every n-cycle, $c \in Z_n(X)$ ($S(c)$ is also an n-cycle and)

$$c - S(c) = \partial b$$

for some $b \in C_{n+1}(X)$, i.e., $c - S(c)$ is a boundary, hence, for all $k \in \mathbb{N}$, $c - S^k(c)$ is also a boundary, indeed

$$c - S^k(c) = c - S(c) + S(c) - S^2(c) + \cdots + S^{k-1}(c) - S^k(c);$$

(4) if the n-cycle c in item (2) is in $C_n(X; U, V)$, then the chain b can be chosen in $C_{n+1}(X; U, V)$, hence, in this case, for all $k \in \mathbb{N}$, $c - S^k(c)$ is also a boundary of the type $\partial b'$ with $b' \in C_{n+1}(X; U, V)$ (do you see it?).

The chain map S basically consists in dividing every singular simplex σ in "smaller" simplexes with vertices in the barycenter of σ, the barycenters of its faces, the barycenters of their faces, and so on, besides the vertices of σ themselves (Figure 5.20). The simplexes obtained in this way are then taken with appropriate coefficients. The fact that $S^k(c)$ belongs to $C_n(X; U, V)$ for a sufficiently large k is guaranteed by continuity and the fact that $\{U, V\}$ is an open cover of X.

Now, suppose that we have S satisfying the properties (1)–(4), and let

$$H(\mathfrak{I}) : H_\bullet(X; U, V) \to H_\bullet(X)$$

be the map induced in homology by the inclusion $\mathfrak{I} : C_\bullet(X; U, V) \to C_\bullet(X)$. We want to show that $H(\mathfrak{I}) : H_n(X; U, V) \to H_n(X)$ is both injective and surjective for all $n \in \mathbb{Z}$. In order not to make confusion, we denote

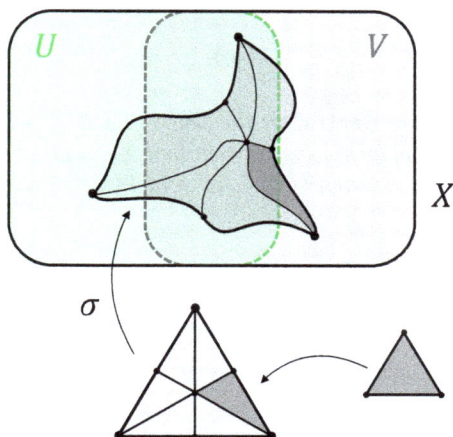

Figure 5.20. Barycentric subdivision of a singular simplex.

by $[c] \in H_n(X)$ the homology class of a cycle $c \in C_n(X)$ and by $[c']_{U,V} \in H_n(X; U, V)$ the homology class of a cycle $c' \in C_n(X; U, V)$. For the surjectivity, let $c \in Z_n(X)$ be an n-cycle in $C_n(X)$ and let $[c] \in H_n(X)$ be its singular homology class. Choose $k \in \mathbb{N}_0$ so that $S^k(c) \in C_n(X; U, V)$. We stress that, as S is a chain map, $S^k(c)$ is a n-cycle as well, i.e., $S^k(c) \in Z_n(X)$. Additionally, from item (3), $S^k(c)$ is homologous to c, hence

$$[c] = [S^k(c)] = [\Im(S^k(c))] = H(\Im)[S^k(c)]_{U,V},$$

i.e., $[c]$ is in the image of $H(\Im)$ as desired. For the injectivity, let $c \in C_n(X; U, V)$ be a cycle such that

$$0 = H(\Im)[c]_{U,V} = [\Im(c)] = [c].$$

So, c is an n-boundary in $C_\bullet(X)$, i.e., $c = \partial b$ for some $b \in C_{n+1}(X)$. Let $k \in \mathbb{N}_0$ be such that $S^k(b) \in C_{n+1}(X; U, V)$. Then, from item (4), $c - S^k(c) = \partial b'$ for some $b' \in C_{n+1}(X; U, V)$. Hence,

$$c = \partial b' + S^k(c) = \partial b' + S^k(\partial b) = \partial b' + \partial S^k(b) = \partial \left(b' + S^k(b) \right).$$

As both b', $S^k(b)$ belong to $C_{n+1}(X; U, V)$, we conclude that c is a boundary in $C_\bullet(X; U, V)$, i.e., $[c]_{U,V} = 0$. From the kernel criterion, $H(\Im)$ is injective as desired. $\qquad\square$

We are now ready to state the main result of this section.

Theorem 5.33 (Mayer–Vietoris Theorem). *Let X be a topological space and let $U, V \subseteq X$ be open subspaces such that $X = U \cup V$. Then, for every $n \in \mathbb{Z}$ there exists a canonical linear map $\Delta : H_n(X) \to H_{n-1}(U \cap V)$ such that the following sequence of linear maps:*

$$\cdots \xleftarrow{H(j_\sharp)} H_{n-1}(U \cap V) \xleftarrow{\Delta} H_n(X) \xleftarrow{H(i_\sharp)} H_n(U) \oplus H_n(V) \xleftarrow{H(j_\sharp)} H_n(U \cap V) \longleftarrow \cdots$$

$$(5.8)$$

is exact. The maps Δ are natural in the sense that if X' is another topological space, $U', V' \subseteq X'$ are open subspaces such that $X' = U' \cup V'$ and $F : X \to X'$ is a continuous map such that $F(U) \subseteq U'$ and $F(V) \subseteq V'$, then the following diagram:

$$(5.9)$$

commutes, where the vertical arrows are the maps induced by F in the obvious way (also explained in the proof). In particular, $\Delta \circ H(F) = H(F|_{U \cap V}) \circ \Delta$.

Proof. As in Proposition 5.32 denote by $\mathfrak{I} : (C_\bullet(X; U, V), \partial) \to (C_\bullet(X), \partial)$ the inclusion. As $\mathfrak{I}$ is a quasi-isomorphism, we can use $H(\mathfrak{I})$ to identify $H_\bullet(X; U, V)$ with $H_\bullet(X)$, and we get Sequence (5.8) from (5.7). We only stress, for future reference, that, after this identification, the connecting homomorphism $\Delta : H_n(X) \to H_{n-1}(U \cap V)$ acts as follows: take an n-cycle $c \in Z_n(X)$ and, by barycentric subdivision or any other method, find an homologous cycle $c' \in Z_n(X) \cap C_n(X; U, V) = Z_n(X; U, V)$. Now, by surjectivity of $i_\sharp : C_n(U) \oplus C_n(V) \to C_n(X; U, V)$, c' can be written in the form $c' = i_\sharp(c_U, c_V)$ with $(c_U, c_V) \in C_n(U) \oplus C_n(V)$. Consider $(\partial c_U, \partial c_V) \in C_{n-1}(U) \oplus C_{n-1}(V)$. Actually, there is a (unique) $(n-1)$-cycle $c_{U \cap V} \in Z_{n-1}(U \cap V)$ such that $j_\sharp(c_{U \cap V}) = (\partial c_U, \partial c_V)$, and we have $\Delta[c] = [c_{U \cap V}] \in H_n(U \cap V)$.

The second part of the statement requires a little explanation. Namely, as F restricts to U, resp. V, in the domain, and to U', resp. V', in the codomain, it also restricts to $U \cap V$ in the domain and to $U' \cap V'$ in the codomain. These restrictions are again continuous maps, hence they induce chain maps $C_\bullet(U) \to C_\bullet(U')$, $C_\bullet(V) \to C_\bullet(V')$, $C_\bullet(U \cap V) \to C_\bullet(U' \cap V')$ that, abusing the notation, we denote $F_\sharp$ again. We also get chain maps

$$F_\sharp : C_\bullet(X; U, V) \to C_\bullet(X'; U', V'), \quad c \mapsto F_\sharp(c),$$

and

$$F_\sharp \oplus F_\sharp : C_\bullet(U) \oplus C_\bullet(V) \to C_\bullet(U') \oplus C_\bullet(V'), \quad (c_U, c_V) \mapsto (F_\sharp(c_U), F_\sharp(c_V))$$

(do you see it?). All these chain maps induce linear maps in homology. The latter will be all denoted $H(F)$ except for the very last one that will be denoted $H(F) \oplus H(F) : H_\bullet(U) \oplus H_\bullet(V) \to H_\bullet(U') \oplus H_\bullet(V')$. It is now easy to see that the diagram

$$
\begin{array}{ccccccccc}
0 & \longrightarrow & (C_\bullet(U \cap V), \partial) & \xrightarrow{j_\sharp} & (C_\bullet(U) \oplus C_\bullet(V), \partial^\oplus) & \xrightarrow{i_\sharp} & (C_\bullet(X; U, V), \partial) & \longrightarrow & 0 \\
& & \downarrow{\scriptstyle F_\sharp} & & \downarrow{\scriptstyle F_\sharp \oplus F_\sharp} & & \downarrow{\scriptstyle F_\sharp} & & \\
0 & \longrightarrow & (C_\bullet(U' \cap V'), \partial) & \xrightarrow{j_\sharp} & (C_\bullet(U') \oplus C_\bullet(V'), \partial^\oplus) & \xrightarrow{i_\sharp} & (C_\bullet(X'; U', V'), \partial) & \longrightarrow & 0
\end{array}
$$

is a morphism of short exact sequences of chain complexes (Definition 2.43, do you see it?). It then follows from Proposition 2.44 that Diagram (5.9)

$$
\begin{array}{ccccccccc}
\cdots \xleftarrow{H(j_\sharp)} & H_{n-1}(U \cap V) & \xleftarrow{\Delta} & H_n(X; U, V) & \xleftarrow{H(i_\sharp)} & H_n(U) \oplus H_n(V) & \xleftarrow{H(j_\sharp)} & H_n(U \cap V) & \xleftarrow{} \cdots \\
& \downarrow{\scriptstyle H(F)} & & \downarrow{\scriptstyle H(F)} & & \downarrow{\scriptstyle H(F) \oplus H(F)} & & \downarrow{\scriptstyle H(F)} & \\
\cdots \xleftarrow{H(j_\sharp)} & H_{n-1}(U' \cap V') & \xleftarrow{\Delta} & H_n(X'; U', V') & \xleftarrow{H(i_\sharp)} & H_n(U') \oplus H_n(V') & \xleftarrow{H(j_\sharp)} & H_n(U' \cap V') & \xleftarrow{} \cdots
\end{array}
$$

commutes. But the isomorphism $H(\mathfrak{I})$ identifies the linear maps

$$H(F) : H_n(X; U, V) \to H_n(X; U', V') \quad \text{and} \quad H(F) : H_n(X) \to H_n(X).$$

Indeed, for every n-cycle c in $C_\bullet(X; U, V)$,

$$H(\mathfrak{I}) \circ H(F)[c]_{U,V} = H(\mathfrak{I})[F_\sharp(c)]_{U',V'} = [\mathfrak{I}(F_\sharp(c))] = [F_\sharp(c)] = H(F)[c]$$
$$= H(F) \circ H(\mathfrak{I})[c]_{U,V}.$$

We conclude that Diagram (5.9) commutes as well. $\qquad\qquad\qquad\square$

Definition 5.34 (Mayer–Vietoris Sequence). The sequence (5.8) is called the *Mayer–Vietoris sequence* (associated to the open cover $\{U, V\}$ of the topological space X).

The Mayer–Vietoris sequence is often useful to compute the singular homology of X from that of $U, V, U \cap V$. The typical example is that of *spheres*.

Example 5.35 (Singular Homology of Spheres). In this example, we compute the singular homology of the sphere

$$S^n = \{x \in \mathbb{R}^{n+1} : \|x\|^2 = 1\} \subseteq \mathbb{R}^{n+1}.$$

We prove that, for every $n \geq 1$,

$$H_i(S^n) = \begin{cases} \mathbb{Z} & \text{if } i = 0, n \\ 0 & \text{otherwise} \end{cases}. \tag{5.10}$$

Note that S^n is path connected for all $n \geq 1$ (do you see it?). Hence $H_0(S^n) = \mathbb{Z}$ from Proposition 5.8. We should actually write $H_0(S^n) \cong \mathbb{Z}$, but being the isomorphism canonical, it is safe to abuse the notation and write $H_0(S^n) = \mathbb{Z}$, meaning that we identify the two abelian groups using our distinguished isomorphism. We will adopt similar abuses also in the sequel. If necessary, we make explicit the canonical isomorphism that we are understanding.

The rest of the proof is by induction on n, and adopts both deformation retraction and Mayer–Vietoris arguments. Consider preliminarily the case $n = 0$ (which is not in the statement but will be useful anyway): the 0-dimensional sphere S^0 is the 2-point space $\{-1, 1\}$ with the discrete topology so that S^0 has exactly 2-path connected components $\{-1\}$ and $\{1\}$ (do you see it?). Hence, the 0-homology of S^0 is canonically isomorphic to $\mathbb{Z}\{-1, 1\} \cong \mathbb{Z}^2$. In a very similar way as for the one point space, one can also show that $H_i(S^0) = 0$ for $i \neq 0$ (and we invite the reader to prove it in details, see also Problem 5.4 in the End-of-Chapter Problem section). Before discussing the base of induction, we further need some general facts about spheres: so let $n \geq 1$, consider the n-sphere $S^n \subseteq \mathbb{R}^{n+1}$, the two points $P_\pm = (0,\ldots,0,\pm 1) \in S^n$ (*north* and *south pole*) and the following two open subsets:

$$U_\pm := S^n \smallsetminus \{P_\pm\} \subseteq S^n,$$

(why are $U_\pm$ open in S^n?). We have $U_+ \cup U_- = S^n$, and $U_+ \cap U_- = S^n \smallsetminus \{P_+, P_-\}$. We claim that U_+, U_- are homeomorphic to $\mathbb{R}^n$, while $U_+ \cap U_-$ is homeomorphic to the punctured space $\mathbb{R}^n \smallsetminus \{0\}$. To see this, consider the *stereographic projection from the north*:

$$\varphi_+ : U_+ \to \mathbb{R}^n$$

$$(x_1,\ldots,x_{n+1}) \mapsto \varphi_+(x_1,\ldots,x_{n+1}) := \left(\frac{x_1}{1 - x_{n+1}}, \ldots, \frac{x_n}{1 - x_{n+1}} \right).$$

The map φ_+ is a well-defined homeomorphism with inverse given by

$$\varphi_+^{-1} : \mathbb{R}^n \to U_+$$

$$(y_1,\ldots,y_n) \mapsto \varphi_+^{-1}(y_1,\ldots,y_n) = \left(\frac{2y_1}{\|y\|^2 + 1}, \ldots, \frac{2y_n}{\|y\|^2 + 1}, \frac{\|y\|^2 - 1}{\|y\|^2 + 1} \right).$$

The open subset U_- is homeomorphic to $\mathbb{R}^n$ as well (via a similar *stereographic projection from the south*). Finally, as φ_+ maps the south pole P_- to $0 \in \mathbb{R}^n$, we also have that $U_+ \cap U_-$ is homeomorphic to the punctured space $\mathbb{R}^n \smallsetminus \{0\}$. We conclude that the singular homologies of $U_\pm$ are the same as those of $\mathbb{R}^n$:

$$H_i(U_\pm) = \begin{cases} \mathbb{Z} & \text{if } i = 0 \\ 0 & \text{otherwise} \end{cases},$$

and the singular homologies of $U_+ \cap U_-$ are the same as those of the punctured space, hence, from Example 5.24, the same as those of the $(n-1)$-sphere S^{n-1}:

$$H_i(U_+ \cap U_-) = H_i(S^{n-1}).$$

More precisely, let $r : \mathbb{R}^n_\times \to S^{n-1}$ be the deformation retraction described in Example 5.24. Then the composition

$$U_+ \cap U_- \xrightarrow{\varphi_+} \mathbb{R}^n_\times \xrightarrow{r} S^{n-1}$$

induces an isomorphism in homology:

$$H(r \circ \varphi_+) : H_i(U_+ \cap U_-) \xrightarrow{\cong} H_i(S^{n-1}). \qquad (5.11)$$

We are now ready to discuss the base of induction: $n = 1$. We already discussed the homologies of the 0-dimensional sphere S^0. We conclude that, when $n = 1$,

$$H_i(U_+ \cap U_-) = H_i(S^0) = \begin{cases} \mathbb{Z}^2 & \text{if } i = 0 \\ 0 & \text{otherwise} \end{cases}.$$

We need to drop one more word on the isomorphisms $H_0(U_\pm) \cong \mathbb{Z}$, $H_0(U_+ \cap U_-) \cong \mathbb{Z}^2$. For the first one, the generator of the free $\mathbb{Z}$-module $H_0(U_\pm) \cong \mathbb{Z}$ is the only path connected component of $U_\pm$, and any singular 0-simplex σ_x, $x \in U_\pm$, is a cycle representing it. We are identifying this generator with $1 \in \mathbb{Z}$. For the second isomorphism $H_0(U_+ \cap U_-) \cong \mathbb{Z}^2$ recall that we are using the stereographic projection to identify $U_+ \cap U_-$ with the punctured line $\mathbb{R}_\times = \mathbb{R} \setminus \{0\}$ and then the homotopy equivalence $\mathbb{R}_\times \to S^0 = \{-1, 1\}$, $t \mapsto t/|t|$. The two generators of the free $\mathbb{Z}$-module $H_0(U_+ \cap U_-) = H_0(\{-1, +1\})$ are the two path connected components $\{+1\}$ and $\{-1\}$ of $\{-1, +1\}$ or, in terms of $U_+ \cap U_-$, the corresponding path connected components

$$\{(x_1, x_2) \in U_+ \cap U_- : x_1 > 0\} \quad \text{and} \quad \{(x_1, x_2) \in U_+ \cap U_- : x_1 < 0\}.$$

Any singular 0-simplex $\sigma_{(x_1, x_2)}$ in $U_+ \cap U_-$ with $x_1 > 0$ is a cycle representing the first one, and any singular 0-simplex $\sigma_{(x_1, x_2)}$ with $x_1 < 0$ is a cycle representing the second one. We are identifying these two generators with $(1, 0) \in \mathbb{Z}^2$ and $(0, 1) \in \mathbb{Z}^2$ respectively (beware that swapping these identifications might change some formulas). Similar considerations hold for the general case $n > 1$ (but beware that, when $n > 1$, $U_+ \cap U_-$ has only one path connected component).

Now, the Mayer–Vietoris sequence associated with the open cover $\{U_+, U_-\}$ of S^1 is

$$0 \longleftarrow H_0(S^1) \xleftarrow{H(i_\sharp)} H_0(U_+) \oplus H_0(U_-) \xleftarrow{H(j_\sharp)} H_0(U_+ \cap U_-) \xleftarrow{\Delta} H_1(S^1) \longleftarrow 0 \longleftarrow \cdots$$

$$\| \qquad\qquad \| \qquad\qquad\qquad \| \qquad\qquad \|$$

$$0 \longleftarrow \mathbb{Z} \longleftarrow \mathbb{Z}^2 \longleftarrow \mathbb{Z}^2 \longleftarrow H_1(S^1) \longleftarrow 0$$

$$(5.12)$$

in low degree, and

$$\cdots \longleftarrow 0 \longleftarrow H_i(S^1) \longleftarrow 0 \longleftarrow \cdots \qquad (5.13)$$

in higher degree $i > 1$.

We leave it to the reader to check that the map $H(j_\sharp) : H_0(U_+ \cap U_-) \to H_0(U_+) \oplus H_0(U_-)$ in (5.12) is given by

$$
\begin{array}{ccc}
H_0(U_+ \cap U_-) & \xrightarrow{\;H(j_\sharp)\;} & H_0(U_+) \oplus H_0(U_-) \\
\| & & \| \\
\mathbb{Z}^2 & \xrightarrow{\hspace{3cm}} & \mathbb{Z}^2
\end{array}
$$

$$(m_1, m_2) \longmapsto (m_1 + m_2, m_1 + m_2)$$

(Exercise 5.12). Therefore, the kernel K of $H(j_\sharp)$ is

$$K := \ker H(j_\sharp) = \{(m, -m) \in \mathbb{Z}^2 : m \in \mathbb{Z}\} \subseteq \mathbb{Z}^2,$$

which is clearly canonically isomorphic to $\mathbb{Z}$ via $\mathbb{Z} \to K$, $m \mapsto (m, -m)$. From exactness (of the Mayer–Vietoris sequence), $\Delta : H_1(S^1) \to H_0(U_+ \cap U_-)$ is an injective linear map whose image is K. We conclude that $H_1(S^1)$ is also (canonically) isomorphic to $\mathbb{Z}$. More precisely, there is a unique isomorphism $H_1(S^1) \cong \mathbb{Z}$ identifying $\Delta : H_1(S^1) \to H_0(U_+ \cap U_-) = \mathbb{Z}^2$ with the homomorphism $\mathbb{Z} \to \mathbb{Z}^2$, $m \mapsto (m, -m)$ (do you see it? See Example 5.36 for a more explicit description of this isomorphism). Finally, from (5.13), $H_i(S^1) = 0$ for higher i. This proves the base of induction.

Next, assume that the claim (5.10) is correct for $1 \le n \le k$ and prove it for $n = k + 1$. In the latter case, $U_+ \cap U_-$ has only one path connected component so that $H_0(U_+ \cap U_-) = \mathbb{Z}$ and the Mayer–Vietoris sequence associated with the open cover $\{U_+, U_-\}$ is

$$
\begin{array}{ccccccccccc}
0 \longleftarrow & H_0(S^{k+1}) & \xleftarrow{H(i_\sharp)} & H_0(U_+) \oplus H_0(U_-) & \xleftarrow{H(j_\sharp)} & H_0(U_+ \cap U_-) & \xleftarrow{\Delta} & H_1(S^{k+1}) & \longleftarrow 0 \longleftarrow \cdots \\
& \| & & \| & & \| & & \| & \\
0 \longleftarrow & \mathbb{Z} & \longleftarrow & \mathbb{Z}^2 & \longleftarrow & \mathbb{Z} & \longleftarrow & H_1(S^{k+1}) & \longleftarrow 0
\end{array}
$$

$$(5.14)$$

in low degree, and

$$
\begin{array}{ccccccc}
\cdots \longleftarrow 0 \longleftarrow & H_{i-1}(U_+ \cap U_-) & \xleftarrow{\Delta} & H_i(S^{k+1}) & \longleftarrow & H_i(U_+) \oplus H_i(U_-) & \longleftarrow \cdots \\
& \| & & \| & & \| & \\
0 \longleftarrow & H_{i-1}(S^k) & \longleftarrow & H_i(S^{k+1}) & \longleftarrow & 0 &
\end{array}
$$

$$(5.15)$$

in higher degree $i > 1$. The map $H(j_\sharp) : H_0(U_+ \cap U_-) \to H_0(U_+) \oplus H_0(U_-)$ in (5.14) is given by

$$
\begin{array}{ccc}
H_0(U_+ \cap U_-) & \xrightarrow{H(j_\sharp)} & H_0(U_+) \oplus H_0(U_-) \\
\| & & \| \\
\mathbb{Z} & \longrightarrow & \mathbb{Z}^2
\end{array}
\qquad ,
$$

$$
m \longmapsto (m, m)
$$

while $H(i_\sharp) : H_0(U_+) \oplus H_0(U_-) \to H_0(S^{k+1})$ is given by

$$
\begin{array}{ccc}
H_0(U_+) \oplus H_0(U_-) & \xrightarrow{H(i_\sharp)} & H_0(S^{k+1}) \\
\| & & \| \\
\mathbb{Z}^2 & \longrightarrow & \mathbb{Z}
\end{array}
$$

$$
(m_1, m_2) \longmapsto m_2 - m_1
$$

(see Exercise 5.12). In particular, $H(j_\sharp)$ is injective and $\ker H(j_\sharp) = 0$. It follows from the exactness of the Mayer–Vietoris sequence that

$$
\operatorname{im}\left(\Delta : H_1(S^{k+1}) \to H_0(U_+ \cap U_-)\right) = 0
$$

as well, i.e., $\Delta : H_1(S^{k+1}) \to H_0(U_+ \cap U_-)$ is the 0 map and $\ker \Delta = H_1(S^{k+1})$. But, from exactness again, Δ is injective, so the only possibility is that $H_1(S^{k+1}) = 0$ (do you see it?). Finally, it follows from (5.15) that $\Delta : H_i(S^{k+1}) \to H_{i-1}(U_+ \cap U_-) = H_{i-1}(S^k)$ is both injective and surjective. We conclude that $H_i(S^{k+1}) \cong H_{i-1}(S^k)$ for all $i > 1$ and from the induction hypothesis we get

$$
H_i(S^{k+1}) \cong H_{i-1}(S^k) = \begin{cases} 0 & \text{if } 1 < i < k \\ \mathbb{Z} & \text{if } i = k+1 \end{cases},
$$

as claimed. Note that, from Example 5.24, the punctured space $\mathbb{R}^{n+1}_\times$ is homotopy equivalent to the sphere S^n. Hence, we also get that

$$
H_i(\mathbb{R}^{n+1}_\times) = H_i(S^n) = \begin{cases} \mathbb{Z} & \text{if } i = 0, n \\ 0 & \text{otherwise} \end{cases}, \tag{5.16}
$$

for all $n > 0$.

The following remark is sometimes useful in applications: for all $n > 0$ the continuous map (reflection with respect to the coordinate hyperplane $x_1 = 0$)

$$T_1 : S^n \to S^n, \quad x = (x_1, x_2 \dots, x_{n+1}) \mapsto T_1(x) := (-x_1, x_2, \dots, x_{n+1})$$

induces the *product by* -1 map in n-homology:

$$\begin{array}{ccc} H_n(S^n) & \xrightarrow{H(T_1)} & H_n(S^n) \\ \| & & \| \\ \mathbb{Z} & \longrightarrow & \mathbb{Z} \end{array}$$

$$m \longmapsto -m$$

To see this, first note that $T_1(U_+) \subseteq U_+$ and $T_1(U_-) \subseteq U_-$. So, according to the Mayer–Vietoris Theorem, T_1 induces a commuting diagram

$$\begin{array}{ccc} H_{n-1}(U_+ \cap U_-) & \xleftarrow{\Delta} & H_n(S^n) \\ \Big\downarrow{H(T_1)} & & \Big\downarrow{H(T_1)} \\ H_{n-1}(U_+ \cap U_-) & \xleftarrow{\Delta} & H_n(S^n) \end{array}$$

Composing the horizontal arrows with the isomorphism (5.11), and using the simple fact that $T_1 \circ r \circ \varphi_+ = r \circ \varphi_+ \circ T_1$, we see that the diagram

$$\begin{array}{ccc} H_{n-1}(S^{n-1}) & \xleftarrow{\Delta} & H_n(S^n) \\ \Big\downarrow{H(T_1)} & & \Big\downarrow{H(T_1)} \\ H_{n-1}(S^{n-1}) & \xleftarrow{\Delta} & H_n(S^n) \end{array} \tag{5.17}$$

commutes as well (do you see it?). For $n = 1$, this diagram boils down to

$$\begin{array}{ccc} (m_1, m_2) & \mathbb{Z}^2 \xleftarrow{\Delta} \mathbb{Z} \\ \Big\uparrow & \Big\downarrow{H(T_1)} \quad \Big\downarrow{H(T_1)} \\ (m_2, m_1) & \mathbb{Z}^2 \xleftarrow{\Delta} \mathbb{Z} \end{array} \tag{5.18}$$

$$(m, -m) \longleftarrow\!\!\shortmid m$$

(Exercise 5.12). It follows that the right vertical arrow is the product by -1 (check the details as an exercise). For higher n, the claim follows by induction from the commutativity of Diagram (5.17) again. One can show in a similar way that actually the reflection $T_i : S^n \to S^n$ with respect to the i-th hyperplane $x_i = 0$ induces the product by -1 in the n-th homology of S^n for all $i = 1, \ldots, n+1$.   ◆

Exercise 5.12. Prove all unproven claims in Example 5.35. Namely, show that

(1) the map $H(j_\sharp) : H_0(U_+ \cap U_-) \to H_0(U_+) \oplus H_0(U_-)$ of Example 5.35 is given by

$$H(j_\sharp) : \mathbb{Z}^2 \to \mathbb{Z}^2, \quad (m_1, m_2) \mapsto (m_1 + m_2, m_1 + m_2),$$

if $n = 1$,

(2) and by

$$H(j_\sharp) : \mathbb{Z} \to \mathbb{Z}^2, \quad m \mapsto (m, m).$$

if $n = k + 1 > 1$;

(3) the map $H(i_\sharp) : H_0(U_+) \oplus H_0(U_-) \to H_0(S^{k+1})$ is given by

$$H(i_\sharp) : \mathbb{Z}^2 \to \mathbb{Z}, \quad (m_1, m_2) \mapsto m_2 - m_1,$$

if $k \geq 0$;

(4) Diagram (5.17) boils down to (5.18) when $n = 1$;

(**Hint:** *When $n = 1$, use the explicit description of the isomorphism $H_0(U_+ \cap U_-) \cong \mathbb{Z}^2$ in the detailed discussion preceding (5.12)*).

Exercise 5.13. Let n, m be non-negative integers. Prove that the spheres S^n, S^m are not homotopy equivalent, unless $n = m$.

Example 5.36 (The Canonical Generator of $H_1(S^1)$). In Example 5.35, we proved, among other things, that $H_1(S^1) = \mathbb{Z}$. More precisely, we proved that there is a unique abelian group isomorphism $H_1(S^1) \cong \mathbb{Z}$ identifying the monomorphism $\Delta : H_1(S^1) \to H_0(U_+ \cap U_-) = \mathbb{Z}^2$ with the linear map $\mathbb{Z} \to \mathbb{Z}^2, m \mapsto (m, -m)$. Here Δ is the connecting homomorphism in the Mayer–Vietoris sequence (5.12) associated to the open cover $\{U_+, U_-\}$ of S^1 (see Example 5.35 for more details). In this example we want to provide a more explicit description of the isomorphism $H_1(S^1) \cong \mathbb{Z}$. We do this finding the generator in $H_1(S^1)$ that corresponds to the canonical generator 1 in $\mathbb{Z}$. In other words, we find a distinguished 1-cycle $c \in Z_1(S^1)$

such that $\Delta[c] = (1, -1) \in \mathbb{Z}^2 = H_0(U_+ \cap U_-)$. When studying $S^1 \subseteq \mathbb{R}^2$, it is often convenient to interpret a pair $(x_1, x_2) \in \mathbb{R}^2$ as a complex number $x_1 + ix_2$ using that $\mathbb{C} = \mathbb{R}^2$ as real vector spaces. From now on, in this example, we adopt this approach. So, S^1 consists of complex numbers of the form $e^{i\theta}$, with $\theta \in \mathbb{R}$. Let $c : \Delta_1 \to S^1$ be the singular 1-simplex given by

$$c(x_0, x_1) := e^{2\pi i x_0}.$$

In other words, c *wraps* the standard 1-simplex Δ_1 once around the circle S^1 counterclockwise, starting from (and ending in) 1 (see Figure 5.21). Clearly, c is a 1-cycle. We want to compute $\Delta[c]$. According to the very definition of the connecting homomorphism in the Mayer–Vietoris sequence we can do this in four steps:

(1) we find a 1-cycle $c' \in C_1(S^1; U_+, U_-)$ such that $c' - c = \partial b$ for some 2-chain $b \in C_2(S^1)$;
(2) we write c' in the form $c' = i_{U_-\sharp}(c_-) - i_{U_+\sharp}(c_+)$ for some $c_\pm \in C_1(U_\pm)$;
(3) we compute $\partial c_\pm$ and notice that $\partial c_\pm = j_{U_\pm\sharp}(\bar{c})$ for some $\bar{c} \in C_0(U_+ \cap U_-)$;
(4) we observe that $\bar{c} \in Z_0(U_+ \cap U_-)$ and $\Delta[c] = [\bar{c}] \in H_0(U_+ \cap U_-)$.

Step (1). This is the only non-trivial step. Consider the singular 1-simplexes $\sigma_\pm : \Delta_1 \to S_1$, $(x_0, x_1) \mapsto \mp e^{\pi i x_0}$ and the 1-chain $c' := \sigma_+ + \sigma_-$ (beware that this is a formal linear combination in the free module spanned by 1-simplexes, it is *not* the sum of the two maps $\sigma_\pm$, so don't be tempted to conclude that $c' = 0$!!). It is clear that $\sigma_\pm$ takes values in $U_\pm$, so that $c' \in C_1(S^1; U_+, U_-)$. Additionally, c' is a cycle, indeed

$$\partial c' = \partial \sigma_+ + \partial \sigma_- = \sigma_{\sigma_+(E_1)} - \sigma_{\sigma_+(E_0)} + \sigma_{\sigma_-(E_1)} - \sigma_{\sigma_-(E_0)}$$

$$= \sigma_{x_-} - \sigma_{x_+} + \sigma_{x_+} - \sigma_{x_-} = 0,$$

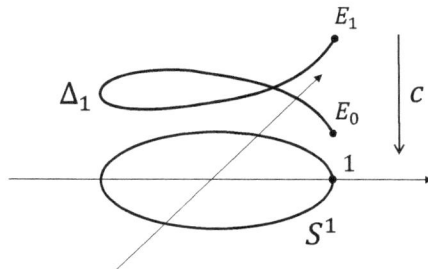

Figure 5.21. The singular 1-simplex c wrapping once around the circle.

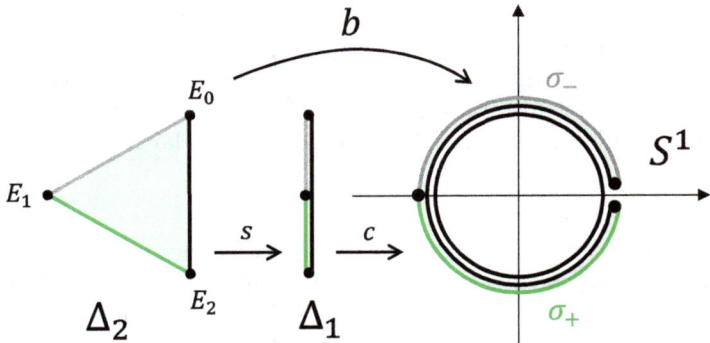

Figure 5.22.   The singular 2-simplex $b = c \circ s$. The boundary of b is $\sigma_+ + \sigma_- - c$.

where, in order to avoid confusion, we denoted $x_\pm = (\pm 1, 0) \in S^1$. Finally, c differs from c' by a boundary. Indeed, consider the singular 2-simplex $b : \Delta_2 \to S^1$, $(x_0, x_1, x_2) \mapsto e^{\pi i(2x_0 + x_1)}$ (b is the composition of the orthogonal projection $s : \Delta_2 \to \Delta_1$, $(x_0, x_1, x_2) \mapsto (x_0 + x_1/2, x_1/2 + x_2)$ onto the first face of Δ_2 followed by c, in other words it first squashes the standard 2-simplex onto its first face, and then wraps it around the circle, see Figure 5.22). Now compute

$$\partial b = b \circ d_0 - b \circ d_1 + b \circ d_2.$$

But

$$b \circ d_0(x_0, x_1) = b(0, x_0, x_1) = e^{\pi i x_0} = \sigma_-(x_0, x_1),$$

$$b \circ d_1(x_0, x_1) = b(x_0, 0, x_1) = e^{2\pi i x_0} = c(x_0, x_1),$$

$$b \circ d_2(x_0, x_1) = b(x_0, x_1, 0) = e^{\pi i(2x_0 + x_1)} = e^{\pi i(x_0 + 1)} = -e^{\pi i x_0} = \sigma_+(x_0, x_1),$$

for all $(x_0, x_1) \in \Delta_1$ (where we used that $x_0 + x_1 = 1$). We conclude that

$$\partial b = \sigma_+ + \sigma_- - c = c' - c.$$

This concludes Step (1).

Step (2). Put $c_\pm = \mp \sigma_\pm$. Then $c_\pm \in C_1(U_\pm)$ and $c' = i_{U_-\sharp}(c_-) - i_{U_+\sharp}(c_+)$ as desired.

Step (3). Compute

$$\partial c_+ = -\partial \sigma_+ = -\sigma_{x_-} + \sigma_{x_+} = \partial \sigma_- = \partial c_-.$$

As $x_\pm \in U_+ \cap U_-$, the 0-chain $\bar{c} := -\sigma_{x_-} + \sigma_{x_+}$ is in $C_0(U_+ \cap U_-)$. The above computation now shows that $\partial c_\pm = j_{U_\pm \#}(\bar{c})$ as desired.

Step (4). Finally, $\Delta[c] = [\bar{c}]$. But the homology class $[\bar{c}]$ is

$$[\bar{c}] = \text{"path connected component of } x_+\text{"}$$
$$- \text{"path connected component of } x_-\text{"}$$

which identifies with $(1, -1)$ under the isomorphism $H_0(U_+ \cap U_-) \cong \mathbb{Z}^2$.

So, the homology class $[c] \in H_1(S^1)$ is exactly the generator that we were looking for. We conclude this example with a remark that will be useful in the following. Besides c, for each $m \in \mathbb{Z}$, consider the singular 1 simplex $c^m : \Delta_1 \to S^1$, defined by

$$c^m(x_0, x_1) := e^{2m\pi i x_0}. \tag{5.19}$$

In other words, c^m wraps Δ_1 around the circle m times counterclockwise (see Figure 5.23 for the case $m = 3$). Clearly, c^m is a 1-cycle for all m. In a very similar way as we did for $c' - c$, it is not difficult to see that $mc - c^m$ is actually a boundary (Exercise 5.14). So c^m is homologous to mc and its homology class is

$$[c^m] = m[c] \tag{5.20}$$

which identifies simply with $m \in \mathbb{Z}$ under the isomorphism $H_1(S^1) \cong \mathbb{Z}$. In this way, we have found a canonical representative for every homology class in $H_1(S^1)$. ◆

Exercise 5.14. Show that the 1-cycle c^m in $C_1(S^1)$ defined by (5.20) is homologous to the 1-cycle mc, where $c = c^1$ (**Hint:** *Consider the singular 2-simplex*

$$b : \Delta_2 \to S^1, \quad (x_0, x_1, x_2) \mapsto e^{2\pi i(mx_0 + (m-1)x_1)}$$

and show that $\partial b = c^{m-1} + c - c^m$, then use induction).

Example 5.37 (Singular Homology of the 2-Punctured Plane). In this example, we use the Mayer–Vietoris sequence together with Example 5.35 to compute the singular homology of the 2-punctured plane $X = \mathbb{R}^2 \setminus \{x_+, x_-\}$, where we set $x_\pm = (\pm 1, 0)$ or, which is the same, the homology of the eight figure $Y = C_- \cup C_+$ (see Example 5.25 for the notation).

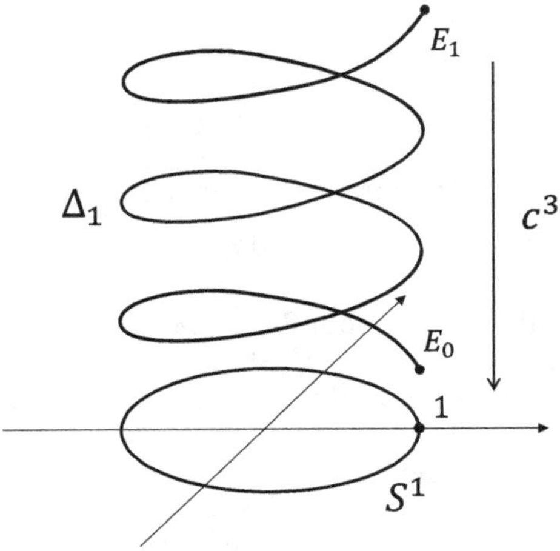

Figure 5.23.   The singular 1-simplex c^3 wraps 3 times around the circle.

We begin remarking that Y is path connected, hence $H_0(Y) = \mathbb{Z}$. Now, consider the following two open subsets in Y:

$$U_\pm = Y \smallsetminus \{(0, \mp 2)\}$$

(why are $U_\pm$ open in Y?). We have $U_+ \cup U_- = Y$ and $U_+ \cap U_- = Y \smallsetminus \{(0,2), (0,-2)\}$. Moreover C_- is a deformation retract in U_-. An explicit deformation retraction $r : U_- \to C_-$ is given by

$$r(x,y) = \begin{cases} (x,y) & \text{if } x \le 0 \\ (0,0) & \text{if } x > 0 \end{cases}.$$

(We give up on presenting a precise homotopy between $i_{C_-} \circ r$ and id_{U_-}, but we hope that the reader has at least an intuition of the fact that such a homotopy exists). Similarly, C_+ is a deformation retract of U_+ and $\{(0,0)\}$ is a deformation retract of $U_+ \cap U_-$. In low degree, the Mayer–Vietoris sequence associated with the open cover $\{U_+, U_-\}$ is

$$\cdots \leftarrow H_0(U_+) \oplus H_0(U_-) \leftarrow H_0(U_+ \cap U_-) \leftarrow H_1(Y) \leftarrow H_1(U_+) \oplus H_1(U_-) \leftarrow H_1(U_+ \cap U_-) \leftarrow \cdots$$
$$\parallel \qquad\qquad \parallel \qquad\qquad \parallel \qquad\qquad \parallel \qquad\qquad \parallel$$
$$\cdots \leftarrow \mathbb{Z}^2 \leftarrow \mathbb{Z} \leftarrow H_1(Y) \leftarrow \mathbb{Z}^2 \leftarrow 0 \quad.$$

$$(m,m) \longleftarrow m$$

As the map $H_0(U_+ \cap U_-) \to H_0(U_+) \oplus H_0(U_-)$ is injective, from exactness, the map $H_1(Y) \to H_0(U_+ \cap U_-)$ must be 0. So the map $H_1(U_+) \oplus$

$H_1(U_-) \to H_1(Y)$ is surjective. But it is also injective, and we conclude that $H_1(Y) = \mathbb{Z}^2$. In higher degree $i > 1$, the Mayer–Vietoris sequence is

$$\cdots \longleftarrow H_{i-1}(U_+ \cap U_-) \longleftarrow H_i(Y) \longleftarrow H_i(U_+) \oplus H_i(U_-) \longleftarrow \cdots$$
$$\| \qquad\qquad \| \qquad\qquad \|$$
$$0 \longleftarrow H_i(Y) \longleftarrow 0$$

So, $H_i(Y) = 0$ for $i > 1$. We conclude that

$$H_i(X) = H_i(Y) = \begin{cases} \mathbb{Z} & \text{if } i = 0 \\ \mathbb{Z}^2 & \text{if } i = 1 \\ 0 & \text{otherwise} \end{cases}.$$

◆

Exercise 5.15. Use appropriate deformation retraction and Mayer–Vietoris sequence arguments to compute the singular homology of the *2-punctured 3D space* $X := \mathbb{R}^3 \setminus \{(0,0,1), (0,0,-1)\}$ (you can be sloppy on homotopy arguments!).

Exercise 5.16. Use appropriate deformation retraction and Mayer–Vietoris sequence arguments to compute the singular homology of the *3-punctured plane*

$$X := \mathbb{R}^2 \setminus \{(1,0), (-1,0), (0,1)\}$$

(you can be sloppy on homotopy arguments!).

Example 5.38 (Singular Homology of the Klein Bottle). The singular homologies that we have computed so far are all free abelian groups (when non-trivial). We now provide an example of a topological space with a non-free singular homology: the *Klein Bottle* (one further example is provided by Problem 5.13 in the End-of-Chapter Problem section). Recall that the Klein Bottle is the topological space obtained from a square by identifying the opposite sides as illustrated in Figure 5.24.

To be more precise, take the *unit square* $[0,1] \times [0,1] \subseteq \mathbb{R}^2$ with its subspaces topology and, for all $x, y \in [0,1]$, identify the point $(x,0)$ with the point $(x,1)$ and the point $(0,y)$ with the point $(1, 1-y)$. The Klein bottle K is the *quotient topological space* under this identification (see Figure 5.25).

We want to compute the singular homology of K. To do this, we use the Mayer–Vietoris Theorem. We begin remarking that K is path connected so that $H_0(K) = \mathbb{Z}$ (do you see it?). Next, denote by $P \in K$ the point

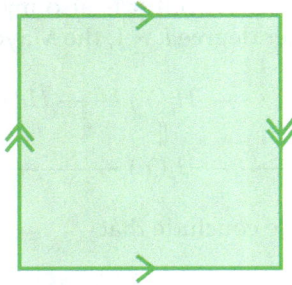

Figure 5.24.   The Klein Bottle.

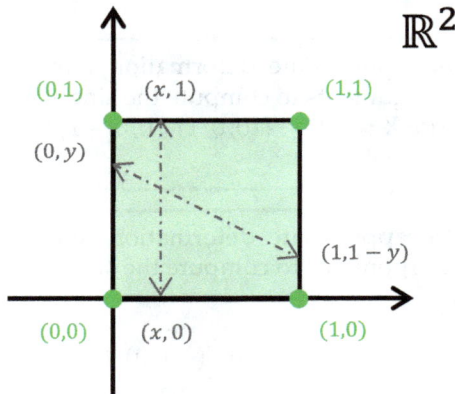

Figure 5.25.   The unit square.

corresponding to the center of the square, and consider the following two open subsets in K:

- $U = K \smallsetminus \{P\}$ (do you see that it is indeed open?);
- the open subspace V corresponding to an open disk around P not touching the boundary of the square;

We have $U \cup V = K$ and $U \cap V = V \smallsetminus \{P\}$. Moreover, the subspace $Y \subseteq U$ corresponding to the boundary of the square is a deformation retract of U. Note that Y is homeomorphic to the eight figure from Example 5.25 (do you see it?) and consists of two circles C_1, C_2 with a common point Q (see Figure 5.26).

Moreover, $U \cap V = V \smallsetminus \{P\}$ is homeomorphic to the punctured plane and, therefore, it is homotopy equivalent to the circle S^1. Hence, from Example 5.37, in low degree, the Mayer–Vietoris sequence associated with

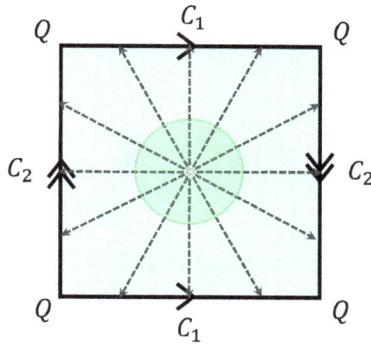

Figure 5.26. An open cover of the Klein Bottle.

the open cover $\{U, V\}$ of K is

$$\cdots \leftarrow H_0(U) \oplus H_0(V) \leftarrow H_0(U \cap V) \leftarrow H_1(K) \leftarrow H_1(U) \oplus H_1(V) \leftarrow H_1(U \cap V) \leftarrow \cdots$$
$$\| \qquad\qquad \| \qquad\quad \| \qquad\qquad \| \qquad\qquad \|$$
$$\cdots \leftarrow \mathbb{Z} \oplus \mathbb{Z} \leftarrow \mathbb{Z} \leftarrow H_1(K) \leftarrow 0 \oplus \mathbb{Z}^2 \leftarrow \mathbb{Z} \leftarrow \cdots .$$

$$(m, m) \longleftarrow m$$

As the map $H_0(U \cap V) \to H_0(U) \oplus H_0(V)$ is injective, from exactness, the connecting homomorphism $H_1(K) \to H_0(U \cap V)$ is zero, hence $H_1(K)$ is isomorphic the cokernel of the map $H_1(U \cap V) \to H_1(U) \oplus H_1(V) \cong H_1(Y)$, which is the map induced in homology by the inclusion $j_V : U \cap V \to V$ followed by the map induced in homology by the retraction $r : V \to Y$. Now, the 1-homology of $U \cap V = U \setminus \{P\}$ is generated by the homology class $[c]$ of a singular 1-simplex $c : \Delta_1 \to U \setminus \{P\}$ wrapping Δ_1 once around P. The image of c under $r \circ j_V$ is a singular 1-simplex (and a 1-cycle) $c' : \Delta_1 \to Y$ wrapping first around C_1, then around C_2, then around C_1 again but in the opposite direction, and finally around C_2 in the same direction as before. Overall, c' is homologous to a singular 1-simplex (and a 1-cycle) $c'' : \Delta_1 \to Y$ wrapping twice around C_2 so that the map $H_1(U \cap V) \to H_1(U) \oplus H_1(V)$ identifies with the linear map $\mathbb{Z} \to \mathbb{Z}^2$, $p \mapsto (0, 2p)$ (see Equation (5.20)) whose cokernel is isomorphic to $\mathbb{Z}^2 / (0 \oplus 2\mathbb{Z}) = \mathbb{Z} \oplus \mathbb{Z}_2$.

The next segment of the Mayer–Vietoris sequence is

$$\cdots \leftarrow H_1(U) \oplus H_1(V) \leftarrow H_1(U \cap V) \leftarrow H_2(K) \leftarrow H_2(U) \oplus H_2(V) \leftarrow \cdots$$
$$\| \qquad\qquad \| \qquad\quad \| \qquad\qquad \|$$
$$\cdots \leftarrow 0 \oplus \mathbb{Z}^2 \leftarrow \mathbb{Z} \leftarrow H_2(K) \leftarrow 0 \leftarrow \cdots .$$

$$(0, 2p) \longleftarrow p$$

As the map $H_1(U \cap V) \rightarrow H_1(U) \oplus H_1(V)$ is injective, it follows from exactness that $H_2(K) = 0$. Similarly, $H_i(K) = 0$ for all $i > 2$. We conclude that

$$H_i(K) = \begin{cases} \mathbb{Z} & \text{if } i = 0 \\ \mathbb{Z} \oplus \mathbb{Z}_2 & \text{if } i = 1 \\ 0 & \text{otherwise} \end{cases}.$$

Clearly, $H_1(K)$ is not a free abelian group.   ◆

We now provide various interesting applications of Example 5.35, including a topological proof of the Fundamental Theorem of Algebra.

Theorem 5.39 (Topological Invariance of Dimension). *Let m, n be non-negative integers. Then the standard Euclidean spaces $\mathbb{R}^m, \mathbb{R}^n$ are not homeomorphic, unless $m = n$. In other words, the topology of $\mathbb{R}^n$ "knows its dimension".*

Proof. The proof is by contradiction. Clearly, we can assume $m, n \neq 0$ (do you see it?). So, let $0 < m < n$ and suppose that there is a homeomorphism $\Phi : \mathbb{R}^n \rightarrow \mathbb{R}^m$. Then there is also a homeomorphism $\Phi_0 : \mathbb{R}^n \rightarrow \mathbb{R}^m$ such that $\Phi_0(0) = 0$. Indeed, let $\tau : \mathbb{R}^m \rightarrow \mathbb{R}^m$, $x \mapsto x - \Phi(0)$ be the translation by the vector $-\Phi(0)$. It is clear that τ is a homeomorphism (do you see it?). Then $\Phi_0 := \tau \circ \Phi_0$ is a homeomorphism as well, and $\Phi_0(0) = \tau(\Phi(0)) = \Phi(0) - \Phi(0) = 0$. By restriction, $\Phi_0 : \mathbb{R}^n_\times \rightarrow \mathbb{R}^m_\times$ is a homeomorphism of the punctured spaces. In particular it induces an isomorphism in homology. Hence, from (5.16),

$$\mathbb{Z} = H_{n-1}(S^{n-1}) \cong H_{n-1}(\mathbb{R}^n_\times) \cong H_{n-1}(\mathbb{R}^m_\times) \cong H_{n-1}(S^{m-1}) = 0,$$

which is a contradiction.   □

Theorem 5.40 (Brouwer Fixed Point Theorem). *Let n be a positive integer and let*

$$D^n = \{x \in \mathbb{R}^n : \|x\| \leq 1\}$$

be the closed n-dimensional disk. Every continuous map $F : D^n \rightarrow D^n$ has a fixed point, i.e., a point $x_0 \in D^n$ such that $F(x_0) = x_0$.

Proof. The proof is by contradiction. The case $n = 1$ follows from the Bolzano's Theorem and does not require homological methods. So let $n > 1$ and let $F : D^n \rightarrow D^n$ be a continuous map. Suppose that $F(x) \neq x$ for all $x \in D^n$. Then we can construct a map $G : D^n \rightarrow S^{n-1}$ defining $G(x)$ as the intersection point with S^{n-1} of the half line departing from $F(x)$

and passing through $x \in D^{n-1}$ (see Figure 5.27). The map G is continuous (the point $G(x)$ can be computed explicitly and shown to have continuous coordinates in the variable x. Do this as an exercise!). Additionally, if $x \in S^{n-1} \subseteq D^n$, then $G(x) = x$ (do you see it?). In other words, if we denote by $i : S^{n-1} \to D^n$ the inclusion, then the diagram

$$
\begin{array}{ccc}
D^n & \xrightarrow{\ G\ } & S^{n-1} \\
{\scriptstyle i}\uparrow & \nearrow & \\
S^{n-1} & {\scriptstyle \text{id}} &
\end{array}
$$

commutes. It follows that the diagram

$$
\begin{array}{ccc}
H_{n-1}(D^n) & \xrightarrow{\ H(G)\ } & H_{n-1}(S^{n-1}) \\
{\scriptstyle H(i)}\uparrow & \nearrow & \\
H_{n-1}(S^{n-1}) & {\scriptstyle \text{id}} &
\end{array}
$$

commutes as well. But, for $n > 1$, from (5.3) we have $H_{n-1}(D^n) = 0$, and from (5.10) we have $H_{n-1}(S^{n-1}) = \mathbb{Z}$. So, we have a commuting diagram

$$
\begin{array}{ccc}
0 & \longrightarrow & \mathbb{Z} \\
\uparrow & \nearrow & \\
\mathbb{Z} & {\scriptstyle \text{id}} &
\end{array}
$$

which is a cotradiction. $\qquad\square$

Let n be a positive integer. A (continuous) *vector field* on the n-dimensional sphere S^n is a continuous map $Z : S^n \to \mathbb{R}^{n+1}$ such that $Z(x)$ is orthogonal to x for all $x \in S^n$ (i.e., for all $x \in S^n$, the image $Z(x)$ is in the tangent space to S^n at x, see Figure 5.28).

Theorem 5.41 (Hairy Ball Theorem). *Let $n = 2k > 0$ be an even positive integer. Then every continuous vector field $Z : S^n \to \mathbb{R}^{n+1}$ on the n-dimensional sphere vanishes at some point, i.e., there exists $x_0 \in S^n$ such that $Z(x_0) = 0$.*

The proof of the hairy ball theorem is based on the following lemma that might have an independent interest.

Lemma 5.42. *Let n be an even non-negative integer. Then the antipodal map $A : S^n \to S^n$, $x \mapsto -x$ is not homotopic to the identity.*

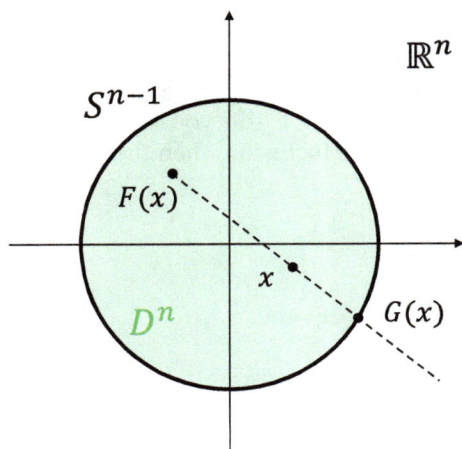

Figure 5.27.   The map G in the proof of Brouwer Theorem.

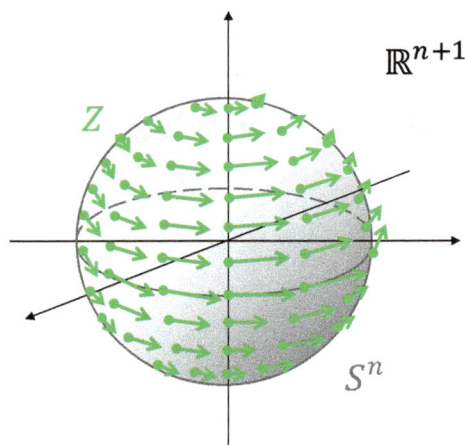

Figure 5.28.   A vector field Z on the sphere S^n.

Proof.   The map A is continuous. Actually, it is the composition of the reflections with respect to all coordinate hyperplanes: $A = T_1 \circ \cdots \circ T_{n+1}$. Every T_i induces the product by -1 in the n-th homology of S^n (see the discussion at the end of Example 5.35). Hence,

$$H_n(A) = H_n(T_1 \circ \cdots \circ T_{n+1}) = H_n(T_1) \circ \cdots \circ H_n(T_{n+1}) = (-1)^{n+1}.$$

So, if n is even, then $n + 1$ is odd and $H_n(A) = -1$. In particular, A does not induce the identity in homology, and cannot be homotopic to id. $\square$

Remark 5.43. As a corollary of Lemma 5.42, we can also prove the following new fixed point theorem: *If n is an even non-negative integer, then any continuous map $F : S^n \to S^n$ homotopic to the identity has a fixed point.* Indeed, suppose by contradiction that F has no fixed points. In this case, $tA(x) + (1-t)F(x) = -tx + (1-t)F(x) \neq 0$ for all $(t,x) \in [0,1] \times S^n$. Indeed, for all $t \in [0,1]$, $-tx + (1-t)F(x)$ is a point of the segment s joining $F(x)$ and $-x$. If 0 belonged to s, then s would be a diameter and its extremal points would be antipodal, i.e., $F(x) = x$. Now consider the map

$$\mathcal{H} : [0,1] \times S^n \to S^n, \quad (t,x) \mapsto \mathcal{H}(t,x) := \frac{-tx + (1-t)F(x)}{\|-tx + (1-t)F(x)\|}.$$

It is a well-defined homotopy between F and A (do you see it?). But "being homotopic" is a transitive relation on continuous maps and A is not homotopic to the identity (from Step 1), while F is by hypothesis. This is a contradiction. $\diamond$

Proof (of Theorem 5.41). Let $Z : S^n \to \mathbb{R}^{n+1}$ be a continuous vector field on S^n. Suppose by contradiction that Z has no zeros, i.e., $Z(x) \neq 0$ for all $x \in S^n$. Then we can define a continuous map $V : S^n \to S^n$ by putting $V(x) = Z(x)/\|Z(x)\|$. In its turn V can be used to define a homotopy $\mathcal{H} : [0,1] \times S^n \to S^n$ between id and A as follows. For all $(t,x) \in [0,1] \times S^n$ put

$$\mathcal{H}(t,x) = (\cos \pi t)\, x + (\sin \pi t)\, V(x).$$

The map $\mathcal{H}$ takes indeed values in S^n:

$$\|\mathcal{H}(t,x)\|^2 = \|(\cos \pi t)\, x + (\sin \pi t)\, V(x)\|^2$$
$$= (\cos^2 \pi t)\|x\|^2 + (\sin^2 \pi t)\|V(x)\|^2 + 2(\cos \pi t \sin \pi t)x \cdot V(x)$$
$$= \cos^2 \pi t + \sin^2 \pi t = 1,$$

where we used that x and $Z(x)$ (hence x and $V(x)$) are orthogonal: $x \cdot Z(x) = x \cdot V(x) = 0$. Finally, for all $x \in S^n$,

$$\mathcal{H}(0,x) = x = \mathrm{id}(x) \quad \text{and} \quad \mathcal{H}(1,x) = -x = A(x),$$

which is a contradiction. This concludes the proof. $\square$

In the case $n = 2$, the hairy ball theorem says, in practice, that it is impossible to comb continuously a hairy (3-dimensional) ball without living out some singular point. This should explain the funny name.

We conclude this chapter by showing that singular homology does even allow to prove the Fundamental Theorem of Algebra.

Theorem 5.44 (Fundamental Theorem of Algebra). *Let $P(z) \in \mathbb{C}[z]$ be a complex polynomial of positive degree in the indeterminate z. Then $P(z)$ possesses a root, i.e., there exists $z_0 \in \mathbb{C}$ such that $P(z_0) = 0$.*

Proof.   Let $m > 0$ be the degree of $P(z)$. We can assume, without loss of generality, that $P(z)$ is a *monic polynomial*:

$$P(z) = z^m + a_{m-1}z^{m-1} + \cdots + a_1 z + a_0,$$

$a_i \in \mathbb{C}$ for all $i = 0, \ldots, m - 1$. Suppose by contradiction that $P(z) \neq 0$ for all $z \in \mathbb{C}$. In this case, we can define the continuous map

$$\mathcal{G} : \mathbb{R} \times S^1 \to S^1, \quad (r, x) \mapsto \mathcal{G}(r, x) := \frac{P(rx)}{|P(rx)|} \frac{|P(r)|}{P(r)}.$$

Consider also the map

$$\mathcal{H} : [0, 1] \times S^1 \to S^1, \quad (t, x) \mapsto \mathcal{H}(t, x) := \begin{cases} \mathcal{G}\left(\dfrac{t}{1-t}, x\right) & \text{if } 0 \le t < 1 \\ x^m & \text{if } t = 1 \end{cases}.$$

The latter map is also continuous. Indeed, it is clearly continuous on $[0, 1) \times S^1$ and, additionally, for all $x \in S^1$,

$$\lim_{t \to 1^-} \mathcal{H}(t, x) = \lim_{t \to 1^-} \mathcal{G}\left(\frac{t}{1-t}, x\right) = \lim_{r \to +\infty} \mathcal{G}(r, x) = \lim_{r \to +\infty} \frac{(rx)^m}{|rx|^m} \frac{r^m}{r^m} = x^m,$$

where we used that $|x| = 1$ (and that the top power term of $P(rx)$ is dominant in the limit $r \to +\infty$). Hence it is a homotopy between $\mathcal{H}_0$ and $\mathcal{H}_1$. Now,

$$\mathcal{H}_0(x) = \mathcal{H}(0, x) = 1 \quad \text{and} \quad \mathcal{H}_1(x) = \mathcal{H}(1, x) = x^m,$$

for all $x \in S^1$. So $\mathcal{H}_1$ is null-homotopic, and must induce the zero map in the first homology $H_1(S^1)$ (Corollary 5.17). However, using the final part of Example 5.36

$$H(\mathcal{H}_1)[c] = [\mathcal{H}_{1\sharp}(c)] = [c^m] = m[c] \neq 0$$

when $m > 0$. This is a contradiction.   □

5.4 End-of-Chapter Problems

Problem 5.1. Repeat, with the appropriate modifications, the discussion following Exercise 5.3 for singular cochains and arbitrary coefficients.

Problem 5.2. State and prove the analog of Theorem 5.16 for singular cochains and arbitrary coefficients.

Problem 5.3. State and prove the analogs of Corollary 5.17, Proposition 2.29 and Proposition 5.22 for singular cochains and arbitrary coefficients.

Problem 5.4. Compute the singular (co)homology of a discrete topological space.

Problem 5.5. Let $(X_i)_{i \in I}$ be a family of topological spaces. Recall that the disjoint union $\coprod_{i \in I} X_i$ possesses a unique topology such that all the inclusions $X_j \hookrightarrow \coprod_{i \in I} X_i$ are continuous maps, $j \in I$. Show that, for each $n \in \mathbb{Z}$, there are canonical abelian group isomorphisms

$$H_n \left(\coprod_{i \in I} X_i \right) \cong \bigoplus_{i \in I} H_n(X_i), \quad \text{and} \quad H^n \left(\coprod_{i \in I} X_i \right) \cong \prod_{i \in I} H^n(X_i).$$

Problem 5.6. Prove that every continuous map with values in a star-shaped subspace of $\mathbb{R}^d$ is null-homotopic.

Problem 5.7. Let $r \subseteq \mathbb{R}^d$ be a straight line, $d > 0$. Prove that the topological space $\mathbb{R}^d \setminus r$ possesses a deformation retract homeomorphic to a *cylinder* $\mathbb{R} \times S^{d-1}$.

Problem 5.8. Let $S \subseteq \mathbb{R}^3$ be a circle and let $r \subseteq \mathbb{R}^3$ be a straight line passing through the center of S and orthogonal to the S's plane (e.g., $S = \{(x,y,z) \in \mathbb{R}^3 \text{ such that } z = 0 \text{ and } x^2 + y^2 = 1\}$ and $r = \{x = y = 0\}$). Prove that the topological space $\mathbb{R} \setminus (S \cup r)$ possesses a deformation retract homeomorphic to a *2-torus* $T^2 = S^1 \times S^1$.

Problem 5.9. Let $(X_i)_{i \in I}$ be a family of topological space. For each $i \in I$, fix a point $x_i \in X_i$. Denote by $\bigvee_{i \in I}(X_i, x_i)$ the topological space obtained from the disjoint union $\coprod_{i \in I} X_i$ by quotienting with respect to the equivalence relation $\sim$ that identifies the points x_i:

$$\bigvee_{i \in I}(X_i, x_i) := \coprod_{i \in I} X_i \Big/ \sim$$

The topological space $\bigvee_{i \in I}(X_i, x_i)$ is also called the *bouquet* or the *wedge sum* of the *pointed topological spaces* (X_i, x_i) (see Figure 5.29).

Figure 5.29.   A bouquet of topological spaces.

For example, the eight figure is homeomorphic to a bouquet of two circles. Prove that for any choice of finitely many distinct points $x_1, \dots, x_p \in \mathbb{R}^{n+1}$ the *punctured space* $\mathbb{R}^{n+1} \smallsetminus \{x_1, \dots, x_p\}$ possesses a deformation retract homeomorphic to the wedge sum of p n-dimensional spheres.

Problem 5.10. Use the Mayer–Vietoris theorem and induction to compute the singular homology of a bouquet of p n-dimensional spheres (hence of the Euclidean Space Punctured p-Times, see Problem 5.9).

Problem 5.11. Use the Mayer–Vietoris theorem to compute the singular homology of the wedge sum of a circle and a 2-sphere.

Problem 5.12. Compute the singular homology of the 2-torus $T^2 = S^1 \times S^1$. (**Hint:** *Remember that the 2-torus can be obtained from a square by identifying the opposite sides as in Figure 5.30. Now use the same strategy as in Example 5.38.*)

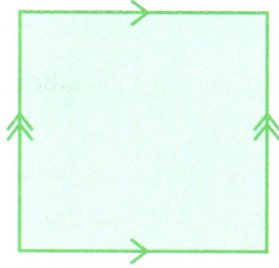

Figure 5.30. The 2-torus.

Problem 5.13. Compute the singular homology of the real projective plane $\mathbb{R}P^2$. (**Hint:** *Remember that the real projective plane can be obtained from a square by identifying the opposite sides as in Figure 5.31. Now use the same strategy as in Example 5.38.*)

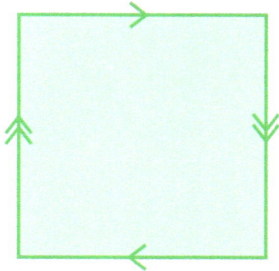

Figure 5.31. The real projective plane.

Problem 5.14. Let X be a topological space. Consider the cochain complex $(C^\bullet(X), \delta)$ of singular cochains in X with coefficients in $\mathbb{Z}$. Recall that a singular n-cochain is a function $c : S_n(X) \to \mathbb{Z}$. In the abelian group,

$$C(X) := \bigoplus_{n \geq 0} C^n(X)$$

take the $\mathbb{Z}$-bilinear composition law

$$\smile \; : C(X) \times C(X) \to C(X)$$

uniquely defined by $C^n(X) \smile C^m(X) \subseteq C^{n+m}(X)$ and

$$c \smile c'(\sigma) = c(\sigma_{n,0})c'(\sigma_{0,m}), \quad c \in C^n(X), c' \in C^m(X), \quad \sigma \in S_{n+m}(X),$$

where $\sigma_{n,0} \in S_n(X)$ and $\sigma_{0,m} \in S_m(X)$ are the singular simplexes given by

$$\sigma_{n,0}(x_0, \ldots, x_n) = \sigma(x_0, \ldots, x_n, 0, \ldots, 0) \quad \text{and}$$
$$\sigma_{0,m}(x_0, \ldots, x_m) = \sigma(0, \ldots, 0, x_0, \ldots, x_m).$$

Prove that $(C(X), +, \smile)$ is a well-defined (non-commutative) ring with unit. Prove also the following identity

$$\delta(c \smile c') = \delta c \smile c' + (-)^n c \smile \delta c',$$

for all $c \in C^n(X)$ and $c' \in C^m(X)$. The product $\smile$ is called the *cup product* in singular cohomology.

Chapter 6

de Rham Cohomology

Numerous (co)chain complexes appear in Differential Geometry as well. In this chapter, we briefly discuss the de Rham complex of an open subset in $\mathbb{R}^n$. Our analysis will mostly parallel that for singular homology in Chapter 5. de Rham cohomology is a diffeomorphism invariant, i.e., two diffeomorphic open subsets have isomorphic de Rham cohomology, but it is also a homotopy invariant, i.e., homotopy equivalent open subsets have isomorphic de Rham cohomology. The advantage of open subsets in the standard Euclidean space over generic topological spaces is that they can be studied via tools from calculus. Actually, open subsets are a special instance of more interesting spaces, namely *smooth manifolds*. de Rham cohomology extends to smooth manifolds and it is usually presented in such generality (see, e.g., Lee, 2013). As the scope of this chapter is mainly illustrative, we do not define smooth manifolds (which will take too much space) and we limit the discussion to open subsets in standard Euclidean spaces. Note however that every smooth manifold is homotopy equivalent to an open subset in some standard Euclidean space, so limiting to the latter case is not tremendously restrictive. We conclude the chapter by sketching the proof of the de Rham theorem stating that de Rham cohomology agrees with singular cohomology. This important result in differential geometry paves the way to a *Calculus based Algebraic Topology* (see Bott and Tu, 1982).

6.1 Differential Forms and de Rham Cohomology

Open subsets in standard Euclidean spaces can be organized in a category **Op** as follows. An object in **Op** is a *non-empty* open subset $U \subseteq \mathbb{R}^n$ for some $n \in \mathbb{N}_0$. The dimension n of the ambient Euclidean space is also called the *dimension* of U, and we write dim $U = n$. If $U \subseteq \mathbb{R}^n$ and $V \subseteq \mathbb{R}^m$ are non-empty open subsets, a morphism in **Op** between U and V is a *smooth map* $F : U \rightarrow V$, i.e., a map that can be differentiated infinitely many times at every point. As the identity $\mathrm{id}_U : U \rightarrow U$ is a smooth map and the composition of smooth maps is a smooth map, we immediately see that **Op** is a category. Isomorphisms in **Op** are *diffeomorphisms*, i.e., smooth maps $F : U \rightarrow U'$ between open subsets $U, U' \subseteq \mathbb{R}^n$ such that F is an invertible map and F^{-1} is a smooth map as well. Note that there are no diffeomorphisms between two open subsets $U \subseteq \mathbb{R}^n$, $V \subseteq \mathbb{R}^m$ if $n \neq m$.

Let us fix our notation about smooth maps. Given open subsets $U \subseteq \mathbb{R}^n$, $V \subseteq \mathbb{R}^m$, the set of smooth maps $F : U \rightarrow V$ will be also denoted as $C^\infty(U, V)$. If $m = 1$ and $V = \mathbb{R}$, then we also simply write $C^\infty(U)$ (instead of $C^\infty(U, \mathbb{R})$). Elements of $C^\infty(U)$ will be also called *smooth functions on U* (reserving the term *smooth maps* to the more general case $C^\infty(U, V)$). With the point-wise operations, smooth functions on U form a real, associative, commutative algebra with unit (do you see it?), i.e., the sum of smooth functions is a smooth function, the product of smooth functions is a smooth function and the product by a real number of a smooth function is a smooth function as well.

In this chapter, following a rather common convention in Differential Geometry, we usually denote $(x^1, \ldots, x^n)$ (with upper indexes) the standard coordinates on an open subset $U \subseteq \mathbb{R}^n$, but we also use $(y^1, \ldots, y^m)$, $(z^1, \ldots, z^p)$, etc. if we deal with more than one open subset in more than one Euclidean space. In this case, for a smooth function $f \in C^\infty(U)$, we also write $f = f(x^1, \ldots, x^n)$ to stress that the coordinates can be promoted to variables. Let $U \subseteq \mathbb{R}^n$ and $V \subseteq \mathbb{R}^m$ be open subsets, let $(x^1, \ldots, x^n)$ be standard coordinates on U, and let $(y^1, \ldots, y^m)$ be standard coordinates on V. A smooth map $F : U \rightarrow V$ can be seen as a vector valued map on U: $F = (F^1, \ldots, F^m)$, where $F^a = F^a(x^1, \ldots, x^n)$ is the smooth function defined by $F^a = y^a \circ F$, $a = 1, \ldots, m$. We also write $F = F(x^1, \ldots, x^n)$.

We now come to *vector fields* and *differential forms*. Let $U \subseteq \mathbb{R}^n$ be a non-empty open subset.

Definition 6.1 (Vector Field). A *vector field* on U is a smooth vector-valued map $X : U \rightarrow \mathbb{R}^n$, i.e., $X \in C^\infty(U, \mathbb{R}^n)$ (we stress that here $n = \dim U$).

Let $X = (X^1, \ldots, X^n)$ be a vector field on U. We interpret X as the assignment of a vector $X(x) = (X^1(x), \ldots, X^n(x))$ applied at the point $x \in U$ for every such x (Figure 6.1). This should explain the terminology "vector field".

A vector field $X = (X^1, \ldots, X^n)$ is completely determined by its *components* $X^i = X^i(x^1, \ldots, x^n) \in C^\infty(U)$, $i = 1, \ldots, n$. Accordingly, it can be

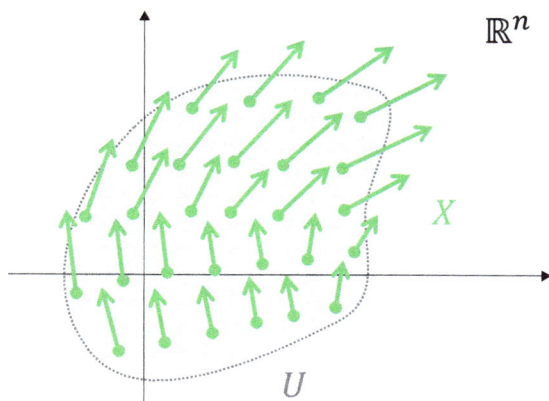

Figure 6.1. A vector field X on an open subset $U \subseteq \mathbb{R}^n$.

identified with the first-order linear differential operator

$$\sum_{i=1}^{n} X^i \frac{\partial}{\partial x^i} : C^\infty(U) \to C^\infty(U)$$

that, abusing the notation, we denote again by X, in other words, we often write

$$X = \sum_{i=1}^{n} X^i \frac{\partial}{\partial x^i} \tag{6.1}$$

instead of $X = (X^1, \ldots, X^n)$. For instance, for a smooth function $f \in C^\infty(U)$, we denote

$$X(f) := \sum_{i=1}^{n} X^i \frac{\partial f}{\partial x^i}.$$

In order not to make confusion between the interpretation of a vector field X as a map $X : U \to \mathbb{R}^n$ and as a differential operator $X : C^\infty(U) \to C^\infty(U)$, from now on the value of a vector field at a point $x \in U$ is denoted as X_x (instead of $X(x)$). Note that the constant vector field

$$E_i := \Big(0, \ldots, \underbrace{1}_{i\text{-th place}}, \ldots, 0 \Big)$$

identifies with the i-th partial derivative $\frac{\partial}{\partial x^i}$.

Exercise 6.1. Prove that, for every vector field X on $U \subseteq \mathbb{R}$, the map

$$X : C^\infty(U) \to C^\infty(U)$$

is a derivation of the associative algebra $C^\infty(U)$.

In the following, we denote by $\mathfrak{X}(U)$ (instead of $C^\infty(U, \mathbb{R}^n)$) the space of vector fields on the open subset $U \subseteq \mathbb{R}^n$. There are various interesting algebraic structures on $\mathfrak{X}(U)$. First of all, if we interpret $C^\infty(U)$ as a ring (forgetting about the vector space structure), then $\mathfrak{X}(U)$ is a $C^\infty(U)$-module: Both the sum and the product by a scalar are defined point-wise, i.e., for all vector fields $X, Y \in \mathfrak{X}(U)$ and all smooth functions $g \in C^\infty(U)$, the sum $X + Y$ is defined by

$$(X + Y)_x := X_x + Y_x, \quad x \in U,$$

and the product gX is defined by

$$(gX)_x := g(x)X_x, \quad x \in U.$$

If we interpret X, Y as differential operators $X, Y : C^\infty(U) \to C^\infty(U)$, then

$$(X + Y)(f) = X(f) + Y(f), \quad f \in C^\infty(U),$$

and

$$(gX)(f) = gX(f), \quad f \in C^\infty(U)$$

(do you see it?).

Exercise 6.2. Prove that the sum and the product by a scalar defined above give to $\mathfrak{X}(U)$ the structure of a module over the ring $C^\infty(U)$.

In particular, the right-hand side of (6.1) is a linear combination in the module $\mathfrak{X}(U)$.

Proposition 6.2. *The $C^\infty(U)$-module $\mathfrak{X}(U)$ is free and finitely generated. Specifically, partial derivatives form a basis in it.*

Proof. We already know that partial derivatives generate $\mathfrak{X}(U)$. It remains to check that they are linearly independent. So, let

$f^1, \ldots, f^n \in C^\infty(U)$ be such that

$$\sum_{i=1}^{n} f^i \frac{\partial}{\partial x^i} = 0.$$

This means that the left-hand side is the 0 differential operator $C^\infty(U) \to C^\infty(U)$. So, it maps every function to the zero function. In particular, for every $j = 1, \ldots, n,$

$$0 = \sum_{i=1}^{n} f^i \frac{\partial}{\partial x^i}(x^j) = \sum_{i=1}^{n} f^i \frac{\partial x^j}{\partial x^i} = \sum_{i=1}^{n} f^i \delta_i^j = f^j.$$

This concludes the proof. $\qquad\square$

Note that constant functions const : $U \to \mathbb{R}$ identify with real numbers (they form a subring in the ring $C^\infty(U)$ isomorphic to the ring $\mathbb{R}$). Restricting the product by a scalar to constant functions, we see that $\mathfrak{X}(U)$ is also a real vector space. This vector space is also equipped with a Lie bracket $[-, -]$ which is easily described in the differential operator language. Namely, let $X, Y : C^\infty(U) \to C^\infty(U)$ be vector fields (seen as differential operators). In particular, they are $\mathbb{R}$-linear endomorphisms of the vector space $C^\infty(U)$. An easy computation that we leave to the reader shows that their commutator is given by

$$[X, Y] = X \circ Y - Y \circ X = \sum_{i=1}^{n} (X(Y^i) - Y(X^i)) \frac{\partial}{\partial x^i}, \qquad (6.2)$$

hence it is a vector field again. It follows that vector fields form a Lie subalgebra in the Lie algebra $\mathrm{End}_{\mathbb{R}} C^\infty(U)$.

Exercise 6.3. Prove Formula (6.2). Prove also that the Lie algebra and the module structures in $\mathfrak{X}(U)$ interact as follows: For any $X, Y \in \mathfrak{X}(U)$ and any $f \in C^\infty(U)$, we have

$$[X, fY] = X(f)Y + f[X, Y]$$

(in particular, the commutator of vector fields is $\mathbb{R}$-bilinear but *not* $C^\infty(U)$-bilinear).

Definition 6.3 (Differential Form). A degree k *differential form* (or simply a *k-form*) on a non-empty open subset $U \subseteq \mathbb{R}^n$ is an alternating

$C^\infty(U)$-multilinear map:

$$\omega : \underbrace{\mathfrak{X}(U) \times \cdots \times \mathfrak{X}(U)}_{k\text{-times}} \to C^\infty(U),$$

i.e., $\omega \in \text{Alt}^k_{C^\infty(U)}(\mathfrak{X}(U); C^\infty(U))$.

Degree k differential forms form a $C^\infty(U)$-module that we denote $\Omega^k(U)$. In particular, $\Omega^1(U)$ is the dual module of $\mathfrak{X}(U)$: $\Omega^1(U) = \mathfrak{X}(U)^* = \text{Hom}_{C^\infty(U)}(\mathfrak{X}(U), C^\infty(U))$. As $\mathfrak{X}(U)$ is free and finitely generated, from Proposition 1.72, $\Omega^1(U)$ is free and finitely generated as well with basis given by the *dual basis*

$$(dx^1, \ldots, dx^n).$$

In other words, $dx^i \in \Omega^1(U)$ is the differential 1-form uniquely defined by

$$dx^i\left(\frac{\partial}{\partial x^j}\right) = \delta^i_j, \quad \text{for all } i = 1, \ldots, n.$$

The differential 1-forms $(dx^1, \ldots, dx^n)$ are linearly independent and generate $\Omega^1(U)$, i.e., every 1-form $\theta \in \Omega^1(U)$ can be uniquely written as

$$\theta = \sum_{i=1}^n \theta_i dx^i,$$

for some smooth functions $\theta_i \in C^\infty(U)$. From Proposition 1.72 again, for every $k \in \mathbb{Z}$, we also have natural $C^\infty(U)$-module isomorphisms

$$\Omega^k(U) \cong \wedge^k \Omega^1(U)$$

that we will always understand in what follows. Additionally, $\Omega^k(U)$ is free and finitely generated as well with basis given by

$$\left(dx^{i_1} \wedge \cdots \wedge dx^{i_k}\right)_{i_1 < \cdots < i_k}.$$

In particular, there are no non-trivial differential forms of degree k for $k > n$. In the following, it is often convenient to expand a differential k-form as follows:

$$\omega = \sum_{i_1, \ldots, i_k} \omega_{i_1 \cdots i_k} dx^{i_1} \wedge \cdots \wedge dx^{i_k},$$

for some smooth functions $\omega_{i_1 \cdots i_k} \in C^\infty(U)$. Note that the latter is *not* a basis expansion, as we are not imposing the ordering $i_1 < \cdots < i_k$ on the sum indexes (hence, there might be repetitions among the generators). However, it is clear that every k-form can be written in this way (and if we assume that the coefficients $\omega_{i_1 \cdots i_k}$ are skew-symmetric in the indexes $i_1, \ldots, i_k$, then they are also unique!).

Next, for all k, we define an $\mathbb{R}$-linear (beware, *not* $C^\infty(U)$-linear) map

$$d : \Omega^k(U) \to \Omega^{k+1}(U)$$

via the following formula:

$$d\omega(X_1, \ldots, X_{k+1}) = \sum_i (-)^{i+1} X_i \big(\omega(X_1, \ldots, \widehat{X_i}, \ldots, X_{k+1}) \big)$$

$$+ \sum_{i<j} (-)^{i+j} \omega \big([X_i, X_j], X_1, \ldots, \widehat{X_i}, \ldots, \widehat{X_j}, \ldots, X_{k+1} \big),$$

for all $X_1, \ldots, X_{k+1} \in \mathfrak{X}(U)$. In order to show that this is well defined, we have to prove various things. First of all that $d\omega$ is a differential $k+1$-form. To do this, we have to show $C^\infty(U)$-linearity in each argument X_i and skew-symmetry. This can be done with a straightforward computation that we omit. We need also to prove that d is $\mathbb{R}$-linear. This is easy and we leave the details to the reader.

Definition 6.4 (de Rham Differential). The operator $d : \Omega^k(U) \to \Omega^{k+1}(U)$ defined above is called the (k-th) *de Rham differential.*

One can actually show that, if ω is given by

$$\omega = \sum_{i_1, \ldots, i_k} \omega_{i_1 \cdots i_k} dx^{i_1} \wedge \cdots \wedge dx^{i_k}$$

for some smooth functions $\omega_{i_1 \cdots i_k} \in C^\infty(U)$, then

$$d\omega = \sum_{i, i_1, \ldots, i_k} \frac{\partial}{\partial x^i} \omega_{i_1 \cdots i_k} dx^i \wedge dx^{i_1} \wedge \cdots \wedge dx^{i_k}. \tag{6.3}$$

If $k = 0$, then $\omega =: f \in \Omega^0(U) = C^\infty(U)$ is a smooth function and

$$df = \sum_{i=1}^n \frac{\partial f}{\partial x^i} dx^i.$$

In particular, the de Rham differential of the coordinate function x^i is exactly dx^i (do you see it?), which explains the notation.

Theorem 6.5. *Let $U \subseteq \mathbb{R}^n$ be a non-empty open subset. Then the sequence of linear maps*

$$0 \longrightarrow C^\infty(U) \overset{d}{\longrightarrow} \Omega^1(U) \overset{d}{\longrightarrow} \cdots \overset{d}{\longrightarrow} \Omega^n(U) \longrightarrow 0$$

is a cochain complex of $\mathbb{R}$-vector spaces.

Proof. A straightforward but long (and a bit intricate) computation that we omit. $\qquad\square$

Definition 6.6 (de Rham Cohomology). The cochain complex $(\Omega^\bullet(U), d)$ is called the *de Rham complex* of U. The cohomology $H_{dR}^\bullet(U) :=$ $H^\bullet(\Omega(U), d)$ is called the *de Rham cohomology* of U. A differential form ω is called *closed* if $d\omega = 0$, i.e., ω is a cocycle, and it is called *exact* if $\omega = d\rho$ for some other differential form ρ, i.e., ω is a coboundary.

Example 6.7. If $U = \mathbb{R}^3$, then $\mathfrak{X}(U)$ is generated by

$$\left(\frac{\partial}{\partial x^1}, \frac{\partial}{\partial x^2}, \frac{\partial}{\partial x^3} \right),$$

$\Omega^1(U)$ is generated by (dx^1, dx^2, dx^3), $\Omega^2(U)$ is generated by

$$\left(dx^2 \wedge dx^3, dx^3 \wedge dx^1, dx^1 \wedge dx^2 \right)$$

and $\Omega^3(U)$ is generated by $dx^1 \wedge dx^2 \wedge dx^3$ (there are no non-trivial higher degree differential forms in this case). Accordingly, there is a $C^\infty(U)$-module isomorphism $\Omega^1(U) \cong C^\infty(\mathbb{R}^3, \mathbb{R}^3)$ (just map a 1-form to its triple of components that can be seen as a vector valued map). Similarly, $\Omega^2(U) \cong C^\infty(\mathbb{R}^3, \mathbb{R}^3)$ and $\Omega^3(U) \cong C^\infty(\mathbb{R}^3)$. A direct computation exploiting Formula (6.3) now reveals that the de Rham complex

$$0 \longrightarrow C^\infty(\mathbb{R}^3) \xrightarrow{d} \Omega^1(\mathbb{R}^3) \xrightarrow{d} \Omega^2(\mathbb{R}^3) \xrightarrow{d} \Omega^3(\mathbb{R}^3) \longrightarrow 0$$

identifies with the $(\mathrm{grad}, \mathrm{curl}, \mathrm{div})$ cochain complex (2.8). Similarly, the de Rham complex of $\mathbb{R}^2$ identifies with the cochain complex (2.9). Check the details as an exercise. $\qquad\blacklozenge$

Example 6.8 (de Rham Cohomology of a Point). When $U = \mathbb{R}^0 = \{0\}$ is 0-dimensional, the de Rham complex reduces to

$$0 \longrightarrow C^\infty(\{0\}) \cong \mathbb{R} \longrightarrow 0.$$

We conclude that

$$H_{dR}^k(\{0\}) = \begin{cases} \mathbb{R} & \text{if } k = 0 \\ 0 & \text{otherwise} \end{cases}.$$

$\qquad\blacklozenge$

Example 6.9 (Degree 0 de Rham Cohomology). Let $U \subseteq \mathbb{R}^n$ be a non-empty open subset. The 0-th de Rham cohomology of U is the kernel of the map

$$d : C^\infty(U) \to \Omega^1(U), \quad f \mapsto df = \sum_{i=1}^n \frac{\partial f}{\partial x^i} dx^i.$$

As the dx^i are linearly independent, $df = 0$ if and only if

$$\frac{\partial f}{\partial x^i} = 0, \quad \text{for all } i = 1, \ldots, n,$$

which in turn implies that f is a locally constant function, i.e., for any $x \in U$ there is an open neighborhood $V \subseteq U$ of x such that $f|_V = \text{const}$. It follows that f is constant on each connected component of U. This means that f descends to a function

$$\overline{f} : \pi_0(U) \to \mathbb{R},$$

on the set $\pi_0(U)$ of connected components of U (note that, for an open subset of $\mathbb{R}^n$, connected components and path connected components are the same thing. This explains why we used the same notation π_0). Specifically, if U_x is the connected components of a point $x \in U$, then $\overline{f}(U_x) := f(x)$ (which is well defined because f is constant on U_x). Conversely, given a function $\overline{f} : \pi_0(U) \to \mathbb{R}$, we can consider the function $f = \overline{f} \circ \pi : U \to \mathbb{R}$, where $\pi : U \to \pi_0(U), x \mapsto \pi(x) := U_x$ is the natural projection. Clearly, f is a locally constant, hence smooth, function. We conclude that there is a bijection

$$H^0_{dR}(U) \to \mathbb{R}^{\pi_0(U)}, \quad f \mapsto \overline{f}.$$

It is clear that such bijection is also $\mathbb{R}$-linear, hence it is a vector space isomorphism. So,

$$H^0_{dR}(U) \cong \mathbb{R}^{\pi_0(U)}.$$

In particular, U is connected if and only if $H^0_{dR}(U) \cong \mathbb{R}$. ◆

6.2 Homotopies and de Rham Cohomology

We begin this section promoting the de Rham complex to a contravariant functor

$$dR : \mathbf{Op} \to \mathbf{CoCh}_\mathbb{R}.$$

For a non-empty open subset $U \subseteq \mathbb{R}^n$, we put $dR(U) := (\Omega^\bullet(U), d)$. It remains to define dR on morphisms. So, let $U \subseteq \mathbb{R}^n$ and $V \subseteq \mathbb{R}^m$

be non-empty open subsets and let $F = (F^1, \ldots, F^m) : U \to V$ be a smooth map. We denote by $(x^1, \ldots, x^n)$ the standard coordinates on U and by $(y^1, \ldots, y^m)$ the standard coordinates on V. Moreover, we denote by $F^a = F^a(x^1, \ldots, x^n) \in C^\infty(U)$ the components of F, $a = 1, \ldots, m$. For every k, define the map

$$F^* : \Omega^k(V) \to \Omega^k(U),$$

mapping the k-form

$$\omega = \sum_{a_1, \ldots, a_k} \omega_{a_1 \cdots a_k} dy^{a_1} \wedge \cdots \wedge dy^{a_k} \in \Omega^k(V) \tag{6.4}$$

to the k-form

$$F^*(\omega) := \sum_{a_1, \ldots, a_k} (\omega_{a_1 \cdots a_k} \circ F) dF^{a_1} \wedge \cdots \wedge dF^{a_k}$$

$$= \sum_{a_1, \ldots, a_k} \sum_{i_1, \ldots, i_k} (\omega_{a_1 \cdots a_k} \circ F) \frac{\partial F^{a_1}}{\partial x^{i_1}} \cdots \frac{\partial F^{a_k}}{\partial x^{i_k}} dx^{i_1} \wedge \cdots \wedge dx^{i_k}. \tag{6.5}$$

It is clear that F^* is $\mathbb{R}$-linear. It is also called the *pull-back* of differential forms along F.

Example 6.10 (Restriction of a Differential Form to an Open Subset). Let $F = i_U : U \to V$ be the inclusion of a non-empty open subset $U \subseteq V \subseteq \mathbb{R}^m$. In this case, the coordinates on U are the restrictions to U of the coordinates on V and we denote both by $(y^1, \ldots, y^m)$. As the composition $f \circ i_U$ of the inclusion with a smooth function $f \in C^\infty(V)$ is just the restriction $f|_U$, in this case Formula (6.5) reduces to

$$F^*(\omega) = i_U^*(\omega) = \sum_{a_1, \ldots, a_k} \omega_{a_1 \cdots a_k}|_U dy^{a_1} \wedge \cdots \wedge dy^{a_k}.$$

The k-form $i_U^*(\omega)$ is called the *restriction* of ω to U and it is also denoted by $\omega|_U$. We conclude that *restricting a differential form to a non-empty open subset amounts to restricting its coefficients.*   ◆

Proposition 6.11. *Let $U \subseteq \mathbb{R}^n$ and $V \subseteq \mathbb{R}^m$ be non-empty open subsets, and let $F : U \to V$ be a smooth map. The family $F^* := (F^* : \Omega^k(V) \to \Omega^k(U))_{k \in \mathbb{Z}}$ is a cochain map $F^* : (\Omega^\bullet(V), d) \to (\Omega^\bullet(U), d)$. Additionally, the assignment $dR : \mathbf{Op} \to \mathbf{CoCh}_\mathbb{R}$ mapping U to its de Rham complex $(\Omega^\bullet(U), d)$ and the smooth function $F : U \to V$ to the pull-back $F^* : (\Omega^\bullet(V), d) \to (\Omega^\bullet(U), d)$ is a contravariant functor.*

Proof. For the first part of the statement, we have to prove that $d \circ F^* = F^* \circ d$. This can be done with a straightforward computation in coordinates exploiting Formulas (6.3) and (6.5). We discuss only the case $k = 1$. So, let $\omega \in \Omega^1(V)$ be given by

$$\omega = \sum_a \omega_a dy^a,$$

then

$$dF^*(\omega) = d \sum_a \sum_i (\omega_a \circ F) \frac{\partial F^a}{\partial x^i} dx^i = \sum_a \sum_{j,i} \frac{\partial}{\partial x^j}\left((\omega_a \circ F)\frac{\partial F^a}{\partial x^i}\right) dx^j \wedge dx^i$$

$$= \sum_a \sum_{j,i} \left(\frac{\partial(\omega_a \circ F)}{\partial x^j}\frac{\partial F^a}{\partial x^i} + (\omega_a \circ F)\frac{\partial^2 F^a}{\partial x^j \partial x^i}\right) dx^j \wedge dx^i.$$

As the wedge product $dx^j \wedge dx^i$ is Skew-symmetric while $\frac{\partial^2 F^a}{\partial x^j \partial x^i}$ is symmetric in the indexes j, i (Schwarz Theorem), the last summand does not contribute (do you see it?) and we get

$$dF^*(\omega) = \sum_a \sum_{j,i} \frac{\partial(\omega_a \circ F)}{\partial x^j}\frac{\partial F^a}{\partial x^i} dx^j \wedge dx^i$$

$$= \sum_{b,a} \sum_{j,i} \left(\frac{\partial \omega_a}{\partial y^b} \circ F\right)\frac{\partial F^b}{\partial x^j}\frac{\partial F^a}{\partial x^i} dx^j \wedge dx^i$$

$$= F^*\left(\sum_{b,a} \frac{\partial \omega_a}{\partial y^b} dy^b \wedge dy^a\right) = F^*(d\omega),$$

as desired. The general case is similar.

For the second part of the statement, it is clear that the pull-back id_U^* along the identity map $\mathrm{id}_U : U \to U$ is the identity: $\mathrm{id}_U^* = \mathrm{id} : \Omega^k(U) \to \Omega^k(U)$. To conclude, consider three non-empty open subsets $U \subseteq \mathbb{R}^n$, $V \subseteq \mathbb{R}^m$, $W \subseteq \mathbb{R}^p$ and two smooth maps

$$U \xrightarrow{F} V \xrightarrow{G} W.$$

We have to prove that $(G \circ F)^* = F^* \circ G^*$. We discuss again only the case $k = 1$. So, denote by $(z^1, \ldots, z^p)$ the standard coordinates on W and consider a differential 1-form

$$\rho = \sum_\alpha \rho_\alpha dz^\alpha.$$

We have

$$(G \circ F)^*(\rho) = \sum_\alpha \sum_i (\rho_\alpha \circ G \circ F) \frac{\partial (G \circ F)^\alpha}{\partial x^i} dx^i$$

$$= \sum_a \sum_\alpha \sum_i (\rho_\alpha \circ G \circ F) \left(\frac{\partial G^\alpha}{\partial y^a} \circ F \right) \frac{\partial F^a}{\partial x^i} dx^i$$

$$= F^* \left(\sum_a \sum_\alpha (\rho_\alpha \circ G) \frac{\partial G^\alpha}{\partial y^a} dy^a \right) = F^* (G^*(\omega)),$$

as desired. The general case is similar. This concludes the proof. ☐

Example 6.12. We already remarked that the de Rham complexes of $\mathbb{R}^3$ and $\mathbb{R}^2$ identify with the cochain complexes $(C_\bullet, d)$ and $(B_\bullet, d_B)$ in Example 2.34. Recall that $(C_\bullet, d)$ and $(B_\bullet, d_B)$ are intertwined by a cochain map $p : (C_\bullet, d) \to (B_\bullet, d_B)$. Now, the map $F : \mathbb{R}^2 \to \mathbb{R}^3$, $(x^1, x^2) \mapsto (x^1, x^2, 0)$ is clearly smooth, and the pull-back $F^* : (\Omega^\bullet(\mathbb{R}^3), d) \to (\Omega^\bullet(\mathbb{R}^2), d)$ identifies with $p : (C_\bullet, d) \to (B_\bullet, d_B)$ (do you see it?).   ◆

Composing the functor $dR : \mathbf{Op} \to \mathbf{CoCh}_\mathbb{R}$ with the k-th cohomology functor $H^k : \mathbf{CoCh}_\mathbb{R} \to \mathbf{Vect}_\mathbb{R}$, we get a new functor denoted as

$$H^k_{dR} : \mathbf{Op} \to \mathbf{Vect}_\mathbb{R},$$

the *k-th de Rham cohomology functor*. Given a morphism $F : U \to V$ between two objects U, V in $\mathbf{Op}$, the linear map $H^k_{dR}(F) : H^k_{dR}(V) \to H^k_{dR}(U)$ associated to it via the functor H^k_{dR} is also called the *map induced by F in the k-th de Rham cohomology*. It immediately follows from the functorial properties of the k-th de Rham cohomology that *diffeomorphic open subsets of $\mathbb{R}^n$ have isomorphic de Rham cohomologies.*

Example 6.13 (Map Induced in de Rham Cohomology by a Constant Map). Let $U \subseteq \mathbb{R}^n$ and $V \subseteq \mathbb{R}^m$ be non-empty open subsets. Take a point $y_0 \in V$ and consider the constant map $c_{y_0} : U \to V$ mapping every point $x \in U$ to y_0. Clearly, c_{y_0} is a smooth map. We want to compute the induced map in de Rham cohomology $H^k_{dR}(c_{y_0}) : H^k_{dR}(V) \to H^k_{dR}(U)$ for all k. First note that, from Formula (6.5), the pull-back along a constant map vanishes in all degrees but the 0-th one. It immediately follows that $H^k_{dR}(\mathsf{c}_{y_0}) = 0$ for all $k \neq 0$. It remains to compute

$$H^0_{dR}(\mathsf{c}_{y_0}) : H^0_{dR}(V) \to H^0_{dR}(U).$$

From Example 6.9, $H^0_{dR}(V) \cong \mathbb{R}^{\pi_0(V)}$ and $H^0_{dR}(V) \cong \mathbb{R}^{\pi_0(U)}$ (where, as usual, $\pi_0(U), \pi_0(V)$ denote the sets of connected components of U, V, respectively). It is then immediate to see that

$$H^0_{dR}(\mathsf{c}_{y_0}) : \mathbb{R}^{\pi_0(V)} \to \mathbb{R}^{\pi_0(U)}$$

maps a function $f : \pi_0(V) \to \mathbb{R}$ to the constant function whose unique value is $f(V_{y_0})$ (where V_{y_0} is the connected component of y_0 in V). ◆

We now come to smooth homotopies. Let $U \subseteq \mathbb{R}^n$ and $V \subseteq \mathbb{R}^m$ be non-empty open subsets and let $F, G : U \to V$ be smooth maps.

Definition 6.14 (Smooth Homotopy). A *smooth homotopy* between the smooth maps $F, G : U \to V$ is a smooth map $\mathcal{H} : [0,1] \times U \to V$ such that

$$\mathcal{H}(0, x) = F(x) \quad \text{and} \quad \mathcal{H}(1, x) = G(x)$$

for all $x \in U$ (in particular, $\mathcal{H}$ is a geometric homotopy). Two smooth maps are said to be *smoothly homotopic* if there exists a smooth homotopy $\mathcal{H}$ between them. In this case, we write $F \sim_{\mathcal{H}} G$.

For smooth homotopies, we adopt the same notation as for geometric homotopies denoting by $\mathcal{H}_t : U \to V$ the map defined by $\mathcal{H}_t(x) := \mathcal{H}(t, x)$ for all $x \in U$. "Being smoothly homotopic" is an equivalence relation on the set of smooth maps between given non-empty open subsets $U \subseteq \mathbb{R}^n$, $V \subseteq \mathbb{R}^m$. Reflexivity and symmetry can be proved exactly as for geometric homotopies while transitivity is a little bit more intricate in this case. We provide just the idea of the proof. Given smooth maps $F, G, L : U \to V$ and smooth homotopies $\mathcal{H}, \mathcal{K}$ such that $F \sim_{\mathcal{H}} G$ and $G \sim_{\mathcal{K}} L$ we first construct the geometric homotopy $\mathcal{H} * \mathcal{K}$ exactly as in the proof of Proposition 5.12. In general, $\mathcal{H} * \mathcal{K}$ is continuous but it is only smooth around points $(t, x) \in [0,1] \times U$ with $t \neq 1/2$. However, it is possible to "*smooth out*" $\mathcal{H} * \mathcal{K}$ in a small neighborhood $\mathcal{U}$ of $\{1/2\} \times U$ leaving it unchanged outside $\mathcal{U}$ so that it is still a (now smooth) homotopy.

Example 6.15. The same exact argument as in Example 5.14 works in the smooth setting and shows that any two smooth maps $F, G : U \to V$ are homotopic if $V \subseteq \mathbb{R}^m$ is a convex non-empty open subset (e.g., $\mathbb{R}^m$ itself). ◆

Proposition 6.16. *Smooth homotopies respect the composition of smooth maps.*

Proof.   The same as for Proposition 5.15. $\square$

Theorem 6.17. *Let $F, G : U \to V$ be smoothly homotopic smooth maps between non-empty open subsets $U \subseteq \mathbb{R}^n$ and $V \subseteq \mathbb{R}^m$. Then F, G induce the same map in de Rham cohomology:*

$$H^k_{dR}(F) = H^k_{dR}(G), \quad \text{for all } k \in \mathbb{Z}.$$

Proof.   Let $\mathcal{H} : [0,1] \times U \to V$ be a smooth homotopy between F and G. This means that $\mathcal{H}_0 = F$ and $\mathcal{H}_1 = G$. Now, take a differential form $\omega \in \Omega^k(V)$. We want to compare the k-forms $G^*(\omega), F^*(\omega)$. To do this, we compute

$$G^*(\omega) - F^*(\omega) = \mathcal{H}_1^*(\omega) - \mathcal{H}_0^*(\omega) = \int_0^1 \frac{d\mathcal{H}_t^*(\omega)}{dt} \, dt. \qquad (6.6)$$

The last equality might be intuitive but actually needs some explanations: Both the integral and the derivative in the last term are computed component-wise. Namely, let $(x^1, \ldots, x^n)$ and $(y^1, \ldots, y^m)$ be standard coordinates on U and V, respectively. Let $(\Omega_t)_{t \in [0,1]}$ be a 1-parameter family of differential k-forms on U of the type

$$\Omega_t = \sum_{i_1, \ldots, i_k} \Omega_{i_1 \cdots i_k}(t, x^1, \ldots, x^n) dx^{i_1} \wedge \cdots \wedge dx^{i_k},$$

where $\Omega_{i_1 \cdots i_k}(t, x^1, \ldots, x^n)$ are smooth functions of both the variables $(x^1, \ldots, x^n)$ and t. Any such family is called a *smooth 1-parameter family of differential forms*. For any such family $(\Omega_t)_{t \in [0,1]}$, it makes sense to consider the families

$$\left(\frac{d}{d\tau} \Big|_{\tau = t} \Omega_\tau \right)_{t \in [0,1]} \quad \text{and} \quad \left(\int_0^t \Omega_\tau \, d\tau \right)_{t \in [0,1]}$$

defined by

$$\frac{d}{d\tau} \Big|_{\tau = t} \Omega_\tau := \sum_{i_1, \ldots, i_k} \left(\frac{\partial}{\partial \tau} \Big|_{\tau = t} \Omega_{i_1 \cdots i_k}(\tau, x^1, \ldots, x^n) \right) dx^{i_1} \wedge \cdots \wedge dx^{i_k}, \quad (6.7)$$

and

$$\int_0^t \Omega_\tau \, d\tau = \sum_{i_1, \ldots, i_k} \left(\int_0^t \Omega_{i_1 \cdots i_k}(\tau, x^1, \ldots, x^n) \, d\tau \right) dx^{i_1} \wedge \cdots \wedge dx^{i_k}, \quad (6.8)$$

respectively, and they are smooth 1-parameter families again. A direct computation also shows that taking "time" derivatives and integrals of smooth 1-parameter families commute with the de Rham differential, i.e., for any smooth 1-parameter family of differential forms $(\Omega_t)_{t\in[0,1]}$, the family $(d\Omega_t)_{t\in[0,1]}$ is a smooth 1-parameter family again and moreover

$$\frac{d}{d\tau}\Big|_{\tau=t} d\Omega_\tau = d\frac{d}{d\tau}\Big|_{\tau=t}\Omega_\tau \quad \text{and} \quad \int_0^t d\Omega_\tau\, d\tau = d\int_0^t \Omega_\tau\, d\tau \quad \text{for all } t \in [0,1].$$

Now, let

$$\omega = \sum_{a_1,\dots,a_k} \omega_{a_1\cdots a_k} dy^{a_1} \wedge \cdots \wedge dy^{a_k}.$$

From Formula (6.5) we easily see that $(\mathcal{H}_t^*(\omega))_{t\in[0,1]}$ is a smooth 1-parameter family of differential forms so that the last term in (6.6) makes sense. The last equality in (6.6) immediately follows from Definitions (6.7) and (6.8) and the Fundamental Theorem of Calculus (do you see it?). Next, for all $t \in [0,1]$ we define an $\mathbb{R}$-linear map

$$i_t^{\mathcal{H}} : \Omega^k(V) \to \Omega^{k-1}(U)$$

by putting

$$i_t^{\mathcal{H}}\omega := \sum_{a_1,\dots,a_k} \sum_{j=1}^k (-)^{j+1}\left(\omega_{a_1\cdots a_k} \circ \mathcal{H}_t\right) \frac{d}{dt}\mathcal{H}^{a_t} d\mathcal{H}_t^{a_1} \wedge \cdots \wedge \widehat{d\mathcal{H}_t^{a_j}} \wedge \cdots \wedge d\mathcal{H}_t^{a_k}.$$

A direct computation that we omit (but the brave reader is invited to try to perform it) shows that, for all $\omega \in \Omega^k(V)$,

$$\frac{d\mathcal{H}_t^*(\omega)}{dt} = di_t^{\mathcal{H}}\omega + i_t^{\mathcal{H}}d\omega.$$

The latter formula is sometimes referred to as the *Infinitesimal Homotopy Formula*. Integrating both sides of the Infinitesimal Homotopy Formula and using (6.6) we find

$$G^*(\omega) - F^*(\omega) = \int_0^1 \left(di_t^{\mathcal{H}}\omega + i_t^{\mathcal{H}}d\omega\right) dt = \int_0^1 di_t^{\mathcal{H}}\omega\, dt + \int_0^1 i_t^{\mathcal{H}}d\omega\, dt$$

$$= d\int_0^1 i_t^{\mathcal{H}}\omega\, dt + \int_0^1 i_t^{\mathcal{H}}d\omega\, dt.$$

This shows that the linear operator

$$h^{\mathcal{H}} : \Omega^k(V) \to \Omega^{k-1}(U),$$

defined by putting

$$h^{\mathcal{H}}(\omega) := \int_0^1 i_t^{\mathcal{H}} \omega \, dt,$$

is an algebraic homotopy between the cochain maps $G^*, F^* : (\Omega^\bullet(V), d) \to (\Omega^\bullet(U), d)$. This concludes the proof. $\square$

Corollary 6.18. *If $F : U \to V$ is a null-homotopic smooth map between non-empty open subsets $U \subseteq \mathbb{R}^n$ and $V \subseteq \mathbb{R}^m$, then $H^k_{dR}(F) = 0$ for all $k \neq 0$.*

Example 6.19. For any smooth map $F : U \to V$ between non-empty open subsets $U \subseteq \mathbb{R}^n$ and $V \subseteq \mathbb{R}^m$, with V convex we have $H^k_{dR}(F) = 0$ for all $k \neq 0$.   ◆

Definition 6.20 (Smooth Homotopy Equivalence). A smooth map $F : U \to U'$ between non-empty open subsets $U \subseteq \mathbb{R}^n$ and $U' \subseteq \mathbb{R}^{n'}$ is a *smooth homotopy equivalence* if there exists a smooth map $G : U' \to U$ in the other direction such that $G \circ F$ is smoothly homotopic to the identity of U and $F \circ G$ is smoothly homotopic to the identity of U'. In this situation, we also say that G is a *smooth homotopy inverse* of F (and vice-versa) or that *G inverts F up to smooth homotopies*. If U, U' are connected by a smooth homotopy equivalence, we say that they are *smoothly homotopy equivalent*.

Proposition 6.21. *Let $F : U \to U'$ be a smooth homotopy equivalence between non-empty open subsets $U \subseteq \mathbb{R}^n$ and $U' \subseteq \mathbb{R}^{n'}$, and let $G : U' \to U$ be a smooth homotopy inverse of F. Then F, G induce mutually inverse vector space isomorphisms in de Rham cohomology, i.e., $H^k_{dR}(F) : H^k_{dR}(U') \to H^k_{dR}(U)$ and $H^k_{dR}(G) : H^k_{dR}(U) \to H^k_{dR}(U')$ are vector space isomorphisms and*

$$H^k_{dR}(F)^{-1} = H^k_{dR}(G) \quad \text{for all } k \in \mathbb{Z}.$$

In particular, smoothly homotopy equivalent open subsets in some standard Euclidean space have isomorphic de Rham cohomologies.

Proof. Formally identical to the proof of Proposition 5.20 (up to some minor changes that we leave to the reader). $\square$

Definition 6.22 (Smoothly Contractible Open Subset). A non-empty open subset $U \subseteq \mathbb{R}^n$ is *smoothly contractible* if there exists a point $x_0 \in U$ such that the constant map $c_{x_0} : U \to U$ is smoothly homotopic to the identity of U.

Proposition 6.23. *Let $U \subseteq \mathbb{R}^n$ be a smoothly contractible open subset. Then*

$$H_{dR}^k(U) \cong \begin{cases} \mathbb{R} & \text{if } k = 0 \\ 0 & \text{otherwise} \end{cases}.$$

Proof. Let $x_0 \in U$ be a point as in Definition 6.22. Then the map $\mathbb{R}^0 \to U$, $0 \mapsto x_0$ is a smooth homotopy equivalence with smooth homotopy inverse given by $U \to \mathbb{R}^0$, $x \mapsto 0$ (do you see it?). It follows that $H_{dR}^k(U) \cong H_{dR}^k(\mathbb{R}^0)$ and the statement follows from Example 6.8. $\square$

Exercise 6.4. Let $U \subseteq \mathbb{R}^n$ and $V \subseteq \mathbb{R}^m$ be non-empty open subsets. Prove that if V is smoothly contractible, then the non-empty open subset $U \times V \subseteq \mathbb{R}^n \times \mathbb{R}^m = \mathbb{R}^{n+m}$ is smoothly homotopy equivalent to U.

Example 6.24 ($\mathbb{R}^n$ is Smoothly Contractible). Recall that an open subset $U \subseteq \mathbb{R}^n$ is *star-shaped* if there exists a point $x_0 \in U$ such that, for all $x \in U$, the segment joining x_0 and x is entirely contained into U. In this case, x_0 is called a *star center* for U. For instance, any (non-empty) convex open subset U is star-shaped and any point in U is a star center. Any star-shaped open subset $U \subseteq \mathbb{R}^n$ is smoothly contractible. Indeed, let $x_0 \in U$ be a star center. Then the map

$$\mathcal{H} : [0,1] \times U \to U, \quad (t,x) \mapsto tx + (1-t)x_0 \tag{6.9}$$

is a well-defined smooth homotopy between the constant map $c_{x_0} : U \to U$ and the identity of U. It immediately follows from Proposition 6.23 that, if $U \subseteq \mathbb{R}^n$ is a star-shaped open subset (for instance, U is a convex open subset), then

$$H_{dR}^k(U) \cong \begin{cases} \mathbb{R} & \text{if } k = 0 \\ 0 & \text{otherwise} \end{cases}.$$

In particular,

$$H^k_{dR}(\mathbb{R}^n) \cong \begin{cases} \mathbb{R} & \text{if } k = 0 \\ 0 & \text{otherwise} \end{cases}.$$

We already remarked that the cochain complex in Example 2.27 is canonically isomorphic to the de Rham complex of $\mathbb{R}^3$. The algebraic homotopy h defined therein does actually agree with the algebraic homotopy $h^{\mathcal{H}}$ from the proof of Theorem 6.17 where

$$\mathcal{H} : [0, 1] \times \mathbb{R}^3 \to \mathbb{R}^3, \quad (t, x) \mapsto tx$$

is the smooth homotopy constructed as in (6.9) with $x_0 = 0$ ($\mathbb{R}^3$ is star-shaped and 0 is a star center for it). ◆

The following theorem is an easy consequence of Example 6.24.

Theorem 6.25 (Poincaré Lemma). *Let $V \subseteq \mathbb{R}^n$ be a non-empty open subset and let k be a positive integer. Every closed differential k-form $\omega \in \Omega^k(V)$ on V is locally exact, i.e., for every $x_0 \in V$, there exists an open neighborhood $U \subseteq V$ of x_0, and a differential $(k-1)$-form $\rho \in \Omega^{k-1}(U)$ such that $\omega|_U = d\rho$.*

Proof.   Let $\omega \in \Omega^k(V)$ be a closed differential form, i.e., $d\omega = 0$, and let $x_0 \in V$. Choose a star-shaped open neighborhood $U \subseteq V$ of x_0. It always exists (we can take, e.g., an open disk centered in x_0 and entirely contained in V). The restriction map $\Omega^k(V) \to \Omega^k(U)$, $\eta \mapsto \eta|_U$ is (the pull-back along the inclusion $i_U : U \to V$, hence it is) a cochain map. So, $\omega|_U \in \Omega^k(U)$ is a closed differential form on U (see the discussion immediately preceding Proposition 6.11 about restricting a differential form to an open subset). But $k > 0$ so, from Example 6.24, $H^k(U) = 0$. This means that $\ker d = \text{im}\, d$ and every closed differential form on U is exact. □

Smooth homotopy equivalence is an equivalence relation. The proof of this fact is very similar to that of Proposition 5.26 and we leave the details to the reader (but take into account the discussion following Definition 6.14).

Remark 6.26. Let $U \subseteq \mathbb{R}^n$ and $V \subseteq \mathbb{R}^m$ be non-empty open subsets and let $F, G : U \to V$ be smooth maps. In particular, F, G are continuous maps and it makes sense to wonder whether there is a continuous homotopy between them (in which case, they induce the same map in singular homology). It can actually be proved that if such a continuous homotopy exists, then a smooth homotopy exists as well. The proof is based on an appropriate *approximation technique* of continuous maps by smooth maps

(Lee, 2013, Chapter 6). We conclude that two smooth maps $F, G : U \to V$ induce the same map in de Rham cohomology provided only they are *continuously homotopic*.

Even more, one can show that any continuous map between non-empty open subsets in some standard Euclidean spaces is homotopic to a smooth map and, using this, we can conclude that two such open subsets are smoothly homotopy equivalent if and only if they are continuously homotopy equivalent. ◇

Remark 6.27. Consider the category **Op** of non-empty open subsets in some standard Euclidean space. Define a new category **hOp** as follows. The objects in **hOp** are the same as in **Op**. In order to define morphisms, recall that "being smoothly homotopic" is an equivalence relation on the set $\mathrm{Hom}_{\mathbf{Op}}(U, V)$ of smooth maps between non-empty open subsets $U \subseteq \mathbb{R}^n$, $V \subseteq \mathbb{R}^m$. Denote by $\sim$ this equivalence relation and, for any two non-empty open subsets $U \subseteq \mathbb{R}^n$, $V \subseteq \mathbb{R}^m$, put

$$\mathrm{Hom}_{\mathbf{hOp}}(U, V) := \mathrm{Hom}_{\mathbf{Op}}(U, V)/\sim ,$$

the set of *smooth homotopy classes* of smooth maps. Given a smooth map $F : U \to V$, we denote by $[F]_\sim \in \mathrm{Hom}_{\mathbf{hOp}}(U, V)$ its smooth homotopy class. The composition law of morphisms in **hOp** is defined as follows. Let

$$U \xrightarrow{F} V \xrightarrow{G} W$$

be smooth maps between non-empty open subsets $U \subseteq \mathbb{R}^n$, $V \subseteq \mathbb{R}^m$, $W \subseteq \mathbb{R}^p$. We put

$$[G]_\sim \circ [F]_\sim := [G \circ F]_\sim.$$

As smooth homotopies respect the composition of continuous maps, this is well defined (do you see it?). The composition law in **hOp** defined in this way is clearly associative. The units are the smooth homotopy classes of the identity maps. The isomorphisms in **hOp** are the (smooth homotopy classes of) smooth homotopy equivalences (do you see it?). The category **hOp** is called the *homotopy category of* **Op**.

It should now be clear that the de Rham complex can also be seen as a contravariant functor

$$dR_\bullet : \mathbf{hOp} \to \mathbf{hCh}_{\mathbb{R}}$$

from the homotopy category of **Op** to the homotopy category of chain complexes (of real vector spaces). Similarly, for all $k \in \mathbb{Z}$, the k-th de Rham cohomology functor can be seen as a functor

$$H^k_{dR} : \mathbf{hOp} \to \mathbf{Vect}_{\mathbb{R}}.$$ ◇

6.3 Mayer–Vietoris Sequence in de Rham Cohomology

Let $W \subseteq \mathbb{R}^n$ be a non-empty open subset and let $\{U, V\}$ be an open cover of W, i.e., $U, V \subseteq X$ are (non-empty) open subsets such that $W = U \cup V$. We assume that $U \cap V \neq \emptyset$. We have a commuting diagram of smooth maps:

$$
\begin{array}{ccc}
 & W & \\
{}^{i_U}\nearrow & & \nwarrow {}^{i_V} \\
U & & V \\
\nwarrow {}_{j_U} & & \nearrow {}_{j_V} \\
 & U \cap V &
\end{array}
\tag{6.10}
$$

where the arrows are the inclusions. Applying the de Rham complex functor to Diagram (6.10), we get a commuting diagram of cochain maps:

$$
\begin{array}{ccc}
 & (\Omega^\bullet(W), d) & \\
{}^{i_U^*}\swarrow & & \searrow {}^{i_V^*} \\
(\Omega^\bullet(U), d) & & (\Omega^\bullet(V), d). \\
{}_{j_U^*}\searrow & & \swarrow {}_{j_V^*} \\
 & (\Omega^\bullet(U \cap V), d) &
\end{array}
$$

We can combine the top cochain maps in a single chain map

$$
i^* : (\Omega^\bullet(W), d) \to (\Omega^\bullet(U) \oplus \Omega^\bullet(V), d^\oplus)
$$
$$
\omega \mapsto i^*\omega := (i_U^*\omega, i_V^*\omega) = (\omega|_U, \omega|_V).
$$

We can also combine the bottom cochain maps in a single chain map

$$
j^* : (\Omega^\bullet(U) \oplus \Omega^\bullet(V), d^\oplus) \to (\Omega^\bullet(U \cap V), d)
$$
$$
(\omega_U, \omega_V) \mapsto j^*(\omega_U, \omega_V) := j_V^*\omega_V - j_U^*\omega_V = \omega_V|_{U \cap V} - \omega_U|_{U \cap V}.
$$

Hence, we get a sequence of cochain maps

$$
0 \longrightarrow (\Omega^\bullet(W), d) \xrightarrow{i^*} (\Omega^\bullet(U) \oplus \Omega^\bullet(V), d^\oplus) \xrightarrow{j^*} (\Omega^\bullet(U \cap V), d) \longrightarrow 0.
\tag{6.11}
$$

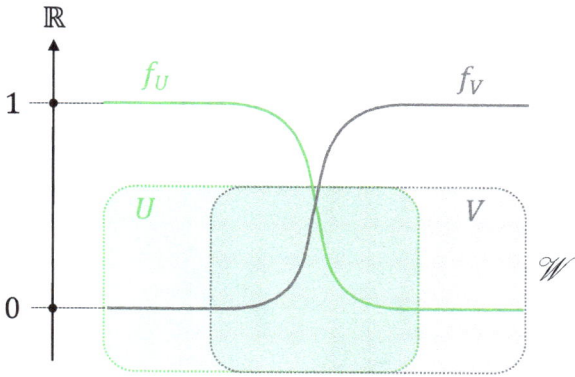

Figure 6.2. A partition of unity.

Lemma 6.28. *The sequence (6.11) is a short exact sequence of cochain complexes.*

Proof. It is clear that i^* is injective. Now, take $\rho \in \Omega^k(\mathcal{W})$ and compute

$$j^*\left(i^*(\omega)\right) = j^*(\omega|_U, \omega|_V) = \omega|_V|_{U\cap V} - \omega|_U|_{U\cap V} = \omega|_{U\cap V} - \omega|_{U\cap V} = 0.$$

This shows that $\operatorname{im} i^* \subseteq \ker j^*$. Next, take $(\omega_U, \omega_V) \in \ker j^*$. So,

$$0 = j^*(\omega_U, \omega_V) = \omega_V|_{U\cap V} - \omega_U|_{U\cap V},$$

i.e., $\omega_U|_{U\cap V} = \omega_V|_{U\cap V}$. In other words, ω_U, ω_V agree on $U \cap V$. This clearly implies that there exists a (necessarily unique) differential form ω on $\mathcal{W}$ such that $\omega|_U = \omega_U$ and $\omega|_V = \omega_V$ (do you see it?). In other words, $(\omega_U, \omega_V) = i^*\omega$. So, $\ker j^* \subseteq \operatorname{im} i^*$, hence $\ker j^* = \operatorname{im} i^*$.

To conclude, we have to show that j^* is surjective. This is done with a technical trick. One can prove that there exist two (non-unique) smooth functions $f_U, f_V \in C^\infty(\mathcal{W})$ such that $f_U + f_V = 1$ and, additionally, the support of f_U is entirely contained in U while the support of f_V is entirely contained into V (Figure 6.2). Recall that the support of a function $f : \mathcal{W} \to \mathbb{R}$ is the topological closure in $\mathcal{W}$ of the subset

$$\{x \in \mathcal{W} : f(x) \neq 0\}.$$

Any pair $\{f_U, f_V\}$ as above is called a *partition of unity* (subordinate to the open cover $\{U, V\}$ of $\mathcal{W}$). The existence of partitions of unity, particularly in our simple setting, is not hard to prove but we prefer to omit the technical details (but see, e.g., Lee, 2013, Chapter 2 where partitions of unity in the general setting of smooth manifolds are discussed).

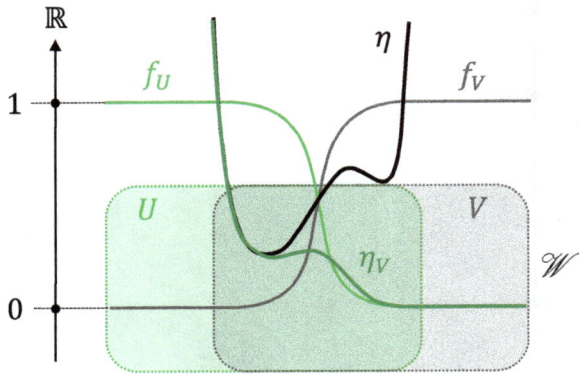

Figure 6.3.   The form η_V in the proof of Lemma 6.28.

So, let $\{f_U, f_V\}$ be a partition of unity. Take a differential form $\eta \in \Omega^k(U \cap V)$ and consider the differential forms

$$\eta_V' := f_U|_{U \cap V} \cdot \eta, \quad \text{and} \quad \eta_U' := -f_V|_{U \cap V} \cdot \eta.$$

We have

$$\eta_V' - \eta_U' = f_U|_{U \cap V} \cdot \eta + f_V|_{U \cap V} \cdot \eta = (f_U|_{U \cap V} + f_V|_{U \cap V}) \eta$$
$$= (f_U + f_V)|_{U \cap V} \cdot \eta = \eta.$$

Moreover, as the support of f_U is contained into U, there exists a unique differential form $\eta_V \in \Omega^k(V)$ such that $\eta_V|_{U \cap V} = \eta_V'$ and whose coefficients all vanish in $V \smallsetminus U$ (Figure 6.3). Similarly, there exists a differential form $\eta_U \in \Omega^k(U)$ such that $\eta_U|_{U \cap V} = \eta_U'$. Consider $(\eta_U, \eta_V) \in \Omega^k(U) \oplus \Omega^k(V)$ and compute

$$j^*(\eta_U, \eta_V) = \eta_V|_{U \cap V} - \eta_U|_{U \cap V} = \eta_V' - \eta_U' = \eta.$$

This shows that j^* is surjective and concludes the proof.   □

We are now ready to state the main result of this section.

Theorem 6.29 (Smooth Mayer–Vietoris Theorem). *Let $\mathcal{W} \subseteq \mathbb{R}^n$ be a non-empty open subset and let $U, V \subseteq \mathcal{W}$ be non-empty open subsets such that $\mathcal{W} = U \cup V$ and $U \cap V \neq \varnothing$. Then, for every $k \in \mathbb{Z}$ there exists a linear map $\Delta : H_{dR}^{k-1}(U \cap V) \to H_{dR}^k(\mathcal{W})$ such that the following sequence of linear maps*

$$\cdots \xrightarrow{H(j^*)} H_{dR}^{k-1}(U \cap V) \xrightarrow{\Delta} H_{dR}^k(\mathcal{W}) \xrightarrow{H(i^*)} H_{dR}^k(U) \oplus H_{dR}^k(V) \xrightarrow{H(j^*)} H_{dR}^k(U \cap V) \longrightarrow \cdots$$

(6.12)

is exact. The maps Δ are natural in the sense that if $\mathcal{W}' \subseteq \mathbb{R}^{n'}$ is another non-empty open subset, $U', V' \subseteq \mathcal{W}'$ are non-empty open subsets such that $\mathcal{W}' = U' \cup V', U' \cap V' \neq \varnothing$, and $F : \mathcal{W} \to \mathcal{W}'$ is a smooth map such that $F(U) \subseteq U'$ and $F(V) \subseteq V'$, then the following diagram

$$\cdots \xrightarrow{H(j^*)} H_{dR}^{k-1}(U \cap V) \xrightarrow{\Delta} H_{dR}^k(\mathcal{W}) \xrightarrow{H(i^*)} H_{dR}^k(U) \oplus H_{dR}^k(V) \xrightarrow{H(j^*)} H_{dR}^k(U \cap V) \longrightarrow \cdots$$

$$\cdots \xrightarrow{H(j^*)} H_{dR}^{k-1}(U' \cap V') \xrightarrow{\Delta} H_{dR}^k(\mathcal{W}') \xrightarrow{H(i^*)} H_{dR}^k(U') \oplus H_{dR}^k(V') \xrightarrow{H(j^*)} H_{dR}^k(U' \cap V') \longrightarrow \cdots$$

commutes, where the vertical arrows are the maps induced by F in the obvious way.

Proof. It is enough to consider the long exact sequence induced in cohomology by the short exact sequence of cochain complexes (6.11). We leave the details to the reader. We only recall how the connecting homomorphism $\Delta : H_{dR}^{k-1}(U \cap V) \to H_{dR}^k(\mathcal{W})$ acts. Take a closed $(k-1)$-form $\eta \in H_{dR}^{k-1}(U \cap V)$. Using, e.g., a partition of unity as in the proof of Lemma 6.28, find a cochain $(\eta_U, \eta_V) \in \Omega^{k-1}(U) \oplus \Omega^{k-1}(V)$ such that $j^*(\eta_U, \eta_V) = \eta$. Take its differential $d^{\oplus}(\eta_U, \eta_V) = (d\eta_U, d\eta_V)$ and note that $(d\eta_U, d\eta_V) \in \ker j^* = \operatorname{im} i^*$. Hence, there exists (a unique) closed k-form $\omega \in \Omega^k(\mathcal{W})$ such that $i^*(\omega) = (d\eta_U, d\eta_V)$, i.e., $\omega|_U = d\eta_U$ and $\omega|_V = d\eta_V$. Finally, $\Delta[\eta] = [\omega]$. $\square$

Definition 6.30 (Mayer–Vietoris Sequence in de Rham Cohomology). The sequence (6.12) is called the *Mayer–Vietoris sequence* in de Rham cohomology (associated with the open cover $\{U, V\}$ of $\mathcal{W}$).

Example 6.31 (de Rham Cohomology of the Punctured Euclidean Space). As an application of the Smooth Mayer–Vietoris Theorem, we compute the de Rham cohomology of the punctured space $\mathbb{R}_\times^{n+1} := \mathbb{R}^{n+1} \smallsetminus \{0\}$. More precisely, we prove that, for all $n > 0$,

$$H_{dR}^k(\mathbb{R}_\times^{n+1}) = \begin{cases} \mathbb{R} & \text{if } k = 0, n \\ 0 & \text{otherwise} \end{cases}. \tag{6.13}$$

First of all, note that, for $n > 0$, $\mathbb{R}_\times^{n+1}$ is connected (and path connected, do you see it?) hence, from Example 6.9, $H_{dR}^0(\mathbb{R}_\times^{n+1}) = \mathbb{R}$. The rest of the proof is by induction on n and exploits both smooth homotopy equivalence and Mayer–Vietoris arguments (actually the present proof closely parallels that in Example 5.35). Consider preliminarily the case $n = 0$ (which is not in the statement but will be useful anyway). The punctured line $\mathbb{R}_\times$ has two connected components: $\mathbb{R}_+ := \{\text{positive reals}\}$ and $\mathbb{R}_- := \{\text{negative reals}\}$. Hence, $H^0_{dR}(\mathbb{R}_\times) = \mathbb{R}^2$. The first de Rham cohomology $H^1_{dR}(\mathbb{R}_\times)$ vanishes in this case. Indeed, a differential 1-form ω on $\mathbb{R}_\times$ is the same as a pair (ω_+, ω_-) with $\omega_\pm \in \Omega^1(\mathbb{R}_\pm)$ (do you see it?). Both $\mathbb{R}_\pm$ are diffeomorphic to $\mathbb{R}$ (for instance, the logarithm $\log : \mathbb{R}_+ \to \mathbb{R}$ is a diffeomorphism with inverse diffeomorphism given by the exponential map). As $H^1_{dR}(\mathbb{R}) = 0$ from Example 6.24, both $\omega_\pm$ are exact 1-forms, i.e., there exist functions $f_\pm \in C^\infty(\mathbb{R}_\pm)$ such that $\omega_\pm = df_\pm$. But the pair (f_+, f_-) can be seen as a smooth function f on $\mathbb{R}_\times$, and from $\omega_\pm = df_\pm$ we get $\omega = df$ as desired.

We now pass to the generic case $n > 0$. Denote by $(x^1, \ldots, x^{n+1})$ the coordinates on $\mathbb{R}^{n+1}$. In $\mathbb{R}^{n+1}_\times$, consider the open subsets

$$U_\pm := \mathbb{R}^{n+1} \setminus \text{half line of non-negative/non-positive } x^{n+1}.$$

We have $U_+ \cup U_- = \mathbb{R}^{n+1}_\times$, and

$$U_+ \cap U_- = \mathbb{R}^{n+1} \setminus x^{n+1} \text{ axis}.$$

The open subsets U_+, U_- are both diffeomorphic to $\mathbb{R}^{n+1}$. For instance, the smooth map

$$U_+ \to \mathbb{R}^{n+1}, \quad x \mapsto \left(\varphi_+(x/\|x\|), \log\|x\| \right)$$

where φ_+ is the stereographic projection from the north (see Example 5.35) is a diffeomorphism inverted by the smooth map

$$\mathbb{R}^{n+1} \to U_+, \quad y = (y^1, \ldots, y^{n+1}) \mapsto \exp(y^{n+1})\varphi_+^{-1}(y^1, \ldots, y^n).$$

The intersection $U_+ \cap U_-$ is homotopy equivalent to the punctured space $\mathbb{R}^n_\times$, indeed $U_+ \cap U_- = \mathbb{R}^n_\times \times \mathbb{R}$ and the claim follows from Exercise 6.4. We conclude that the de Rham cohomologies of $U_\pm$ are the same as those of $\mathbb{R}^{n+1}$:

$$H^k_{dR}(U_\pm) = \begin{cases} \mathbb{R} & \text{if } k = 0 \\ 0 & \text{otherwise} \end{cases},$$

and the de Rham cohomologies of $U_+ \cap U_-$ are the same as those of the punctured space $\mathbb{R}^n_\times$:

$$H^k_{dR}(U_+ \cap U_-) = H^k(\mathbb{R}^n_\times), \quad k \geq 0.$$

We are now ready to discuss the base of induction: $n = 1$. We already discussed the de Rham cohomologies of the punctured line $\mathbb{R}_\times$. We conclude that, when $n = 1$,

$$H^k_{dR}(U_+ \cap U_-) = H^k_{dR}(\mathbb{R}_\times) = \begin{cases} \mathbb{R}^2 & \text{if } k = 0 \\ 0 & \text{otherwise} \end{cases}.$$

We need to drop one more word on the isomorphisms $H^0_{dR}(U_\pm) \cong \mathbb{R}$ and $H^0_{dR}(U_+ \cap U_-) \cong \mathbb{R}^2$. For the first one, $H^0_{dR}(U_\pm) = \ker(d : C^\infty(U_\pm) \to \Omega^1(U_\pm))$ is generated by the constant function 1, which we identify with the generator $1 \in \mathbb{R}$ (as we always do in the connected case). As for the isomorphism $H^0_{dR}(U_+ \cap U_-) \cong \mathbb{R}^2$, the two generators of the real vector space $H^0_{dR}(U_+ \cap U_-) \cong \mathbb{R}^2$ are the two functions $f_\pm$, where $f_\pm$ is 1 where $\pm x^1 > 0$ and 0 elsewhere. We are identifying this two generators with $(1, 0) \in \mathbb{R}^2$ and $(0, 1) \in \mathbb{R}^2$, respectively (beware that changing these identifications might change some formulas).

Now, the Mayer–Vietoris sequence associated to the open cover $\{U_+, U_-\}$ of $\mathbb{R}^2_\times$ is

$$0 \longrightarrow H^0_{dR}(\mathbb{R}^2_\times) \xrightarrow{H(i^*)} H^0_{dR}(U_+) \oplus H^0_{dR}(U_-) \xrightarrow{H(j^*)} H^0_{dR}(U_+ \cap U_-) \xrightarrow{\Delta} H^1_{dR}(\mathbb{R}^2_\times) \longrightarrow 0 \longrightarrow \cdots$$
$$0 \longrightarrow \mathbb{R} \longrightarrow \mathbb{R}^2 \longrightarrow \mathbb{R}^2 \longrightarrow H^1_{dR}(\mathbb{R}^2_\times) \longrightarrow 0$$

$$(6.14)$$

in low degree and

$$\cdots \longrightarrow 0 \longrightarrow H^k_{dR}(\mathbb{R}^2_\times) \longrightarrow 0 \longrightarrow \cdots \qquad (6.15)$$

in higher degree $k > 1$.

The map

$$H(j^*) : H^0_{dR}(U_+) \oplus H^0_{dR}(U_-) \to H^0_{dR}(U_+ \cap U_-)$$

in (6.14) is given by

$$H^0_{dR}(U_+) \oplus H^0_{dR}(U_-) \xrightarrow{H(j^*)} H^0_{dR}(U_+ \cap U_-)$$
$$\mathbb{R}^2 \longrightarrow \mathbb{R}^2$$

$$(a_1, a_2) \longmapsto (a_2 - a_1, a_2 - a_1).$$

Therefore, the image I of $H(j^*)$ is

$$I := \operatorname{im} H(j^*) = \left\{ (a, a) \in \mathbb{R}^2 : a \in \mathbb{R} \right\} \subseteq \mathbb{R}^2,$$

and the cokernel $H^0_{dR}(U_+ \cap U_-)/\operatorname{im} H(j^*)$ is canonically isomorphic to $\mathbb{R}$ via the map $(a, b) \bmod \operatorname{im} H(j^*) \mapsto b - a$. From exactness (of the Mayer–Vietoris sequence), $H^1_{dR}(\mathbb{R}^2_\times)$ is isomorphic to the latter cokernel. More precisely, there is a unique isomorphism $H^1_{dR}(\mathbb{R}^2_\times) \cong \mathbb{R}$ identifying

$\Delta : H^0_{dR}(U_+ \cap U_-) = \mathbb{R}^2 \to H^1_{dR}(\mathbb{R}^2_\times)$ with the linear map $\mathbb{R}^2 \to \mathbb{R}$, $(a, b) \mapsto b - a$. Finally, from (6.15), $H^k_{dR}(\mathbb{R}^2_\times) = 0$ for higher k. This proves the base of induction.

Next, assume that claim (6.13) is correct for $n = m$ and prove it for $n = m + 1$. In the latter case, $U_+ \cap U_-$ has only one path connected component so that $H^0_{dR}(U_+ \cap U_-) = \mathbb{R}$ and the Mayer–Vietoris sequence associated with the open cover $\{U_+, U_-\}$ is

$$0 \longrightarrow H^0_{dR}(\mathbb{R}^{m+2}_\times) \xrightarrow{H(i^*)} H^0_{dR}(U_+) \oplus H^0_{dR}(U_-) \xrightarrow{H(j^*)} H^0_{dR}(U_+ \cap U_-) \xrightarrow{\Delta} H^1_{dR}(\mathbb{R}^{m+2}_\times) \longrightarrow 0 \longrightarrow \cdots$$
$$\shortparallel \qquad\qquad \shortparallel \qquad\qquad \shortparallel \qquad\qquad \shortparallel$$
$$0 \longrightarrow \mathbb{R} \longrightarrow \mathbb{R}^2 \longrightarrow \mathbb{R} \longrightarrow H^1_{dR}(\mathbb{R}^{m+2}_\times) \longrightarrow 0 \tag{6.16}$$

in low degree and

$$\cdots \longrightarrow 0 \longrightarrow H^{k-1}_{dR}(U_+ \cap U_-) \xrightarrow{\Delta} H^k_{dR}(\mathbb{R}^{m+2}_\times) \longrightarrow H^k_{dR}(U_+) \oplus H^k_{dR}(U_-) \longrightarrow \cdots$$
$$\shortparallel \qquad\qquad \shortparallel$$
$$0 \longrightarrow H^{k-1}_{dR}(\mathbb{R}^{m+1}_\times) \longrightarrow H^k_{dR}(\mathbb{R}^{m+2}_\times) \longrightarrow 0 \tag{6.17}$$

in higher degree $k > 1$.

The map $H(j^*) : H^0_{dR}(U_+) \oplus H^0_{dR}(U_-) \to H^0_{dR}(U_+ \cap U_-)$ in (6.16) is given by

$$H^0_{dR}(U_+) \oplus H^0_{dR}(U_-) \xrightarrow{H(j^*)} H^0_{dR}(U_+ \cap U_-)$$
$$\shortparallel \qquad\qquad\qquad\qquad \shortparallel$$
$$\mathbb{R}^2 \longrightarrow \mathbb{R}$$

$$(a_1, a_2) \longmapsto a_2 - a_1,$$

while $H(i^*) : H^0_{dR}(\mathbb{R}^{m+2}_\times) \to H^0_{dR}(U_+) \oplus H^0_{dR}(U_-)$ is given by

$$H^0_{dR}(\mathbb{R}^{m+2}_\times) \xrightarrow{H(i^*)} H^0_{dR}(U_+) \oplus H^0_{dR}(U_-)$$
$$\shortparallel \qquad\qquad\qquad\qquad \shortparallel$$
$$\mathbb{R} \longrightarrow \mathbb{R}^2$$

$$a \longmapsto (a, a).$$

In particular, $H(j^*)$ is surjective and $\operatorname{im} H(j^*) = H^0_{dR}(U_+ \cap U_-)$. It follows from the exactness of the Mayer–Vietoris sequence that

$$\Delta : H^0_{dR}(U_+ \cap U_-) \to H^1_{dR}(\mathbb{R}^{m+2}_\times)$$

is the zero map and, from exactness again, $H^1_{dR}(\mathbb{R}^{m+2}_\times) = 0$ (do you see it?). Finally, it follows from (6.17) that $\Delta \; : \; H^{k-1}_{dR}(U_+ \cap U_-) = H^{k-1}_{dR}(\mathbb{R}^{m+1}_\times) \to H^k_{dR}(\mathbb{R}^{m+2}_\times)$ is both injective and surjective. We conclude that $H^k_{dR}(\mathbb{R}^{m+2}_\times) \cong H^{k-1}_{dR}(\mathbb{R}^{m+1}_\times)$ for all $k > 1$, and from the induction hypothesis, we get

$$H^k_{dR}(\mathbb{R}^{m+2}_\times) \cong H^{k-1}_{dR}(\mathbb{R}^{m+1}_\times) = \begin{cases} 0 & \text{if } 1 < k < m \\ \mathbb{R} & \text{if } k = m+1 \end{cases},$$

as claimed. $\blacklozenge$

Exercise 6.5. Compute the de Rham cohomology of the 2-punctured plane and the 2-punctured 3D-space.

We conclude this chapter and this book briefly discussing the relationship between singular homology and de Rham cohomology.

Theorem 6.32 (de Rham Theorem). *Let $U \subseteq \mathbb{R}^n$ be a non-empty open subset. For every $k \in \mathbb{Z}$, there exists a natural real vector space isomorphism*

$$\tau_{dR} : H^k_{dR}(U) \to H^k(U, \mathbb{R}),$$

where $H^k(U, \mathbb{R})$ is the singular cohomology of U with coefficients in $\mathbb{R}$.

Proof. The proof of the de Rham theorem is extremely technical. Here we only discuss how the isomorphism τ_{dR} roughly works. First of all, for all k, the singular cohomology with coefficients in $\mathbb{R}$ is naturally isomorphic to the dual of the singular homology with coefficients in $\mathbb{R}$:

$$H^k(U, \mathbb{R}) \cong H_k(U, \mathbb{R})^*$$

(this is true for every topological space whenever the ring of coefficients is a field, see Problem 3.38 and Remark 5.5). The latter isomorphism identifies the cohomology class $[\phi]$ of a k-cocycle $\phi \in Z^k(U, \mathbb{R})$ with the linear map

$$\bar{\phi} : H^k(U, \mathbb{R}) \to \mathbb{R}, \quad [c] \mapsto \bar{\phi}[c] := \phi(c),$$

where we used that a singular k-cochain $\phi \in C^k(U, \mathbb{R}) = \mathbb{R}^{S_k(U)}$ can be seen as a linear map $\phi : C_k(U, \mathbb{R}) = \mathbb{R}S_k(U) \to \mathbb{R}$. The real number $\bar{\phi}(c)$ does only depend on the homology class of c and the cohomology class of ϕ. So, we get a well defined map

$$H^k(U, \mathbb{R}) \to H_k(U, \mathbb{R})^*, \quad [\phi] \mapsto \bar{\phi}$$

as desired, and one can show that this map is an isomorphism using that $\mathbb{R}$ is a field (see Problem 3.38).

It remains to show that $H^k_{dR}(U)$ is also dual to $H_k(U, \mathbb{R})$. To do this, one first defines a smooth version of singular homology. Namely, a *smooth singular k-simplex* in U is a smooth map $s : \Delta_k \to U$ (this means that s can be extended to a smooth map on some open neighborhood of Δ_k in $\mathbb{R}^{k+1}$). A smooth singular k-simplex is, in particular, a standard singular k-simplex. As all the face maps of the standard simplex are smooth maps, smooth singular k-simplexes form a semi-simplicial subset in standard singular simplexes (this means that smooth singular simplexes are preserved by the face maps). As a consequence, the real vector spaces spanned by smooth singular simplexes form a subcomplex in the chain complex $(C_\bullet(U, \mathbb{R}), \partial)$ that we denote $(C^\infty_\bullet(U, \mathbb{R}), \partial)$ and call the complex of *smooth singular chains*. A smooth singular chain, i.e., a chain in $(C^\infty_\bullet(U, \mathbb{R}), \partial)$, is a formal linear combination of smooth singular simplexes with real coefficients. The homology $H^\infty_\bullet(U) := H_\bullet(C^\infty(U, \mathbb{R}), \partial)$ is the *smooth singular homology* of U. Using approximation techniques of continuous maps by smooth maps, one can show that the inclusion $(C^\infty_\bullet(U, \mathbb{R}), \partial) \to (C_\bullet(U, \mathbb{R}), \partial)$ is actually a quasi-isomorphism, so we get natural vector space isomorphisms

$$H^\infty_k(U) \cong H_k(U, \mathbb{R}).$$

In other words, every singular k-cycle is homologous to a smooth singular k-cycle and if two smooth singular k-cycles are homologous as standard singular k-cycles, they are also *smoothly homologous*, i.e., they are homologous as smooth singular k-cycles. We are now ready to defined a linear map

$$\tau_{dR} : H^k_{dR}(U) \to H_k(U, \mathbb{R})^*.$$

So, let $\omega \in \Omega^k(U)$ be a closed differential k-form, and let $[c] \in H_k(U, \mathbb{R})$ be a singular k-homology class. We can choose a smooth representative $\tilde{c} \in C^\infty_k(U, \mathbb{R})$ in $[c]$. This means that $\tilde{c}$ is a formal linear combination of smooth singular k-simplexes with real coefficients:

$$\tilde{c} = \sum_i a_i s_i, \quad s_i : \Delta_k \to \mathbb{R}.$$

Now define

$$\bar{\omega}[c] := \sum_i a_i \int_{\Delta_k} s_i^* \omega,$$

the latter integral being just the usual integral of a differential k-form on a measurable domain in a (oriented) hypersurface of $\mathbb{R}^{k+1}$. One can prove, using the (higher dimensional) Stokes Theorem, that the real number $\bar{\omega}[c]$ does only depend on the cohomology class of ω and the homology class of c. Moreover, $\bar{\omega}$ is clearly a linear map. So, we get a well-defined map

$$\tau_{dR} : H^k_{dR}(U, \mathbb{R}) \to H_k(U, \mathbb{R})^*, \quad [\omega] \mapsto \tau_{dR}[\omega] := \bar{\omega},$$

and one can show that this is an isomorphism concluding the proof. $\qquad\square$

6.4 End-of-Chapter Problems

Problem 6.1. Let $U \subseteq \mathbb{R}^n$ be a non-empty open subset in the standard Euclidean space. Consider the vector space

$$\Omega(U) := \bigoplus_{k \in \mathbb{Z}} \Omega^k(U)$$

and the $\mathbb{R}$-bilinear composition law

$$\wedge : \Omega(U) \times \Omega(U) \to \Omega(U) \qquad\qquad (6.18)$$

uniquely defined by $\Omega^k(U) \wedge \Omega^l(U) \subseteq \Omega^{k+l}(U)$ and

$$(\omega \wedge \rho)(X_1, \dots, X_{k+l})$$
$$:= \frac{1}{k!l!} \sum_{\sigma \in S_{k+l}} (-)^\sigma \omega(X_{\sigma(1)}, \dots, X_{\sigma(k)}) \rho(X_{\sigma(k+1)}, \dots, X_{\sigma(k+l)}),$$

$\omega \in \Omega^k(U)$ and $\rho \in \Omega^l(U)$, $X_1, \dots, X_{k+l} \in \mathfrak{X}(U)$, $k, l \geq 0$. Prove that

- $(\Omega(U), \wedge)$ is a well-defined associative $\mathbb{R}$-algebra,
- the composition law (6.18) agrees with the usual wedge product under the isomorphisms $\Omega^p(U) \cong \wedge^p \Omega^1(U)$,
- the product $\wedge$ is *graded commutative*, i.e.,

$$\omega \wedge \rho = (-)^{kl} \rho \wedge \omega, \quad \omega \in \Omega^k(U), \quad \rho \in \Omega^l(U),$$

- the following *graded Leibniz rule* holds:

$$d(\omega \wedge \rho) = d\omega \wedge \rho + (-)^k \omega \wedge d\rho, \quad \omega \in \Omega^k(U), \quad \rho \in \Omega^l(U).$$

Problem 6.2. Let $(U_i)_{i\in I}$ be a family of pair-wise disjoint open subsets in $\mathbb{R}^n$ so that $\coprod_{i\in I} U_i$ is also an open subset in $\mathbb{R}^n$. Prove that, for every $k \in \mathbb{Z}$, there is a canonical isomorphism of vector spaces:

$$H_{dR}^k\left(\coprod_{i\in I} U_i\right) \cong \prod_{i\in I} H_{dR}(U_i).$$

Problem 6.3. Let $U \subseteq \mathbb{R}^n$ be a non-empty open subset, and let γ be a *smooth curve* in U, i.e., a smooth map $\gamma = (\gamma^1, \ldots, \gamma^n) : [a, b] \to U$ from a closed interval $[a, b] \subseteq \mathbb{R}$, $a, b \in \mathbb{R}$, $a < b$. The *integral of a 1-form* $\omega = \sum_{i=1}^n f_i dx^i \in \Omega^1(U)$ along γ is

$$\int_\gamma \omega := \sum_{i=1}^n \int_a^b f_i(\gamma(s))\frac{d\gamma^i(s)}{ds}ds.$$

Prove that

- the integral of a 1-form $\omega \in \Omega^1(U)$ along a curve $\gamma : [a, b] \to U$ is reparameterization invariant, i.e., if $\Phi : [a', b'] \to [a, b]$ is an increasing smooth map between closed intervals, then

$$\int_\gamma \omega = \int_{\gamma\circ\Phi} \omega;$$

- the integral of an exact 1-form $\omega \in \Omega^1(U)$ along a *closed curve* $\gamma : [a, b] \to U$, i.e., $\gamma(a) = \gamma(b)$, is zero;
- the integral of a closed 1-form $\omega \in \Omega^1(U)$ along a curve $\gamma : [0, 1] \to U$ does only depend on the smooth homotopy class of γ, i.e., if $\gamma' : [0, 1] \to U$ is another curve such that

(1) $\gamma'(0) = \gamma(0)$ and $\gamma'(1) = \gamma(1)$, and, additionally,

(2) there exists a smooth homotopy $\mathcal{H} : [0, 1] \times [0, 1] \to U$ between γ and γ' (i.e., $\mathcal{H}$ is a smooth map such that $\mathcal{H}(0, s) = \gamma(s)$ and $\mathcal{H}(1, s) = \gamma'(s)$ for all $s \in [0, 1]$) with $\mathcal{H}(t, 0) = \gamma(0) = \gamma'(0)$ and $\mathcal{H}(t, 1) = \gamma(1) = \gamma'(1)$ for all $t \in [0, 1]$,

then

$$\int_\gamma \omega = \int_{\gamma'} \omega.$$

Problem 6.4. Find a closed but not exact 1-form ω on the punctured plane $\mathbb{R}_{\times}^2 = \mathbb{R}^2 \setminus \{(0,0)\}$. For any $x_0 \in \mathbb{R}_{\times}^2$, find an open neighborhood $U \subseteq \mathbb{R}_{\times}^2$ of x_0 and a function $f \in C^\infty(U)$ such that $\omega|_U = df$.

Problem 6.5. Compute explicitly the operators $i_t^{\mathcal{H}}$ from the proof of Theorem 6.17 in the case when $\mathcal{H}$ is the smooth homotopy:

$$\mathcal{H} : [0,1] \times \mathbb{R}^n \to \mathbb{R}^n, \quad (t,x) \mapsto \mathcal{H}(t,x) := tx.$$

Problem 6.6. Prove that, for any $0 \leq k < n$, and any codimension k affine subspace $S \subseteq \mathbb{R}^n$, the open subspace $\mathbb{R}^n \setminus S$ is smoothly homotopy equivalent to $\mathbb{R}^k \setminus \{0\}$.

Problem 6.7. Let $S \subseteq \mathbb{R}^n$ be an m-dimensional affine subspace, $m < n$. Compute the de Rham cohomology of the open subspace $\mathbb{R}^n \setminus S$.

Problem 6.8. Let $r \subseteq \mathbb{R}^3$ be a straight line and let $x_0, x_1 \in \mathbb{R}^3 \setminus r$ be distinct points. Compute the de Rham cohomology of the open subspaces $\mathbb{R}^3 \setminus (r \cup \{x_0\})$ and $\mathbb{R}^3 \setminus (r \cup \{x_0, x_1\})$.

References

Atiyah M. F. and Macdonald I. G., *Introduction to Commutative Algebra* (4th edn.). Undergraduate Texts in Mathematics, Springer, New York, 2024.

Axler S., *Linear Algebra Done Right* (student economy edn.). Addison-Wesley Series in Mathematics, Westview Press, Boulder, CO, 2016.

Bott R. and Tu L. W., *Differential Forms in Algebraic Topology*. Graduate Text in Mathematics, Vol. 82, Springer-Verlag, New York-Berlin, 1982.

Hatcher A., *Algebraic Topology*, Cambridge University Press, Cambridge, 2002.

Hilgert J. and Neeb K.-H., *Structure and Geometry of Lie Groups*. Springer Monographs in Mathematics, Springer, New York, 2012.

Kosniowski C., *A First Course in Algebraic Topology*, Cambridge University Press, Cambridge-New York, 1980.

Lee J. M., *Introduction to Topological Manifolds* (2nd edn.). Graduate Texts in Mathematics, Vol. 202, Springer, New York, 2011.

Lee J. M., *Introduction to Smooth Manifolds* (2nd edn.). Graduate Texts in Mathematics, Vol. 218, Springer, New York, 2013.

Mac L. S., *Homology*. Classics in Mathematics, Springer-Verlag, Berlin-Göttingen-Heidelberg, 1975.

Mac Lane S., *Categories for the Working Mathematician* (2nd edn.). Graduate Texts in Mathematics, Vol. 5, Springer, New York, 1978.

Manetti M., *Lie Methods in Deformation Theory*. Springer Monographs in Mathematics Unitext, Springer, Singapore, 2022.

May J. P., *Simplicial Objects in Algebraic Topology*. The University of Chicago Press, Chicago-London, 1967.

Poincaré H., Analysis Situs, *Journal de l'École Polytechnique*, **1**(2), 1–123 (1895).

Rotman J. J., *An Introduction to Algebraic Topology*. Graduate Texts in Mathematics, Vol. 119, Springer, New York, 1998.

Rotman J. J., *An Introduction to Homological Algebra* (2nd edn.). Universitext, Springer, New York, 2009.

Vitagliano L., *A Primer on Smooth Manifolds*, World Scientific, Singapore, 2024.

Weibel C. A., *An Introduction to Homological Algebra*. Cambridge Studies in Mathematics, Vol. 38, Cambridge University Press, Cambridge, 1994.

Index

www.ingramcontent.com/pod-product-compliance
Lightning Source LLC
Chambersburg PA
CBHW060240220326
41598CB00027B/3993